SolidWorks Tutorial 2001Plus

A Competency Project Based Approach Utilizing 3D Solid Modeling

David C. Planchard & Marie P. Planchard

SDC PUBLICATIONS

www.schroff.com
www.schroff-europe.com

About the Authors

Marie Planchard is CAD department manager at Mass Bay College in Wellesley Hills, MA. Before developing the CAD engineering design program, she spent 13 years in industry and held a variety of High Technology management positions including Beta Test Manager for CAD software at Computervision Corporation. She has written and presented numerous technical papers on 3D modeling. She is Vice-President of the New England Pro/Users Group, an active member of the SolidWorks Educational Advisory Board, a SolidWorks Research Partner and coordinator for the New England SolidWorks Users Group.

David Planchard is the Director of Corporate Technology Programs at Middlesex Community College in Bedford, MA. Before entering academia, he started his own Semiconductor Capital equipment company and spent over 19 years in the Semiconductor industry in various Engineering and Marketing positions. He holds five U.S. and one International patent. He has published and authored numerous papers on equipment design. He is a member of the New England Pro/Users Group, New England SolidWorks Users Group and the Cisco CCNA Regional Academy Users Group.

David and Marie Planchard are co-founders of D&M Engineering and are active industry and education consultants. They are co-authors of the following SDC Publication books:

- **Engineering Design with SolidWorks 1999, 2000, 2001, 2001Plus**
- **Drawing and Detailing with SolidWorks 2001/2001Plus**
- **SolidWorks Tutorial 2001**
- **Applications in Sheet Metal Using Pro/SHEETMETAL & Pro/ENGINEER**
- **An Introduction to Pro/SHEETMETAL**

INTRODUCTION

Preface

SolidWorks 2001Plus Tutorial was written to assist students, designers, engineers and professionals. SolidWorks 2001Plus Tutorial is a subset of Engineering Design with SolidWorks. The book provides a quick introduction to SolidWorks. The book is focused on providing a solid foundation in SolidWorks using competency-based projects. Desired outcomes and usage competencies are listed for each project.

Commands are presented in a step-by-step progressive approach.

A competency project based curriculum is provided. The learning process is explored through a series of design situations, industry scenarios, projects and objectives.

A progressive learning approach is addressed in each chapter. Each chapter identifies a project with reflective information from the previous project situation. The book is designed to compliment the on-line tutorials contained within SolidWorks.

The authors developed the industry scenarios by combining their own industry experience with the knowledge of engineers, department managers, vendors and manufacturers. These professionals are directly involved with SolidWorks everyday. Their responsibilities go far beyond the creation of a 3D model.

Please contact us with any comments, questions or suggestions on this book.

Marie P. Planchard
Engineering Department Chair
Mass Bay Community College
planchar@massbay.edu

David C. Planchard
Director of Corporate Technology Programs
Middlesex Community College
planchardd@middlesex.cc.ma.us

Trademarks, Disclaimer and Copyrighted Material

SolidWorks and its family of products are registered trademarks of the SolidWorks Corporation. Microsoft Windows, Microsoft Office and its family of products are registered trademarks of the Microsoft Corporation. Other software applications and parts described in this book are trademarks or registered trademarks of their respective owners.

Dimensions of parts are modified for illustration purposes. Every effort is made to provide an accurate text. The authors and the manufacturers shall not be held liable for any parts or drawings developed or designed with this book or any responsibility for inaccuracies that appear in the book. World Wide Web and company information was valid at the time of this printing.

Information in this text is provided from the ASME Engineering Drawing and Related Documentation Publications:

ASME Y14.1 1995
ASME Y14.2M-1992 (R1998)
ASME Y14.3M-1994 (R1999)
ASME Y14.5M-1994

References:

References used in this text:

- SolidWorks Users Guide, SolidWorks Corporation, 2001Plus
- SolidWorks Tutorial, SolidWorks Corporations, 2001Plus
- ASME Y14 Engineering Drawing and Related Documentation Practices
- NBS Handbook 71, Specifications for Dry Cells and Batteries
- Betoline, Wiebe, Miller, Fundamentals of Graphics Communication, Irwin, 1995
- Earle, James, Engineering Design Graphics, Addison Wesley, 1999
- Earle, James, Engineering Drafting, Creative Press, 1995
- Hoelscher, Springer, Dobrovolny, Graphics for Engineers, John Wiley, 1968
- Jensen, Cecil, Interpreting Engineering Drawings, Glencoe 2002.
- Jensen & Helsel, Engineering Drawing and Design, Glencoe, 1990
- Ladouceur and McKeen, Pro/E Solutions and Plastic Design, Onward Press, 1999
- Lockhart & Johnson, Engineering Design Communications, Addison Wesley, 1999
- Meyer, Leo A., Sheet Metal, American Technical Publishers, Homewood, IL, 1995
- Olivo C., Payne, Olivo, T, Basic Blueprint Reading and Sketching, Delmar 1988
- Planchard & Planchard, Drawing and Detailing with SolidWorks, SDC Pub., Mission, KS 2002
- Planchard & Planchard, An Introduction to Pro/SHEETMETAL, SDC Pub., Mission, KS 1999
- Planchard & Planchard, Apps in SheetMetal Using Pro/ENGINEER, SDC Pub., Mission, KS 2000
- Toogood, Roger, Pro/ENGINEER Tutorial, SDC Publications, Mission, KS 1999
- Walker, James, Machining Fundamentals, Goodheart Wilcox, 1999
- 80/20 Product Manual, 80/20, Inc., Columbia City, IN, 2002
- GE Plastics Product Data Sheets, GE Plastics, Pittsfield, MA. 2000
- Reid Tool Supply Product Manual, Reid Tool Supply Co., Muskegon, MI, 2002
- Simpson Strong Tie Product Manual, Simpson Strong Tie, CA, 1998
- Ticona Designing with Plastics – The Fundamentals, Summit, NJ, 2000

Notes:

Table of Contents

What is SolidWorks?

SolidWorks is a design automation software package used to produce parts, assemblies and drawings. SolidWorks is a Windows native 3D solid modeling CAD program. SolidWorks provides easy to use, highest quality design software for engineers and designers who create 3D models and 2D drawings ranging from individual parts to assemblies with thousands of parts.

SolidWorks Corporation, headquartered in Concord, Massachusetts USA, develops and markets innovative design solutions for the Microsoft Windows platform. More information on SolidWorks and its family of products can be found at their URL, **www.SolidWorks.com.**

In SolidWorks, you create 3D parts, assemblies and 2D drawings. The part, assembly and drawing documents are all related.

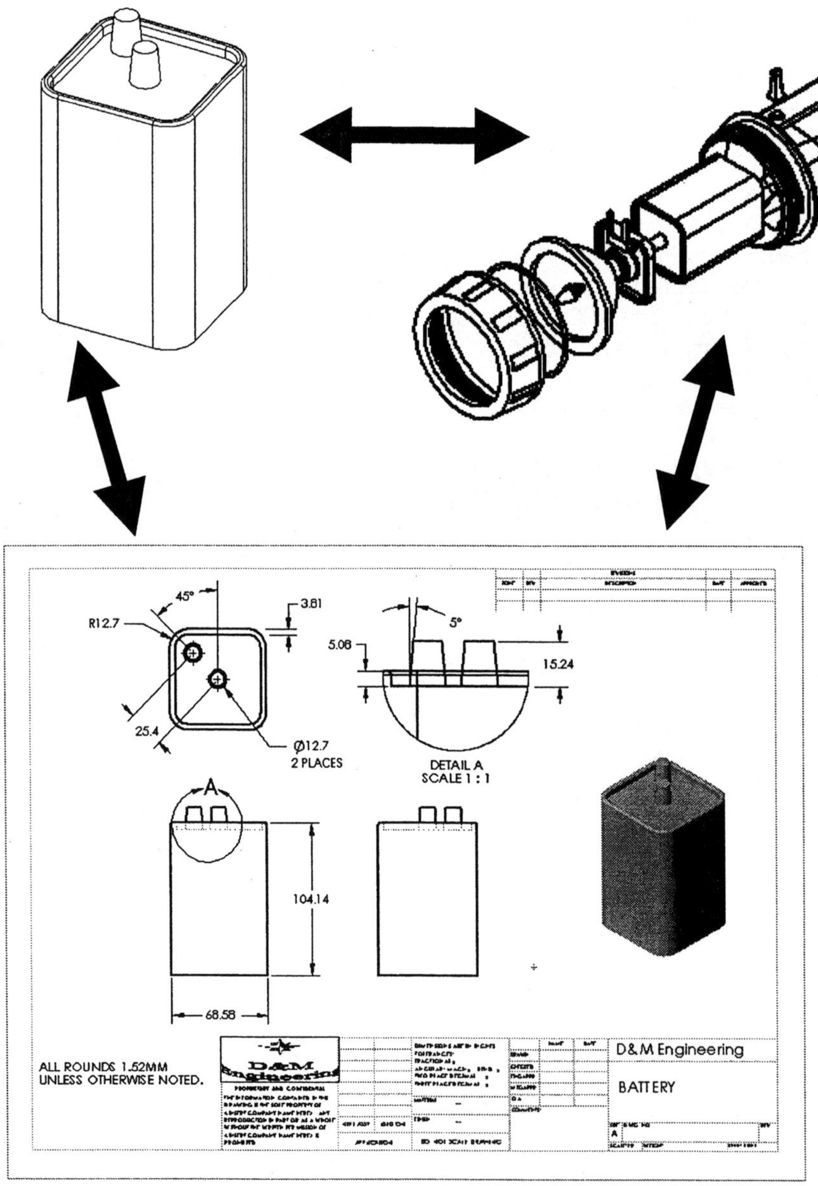

The building blocks of parts are called features. Features such as the Extruded-Boss, Cut, Hole, Fillet, Chamfer and others are used to create parts. Some features are sketched, such as an Extruded-Boss.

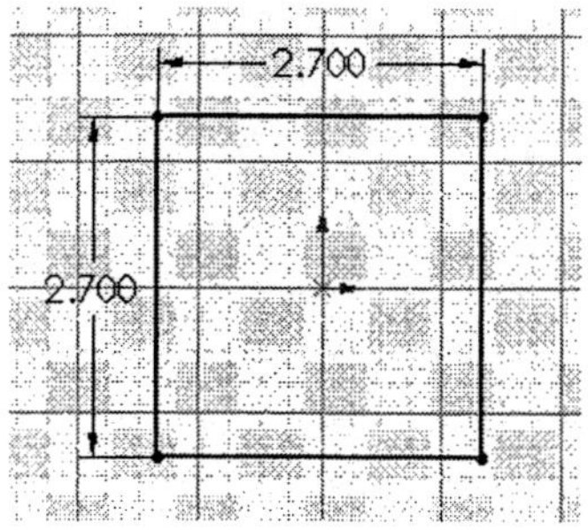

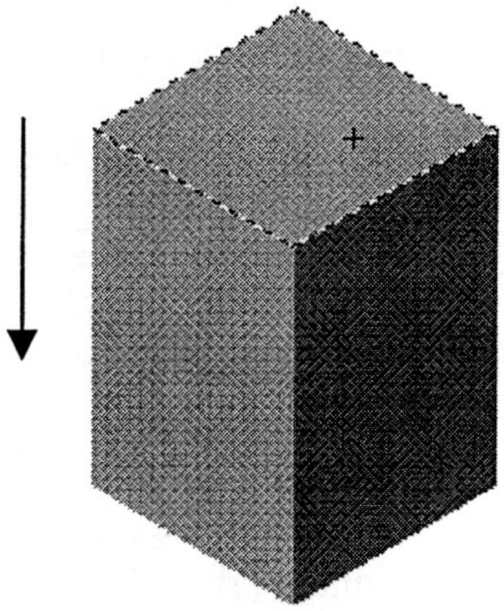

Other features are created by selecting edges or faces of existing features,

such as a Fillet.

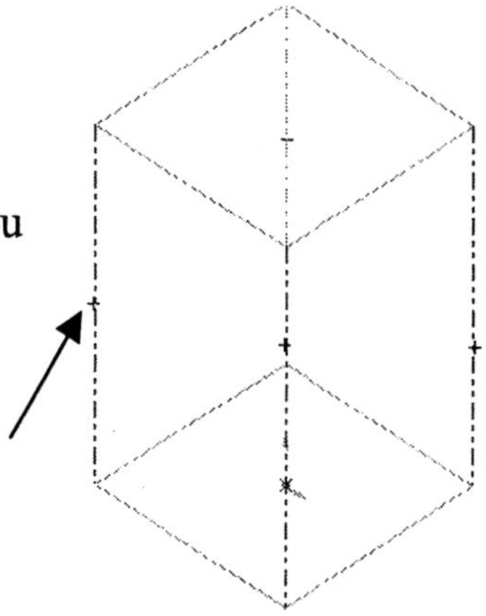

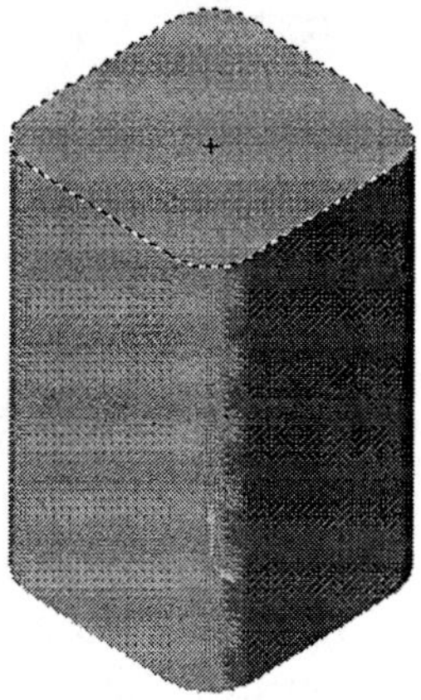

Dimensions drive features. Change a dimension and you change the size of the part.

Geometric relationships are used to maintain the intent of the design.

Create a hole that penetrates through a part. SolidWorks maintains relationships through the change.

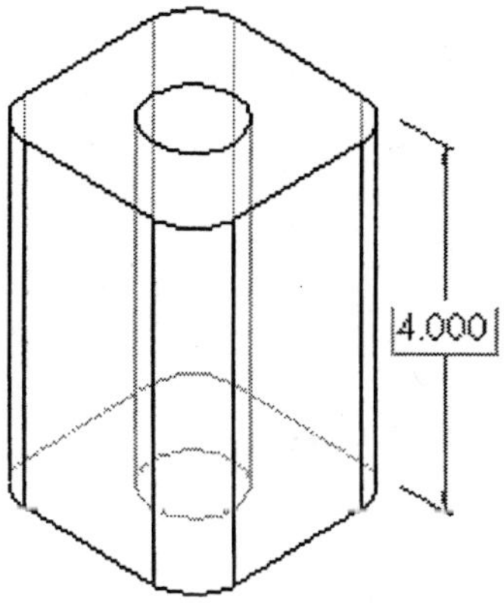

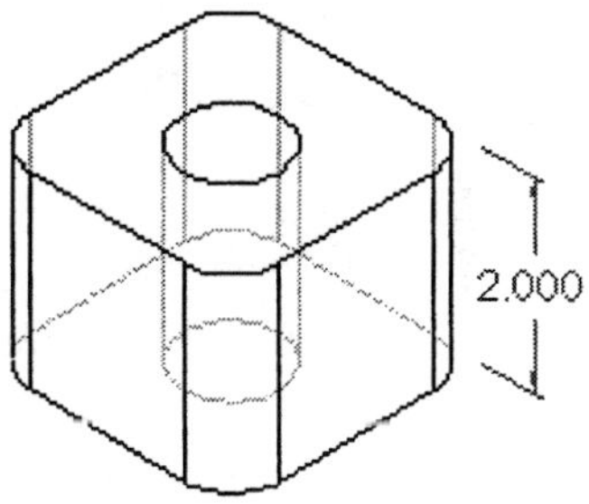

The step-by-step approach used in this text allows you to create parts, assemblies and drawings. The text also allows you to modify and change all components of the model. Change is an integral part of design. Let's begin.

Overview of Projects

Project 1: Extrude Features

How do you start a design in SolidWorks? What is the design intent? How do you take a customer's requirements and covert them into a model? Project 1 introduces the basic concepts behind SolidWorks. In Project 1 you create two parts: BATTERY and BATTERY PLATE. You are exposed to the following features: Extrude Base, Extrude Boss, Extrude Cut and Fillet.

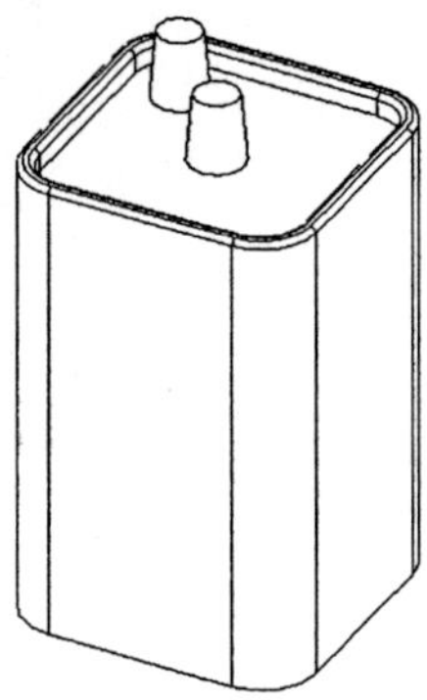

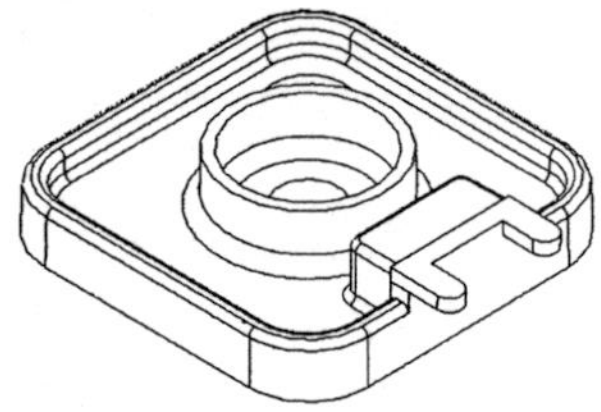

Project 2: Revolve Features

Project 2 introduces the Revolve features. The LENS and BULB are created with the Revolve Base. Explore the Revolve Cut and Revolve Boss features. Circular Pattern, Dome feature and Transparent Feature Properties are introduced.

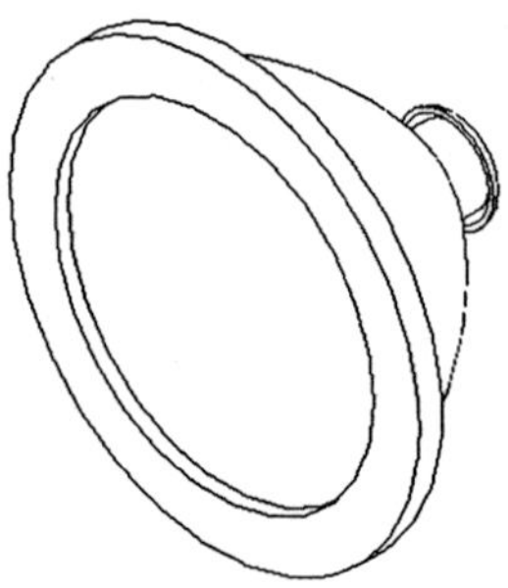

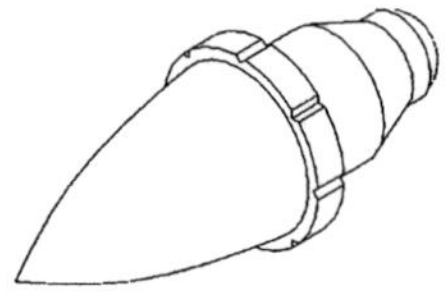

Project 3: Sweep and Loft Features

Project 3 covers the development of the Sweep and Loft features and strengthens the use of previously defined features. The O-RING utilizes a Sweep Base feature. The SWITCH utilizes the Loft Base feature. The LENSCAP and HOUSING are developed with numerous features. A helical curve creates the thread for the LENSCAP and HOUSING.

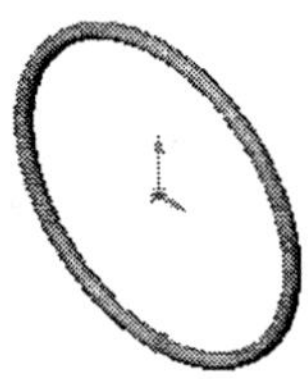

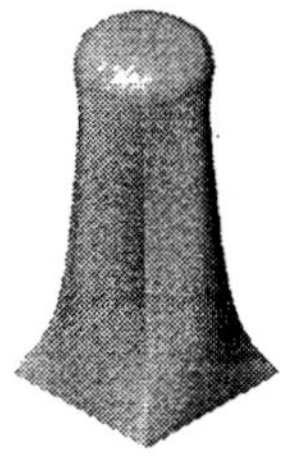

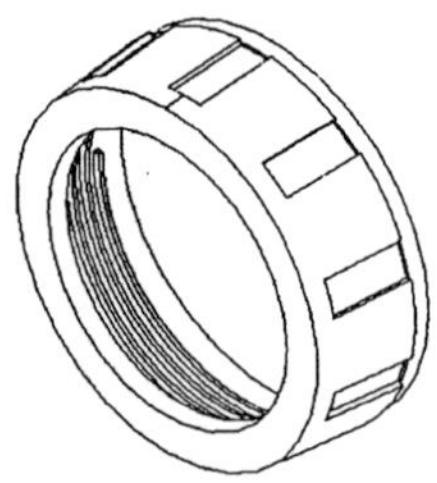

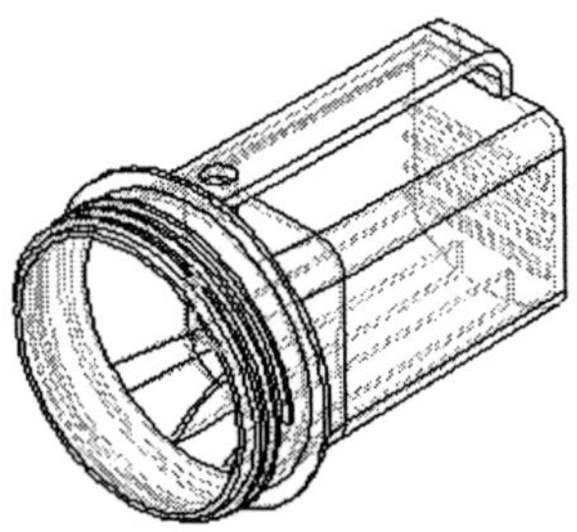

Project 4: Assembly Fundamentals

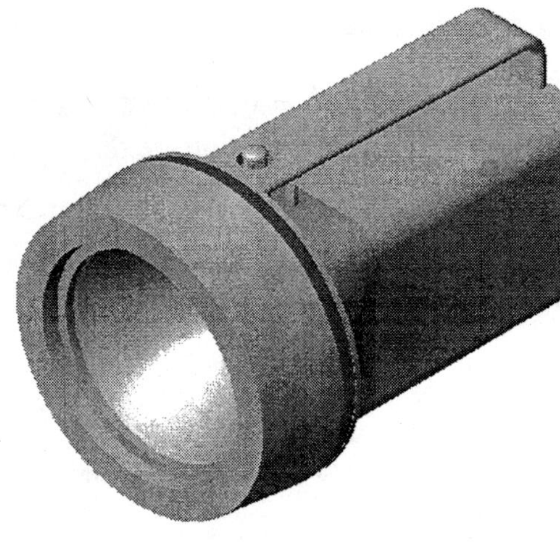

Project 4 develops the LENSAND BULB, BATTERYANDPLATE, and LENSANDCAP assemblies. These assemblies combined with the SWITCH and HOUSING components to complete the FLASHLIGHT assembly.

Project 5: Drawing Fundamentals

Project 5 covers the development of a customized drawing template. Develop a company logo. Create a BATTERY drawing with five views. Develop and incorporate a Bill of Materials into the FLASHLIGHT assembly drawing.

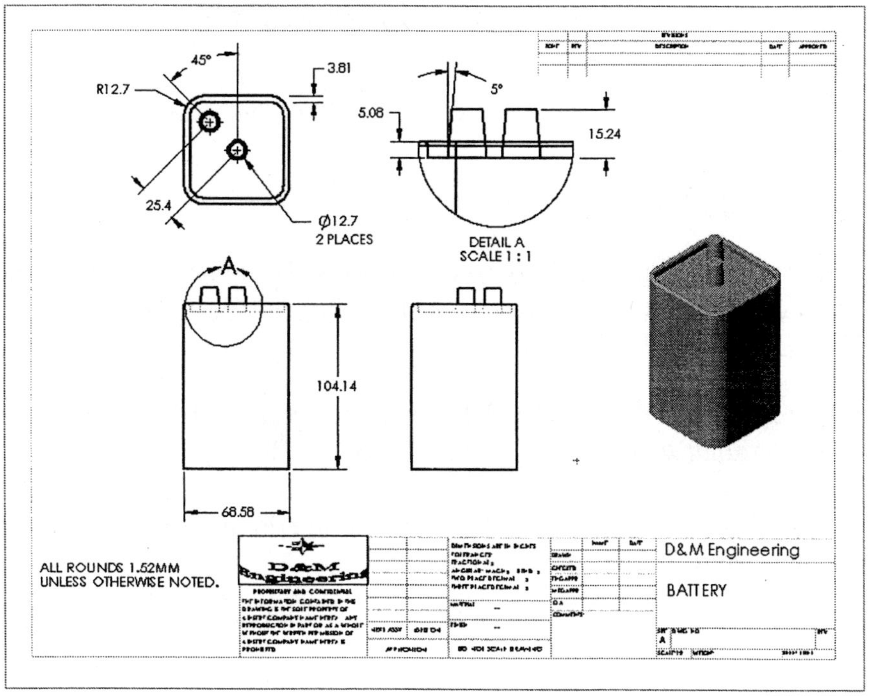

Command Syntax

The following command syntax is used throughout the text. Commands that require you to perform an action are displayed in **Bold** text.

Format	Convention	Example
Bold	All commands actions.	Click **Save**. Click **Tools, Options** from the Main menu.
	Selected icon button.	Click **Rectangle** from the Sketch Tools toolbar.
	Selected geometry: line, circle, arc, point and text	Select the **centerpoint**. Drag the **circle** downward. Click the **arc**.
	Value entries.	Enter **3.0** for Radius. Click **60mm** from the Depth spin box.
Capitalized	Filenames, part names, assembly, component and drawing names.	The BATTERY is contained inside the FLASHLIGHT assembly.
	First letter in a feature name.	Click the **Fillet** feature. Click the **Extrude Base** feature.

Project dimensions are provided in both inches and millimeters. Inches are the primary design units and millimeters are the secondary design units. Millimeters are displayed in brackets [x].

Part dimensions are displayed with three decimal places for inches and two decimal places for millimeters. Control individual decimal display in your drawing to adhere to the following drawing standards: ANSI or ISO. Use PartEnglishTemplate and AssemblyEnglishTemplate for inch units. Use the [PartMetricTemplate] and [AssemblyMetricTemplate] for millimeter units.

This text follows the ASME Y14 Engineering Drawing and Related Documentation Practices for drawings. Display of dimensions and tolerances are as follows:

TYPES of DECIMAL DIMENSIONS (ASME Y14.5M)			
Description	**Example MM**	**Description**	**Example INCH**
Dimension is less than 1mm. Zero precedes the decimal point.	0.9 0.95	Dimension is less than 1 inch. Zero is not used before the decimal point.	.5 .56
Dimension is a whole number. Display no decimal point. Display no zero after decimal point.	19	Express dimension to the same number of decimal places as its tolerance. Add zeros to the right of the decimal point. If the tolerance is expressed to 3 places, then the dimension contains 3 places to the right of the decimal point.	1.750
Dimension exceeds a whole number by a decimal fraction of a millimeter. Display no zero to the right of the decimal.	11.5 11.51		

TABLE 1 **TOLERANCE DISPLAY FOR INCH AND METRIC DIMENSIONS (ASME Y14.5M)**		
DISPLAY	**INCH**	**METRIC**
Dimensions less than 1	.5	0.5
Unilateral Tolerance	$1.417^{+.005}_{-.000}$	$36^{0}_{-0.5}$
Bilateral Tolerance	$1.417^{+.010}_{-.020}$	$36^{+0.25}_{-0.50}$
Limit Tolerance	.571 .463	14.50 11.50

Windows 95/98/2000/NT4.0 Terminology

The mouse pointer provides an integral role in executing SolidWorks commands. The mouse pointer executes commands, selects geometry, displays Pop-Up menus and provides information feedback. A summary of mouse pointer terminology is displayed below:

Item	Description
Click	Press and release the left mouse button.
Double-click	Double press and release the left mouse button.
Click inside	Press the left mouse button. Wait a second and then press the left mouse button inside the text box. This technique is used to modify Feature names in the FeatureManager design tree.
Drag	Point to an object, press and hold down the left mouse button. Move the mouse pointer to a new location, release the left mouse button.
Right-click	Press and release the right mouse button. A Pop-up menu is displayed. Use the left mouse button to select a menu command.
Tool Tip	Position the mouse pointer over an Icon (button). The command is displayed below the mouse pointer.
Mouse pointer feedback	Position the mouse pointer over various areas of the sketch: part, assembly or drawing. The cursor provides feedback depending upon the geometry.

Let's review various Windows terminology that describes: menus, toolbars and commands that constitute the graphical user interface in SolidWorks.

The following Windows terminology is used throughout the text:

Item	Description
Dialog box name	Name of a window to enter information in order to carry out a command.
Box name	Name of a sub-window area inside the dialog box.
Check box	Square box, click to turn on/off an option.
Spin box	Box containing up/down arrow to scroll or type by numerical increments.
Dimmed command	Menu command not currently available (light gray).
Tab	Dialog box sub-headings to simplify complex menus.
Option button	Small circle to activate/deactivate a single dialog box option.
List box	Box containing a list of items. Click the list drop down arrow. Click the desired option.
Text box	Box to type text.
Drop down arrow	Opens a cascading list containing additional options.
OK	Executes the command and closes the dialog box.
CANCEL	Closes the dialog box and leaves the original dialog box settings.
APPLY	Executes the command. The dialog box remains open.

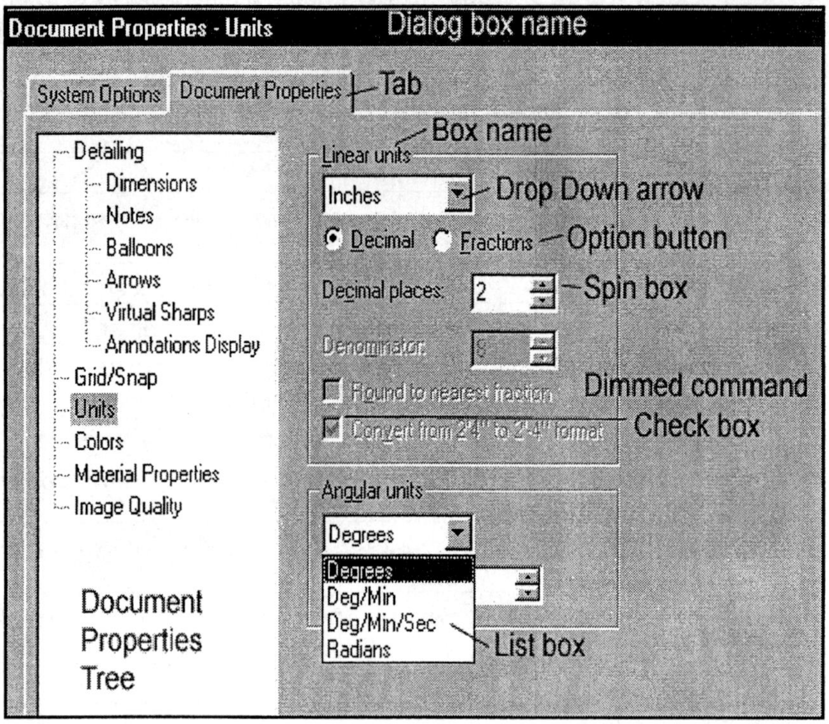
Document Properties - Units
Dialog box name
System Options
Document Properties
Tab
Box name
Linear units
Inches
Drop Down arrow
Decimal
Fractions
Option button
Decimal places:
2
Spin box
Denominator:
Round to nearest fraction
Dimmed command
Convert from 2'4" to 2'-4" format
Check box
Detailing
Dimensions
Notes
Balloons
Arrows
Virtual Sharps
Annotations Display
Grid/Snap
Units
Colors
Material Properties
Image Quality
Angular units
Degrees
Degrees
Deg/Min
Deg/Min/Sec
Radians
List box
Document Properties Tree

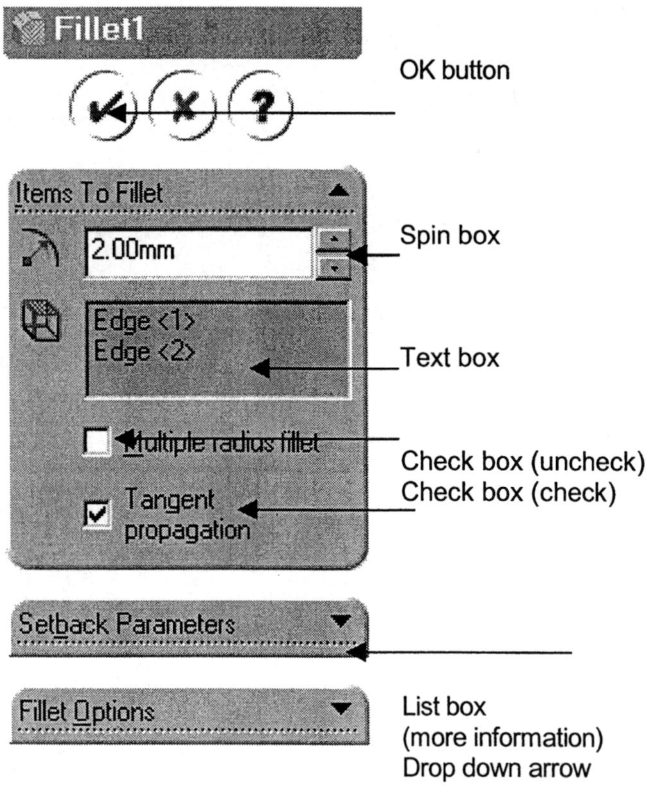
Fillet1
OK button
Items To Fillet
2.00mm
Spin box
Edge <1>
Edge <2>
Text box
Multiple radius fillet
Check box (uncheck)
Tangent propagation
Check box (check)
Setback Parameters
Fillet Options
List box
(more information)
Drop down arrow

Project 1

Extrude Features

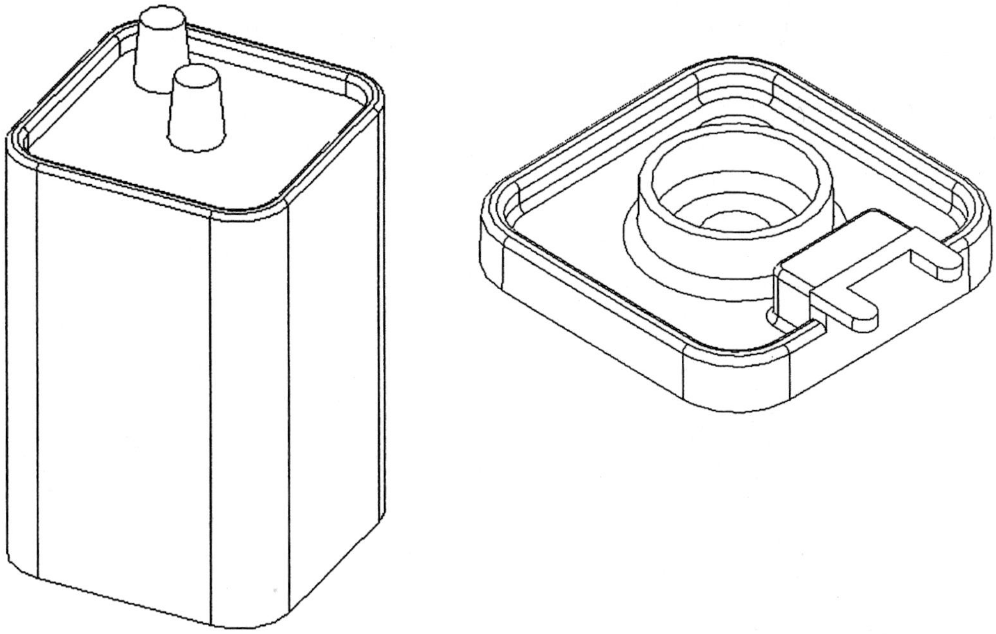

Below are the desired outcomes and usage competencies based upon the completion of Project 1.

Project Desired Outcomes	Usage Competencies
A comprehensive understanding of the customer's design requirements and desires.	To comprehend the fundamental definitions and process of Feature-Based 3D Solid Modeling.
A product design that is cost effective, serviceable and flexible for future manufacturing revisions.	Ability to translate an initial design request and sketches into SolidWorks features: Extruded Base, Cut, Boss, Fillet and Mirror.
Two key FLASHLIGHT parts: • BATTERY • BATTERYPLATE	Specific knowledge and understanding of the Extrude Base feature.

Notes:

Project 1 – Extrude Features

Project Objective

Create two parts for the FLASHLIGHT. Create the BATTERY and BATTERY PLATE parts.

Project Situation

You are employed by a company that specializes in providing promotional trade show products. Your company is expecting a sales order for 100,000 flashlights with a potential of 500,000 units next year. Prototype drawings of the flashlight are required in three weeks.

You are the design engineer responsible for the project. You contact the customer to discuss design options and product specifications. The customer informs you that the flashlights will be used in an international marketing promotional campaign. Key customer requirements:

- Inexpensive reliable flashlight.
- Available advertising space of 10 square inches, 65 square centimeters.
- Light weight, semi indestructible body.
- Self-standing with a handle.

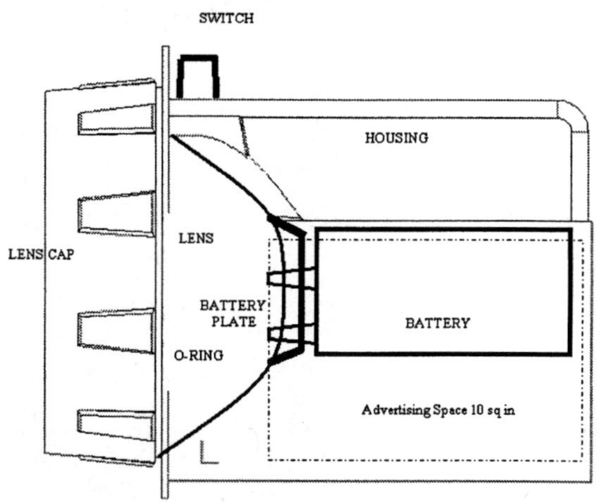

Figure 1.1

Your company's standard product line does not address the above key customer requirements. The customer made it clear that there is no room for negotiation on these key product requirements.

You contact the salesperson and obtain additional information on the customer and product. This is a very valuable customer with a long history of last minute product changes. The job has high visibility with great future potential.

In a design review meeting, you present a conceptual sketch. Your colleagues review the sketch. The team's consensus is to proceed with the conceptual design, Figure 1.1.

The first key design decision is the battery. The battery type will directly affect the flashlight body size, bulb intensity, case structure integrity, weight, manufacturing complexity and cost.

You review two potential battery options:

- A single 6-volt lantern battery.
- Four, 1.5 volt D cell batteries.

The two options affect the product design and specification. Think about it.

A single 6-volt lantern battery is approximately 25% higher in cost and 35% more in weight. The 6-volt lantern battery does provide higher current capabilities and longer battery life.

A special battery holder is required to incorporate the four 1.5 volt D cell configuration. This would directly add to the cost and design time of the flashlight, Figure 1.2.

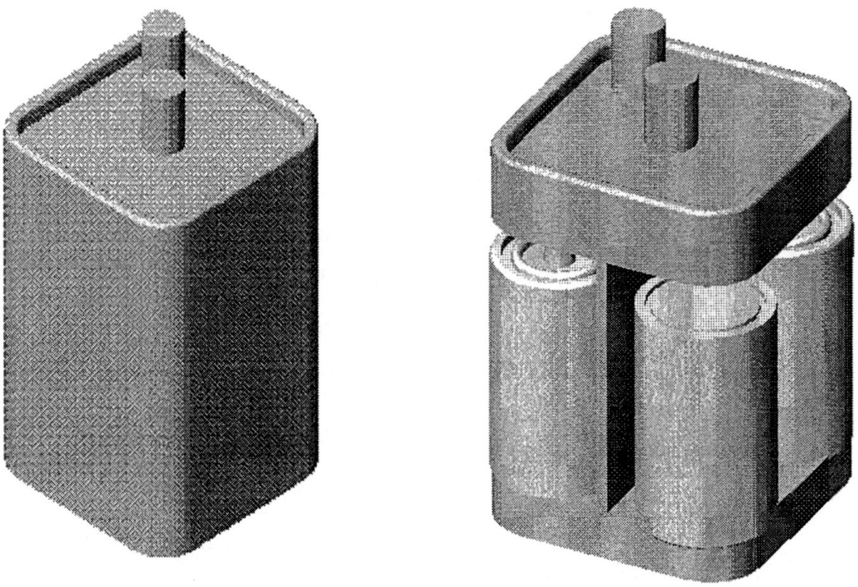

Figure 1.2

Time is critical. For the prototype, you decide to use a standard 6-volt lantern battery. This eliminates the requirement to design and procure a special battery holder. However, you envision the 4-D cell battery model for the next product revision. You design the flashlight to accommodate both battery design options.

Battery dimensional information is required for the design. Where do you go? Potential sources: product catalogs, company web sites, professional standards organizations, design handbooks and colleagues.

The design team decides to purchase the following components: 6-volt BATTERY, LENS, BULB, SWITCH and an O-RING. Your company will design and manufacture the following components: BATTERY PLATE, LENSCAP and HOUSING.

Purchased Parts:	**Designed Parts:**
BATTERY	BATTERY PLATE
LENS	LENS CAP
BULB	HOUSING
SWITCH	
O-RING	

Project Overview

In this project, create two parts of the FLASHLIGHT, Figure 1.3a:

- BATTERY
- BATTERY PLATE

Note: Dimensions and features are used to illustrate the SolidWorks functionality in a design situation. Wall thickness and thread size have been increased for improved picture illustration. Parts have been simplified.

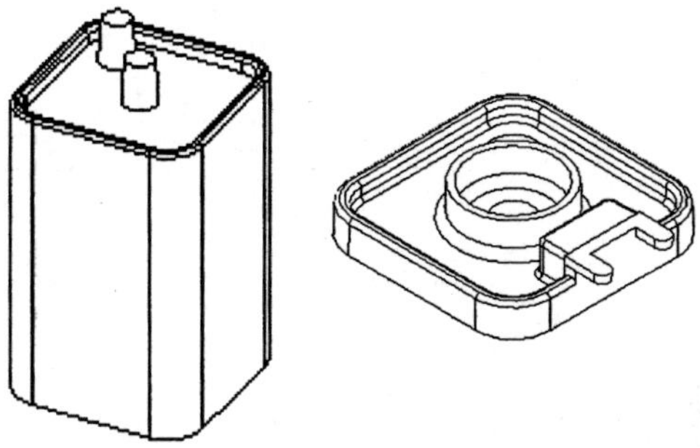

Figure 1.3a

In Project 2, you will create two additional parts of the FLASHLIGHT, Figure 1.3b:

- LENS
- BULB

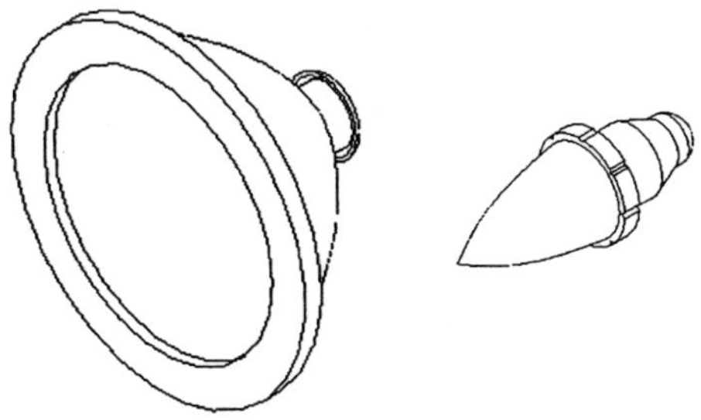

Figure 1.3b

In Project 3, you will create four additional parts to complete the FLASHLIGHT parts, Figure 1.3c.

- O-RING
- SWITCH
- LENSCAP
- HOUSING

Create the FLASHLIGHT assembly in Project 4, and a Parts list in Project 5, Figure 1.3d.

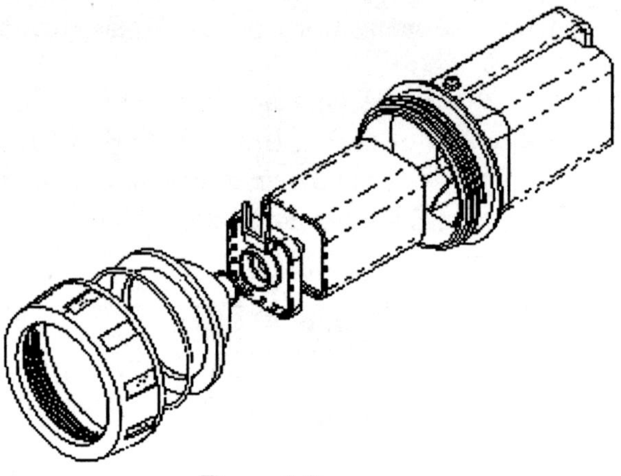

Figure 1.3c

ITEM NO	QTY.	PART NO.	DWG NO
1	1	HOUSING	44-22224
2	1	SWITCH	99-3567
3	1	LENSCAP	44-22223
4	1	O-RING	99-3566
5	1	LENS	99-3565
6	1	BATTERY	99-3564
7	1	BATTERYPLATE	44-22222

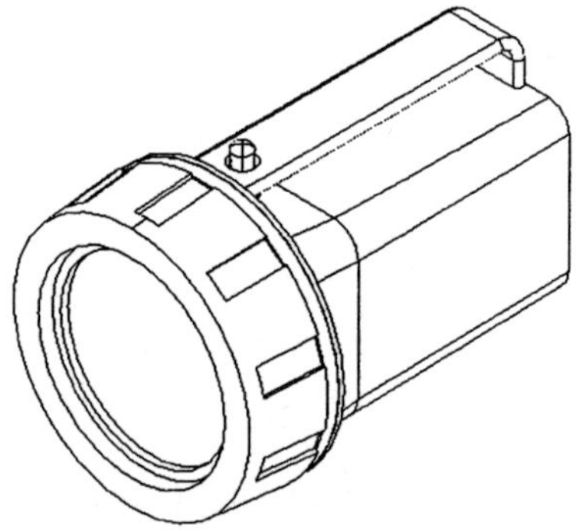

Figure 1.3d

BATTERY

The BATTERY is a simplified representation of an OEM component. The BATTERY consists of the following features:

- Extruded Base
- Extruded Cut
- Fillets
- Extruded Boss

The battery terminals are represented as cylindrical extrusions. The battery dimension is obtained from the ANSI standard 908D.

Note: A 6-volt lantern battery weighs approximately 1.38 pounds, (0.62kg). Locate the center of gravity closest to the center of the battery.

BATTERY Feature Overview

Create the BATTERY, Figure 1.4a. Identify the required BATTERY features.

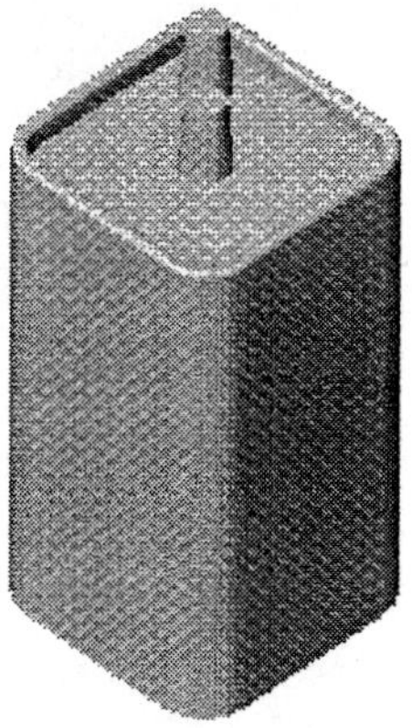

Figure 1.4a

- Extruded Base: The Extruded Base feature is created from a symmetrical square sketch, Figure 1.4b.
- Fillet: The Fillet feature is created by selecting the vertical edges and the top face, Figure 1.4c and Figure 1.4e.
- Extruded Cut: The Extruded Cut feature is created from the top face offset, Figure 1.4d.
- Extruded Boss: The Extruded Boss feature is created to represent the battery terminals, Figure 1.4f.

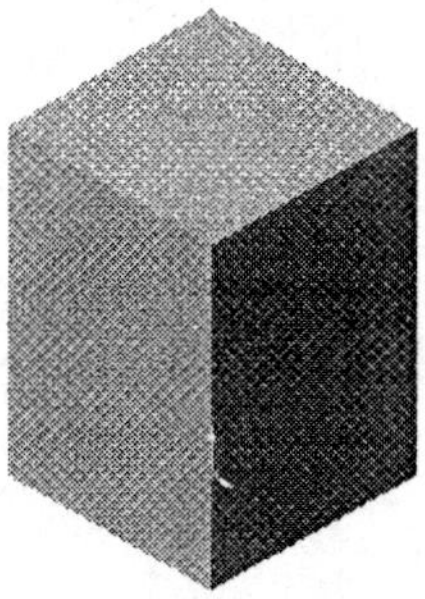

Figure 1.4b

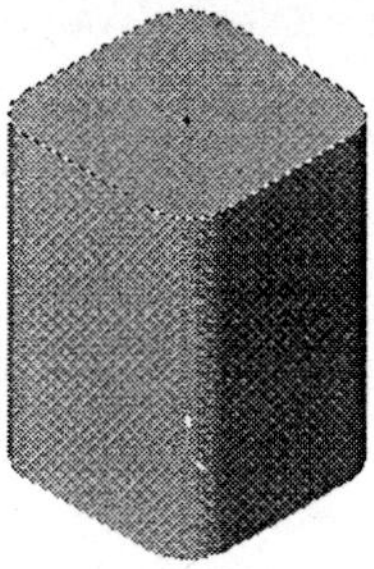

Figure 1.4c

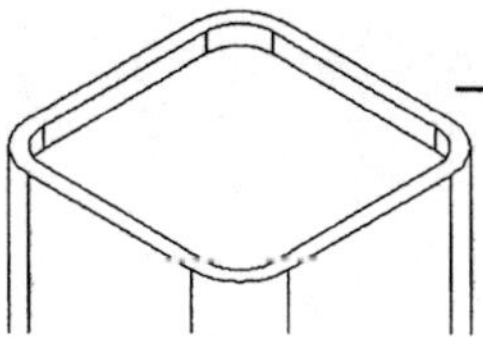

Figure 1.4d

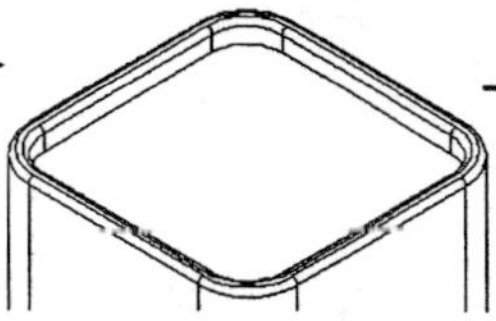

Figure 1.4e

Figure 1.4f

Let's create the BATTERY.

Start SolidWorks

Start a session of SolidWorks. Create a new part called BATTERY. The BATTERY is comprised of numerous features. First, set the Document Properties before you create any features.

Start SolidWorks.

1) Click **Start** on the Windows Taskbar, Start. Click **Programs**. Click the **SolidWorks 2001Plus** SolidWorks 2001Plus folder.

2) Click the **SolidWorks 2001Plus** SolidWorks 2001Plus application. The SolidWorks program window opens.

Create a new part.

3) Click **New** from the Standard toolbar. Part is the default template from the New SolidWorks Document dialog box. Click **OK**.

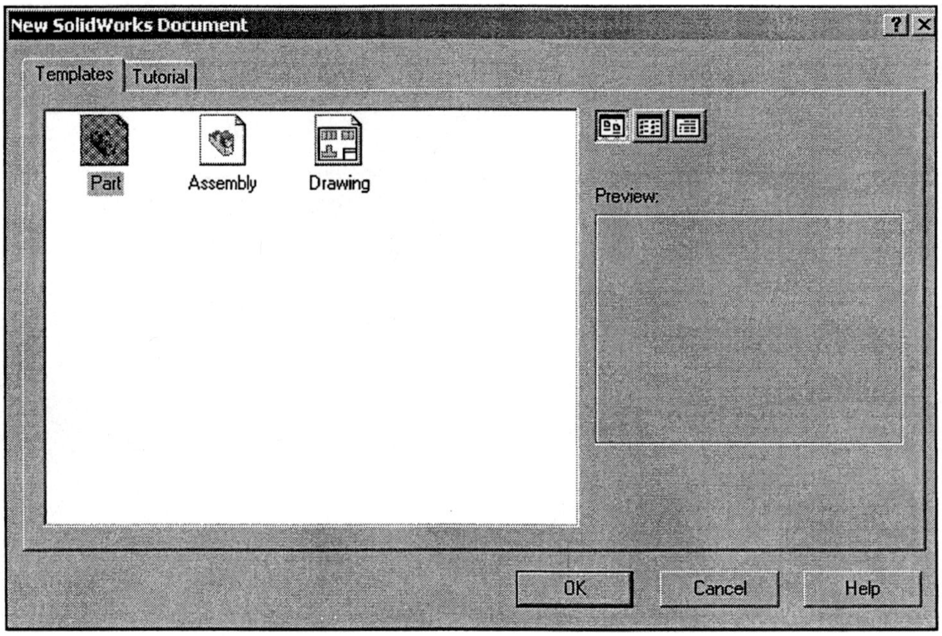

Note: The Document Properties are based on the default document template.

Part1 is displayed. Part1 is the new default part window name.

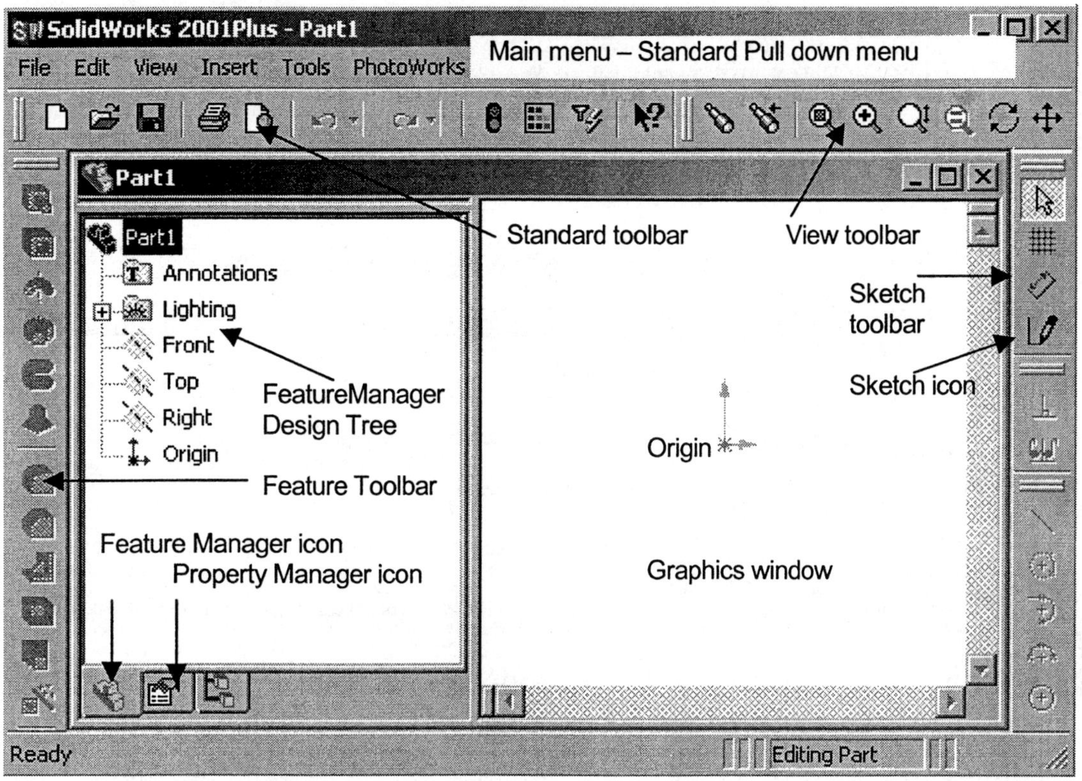

4) Expand the SolidWorks window to full screen. Click the **Maximize** icon in the top right hand corner of the Graphics window.

5) Display the toolbars. Click **View** from the Main menu. Click **Toolbars**. The system places a checkmark in front of the displayed toolbars. Click the **Standard Views**. The Standard Views toolbar is displayed below the Main menu.

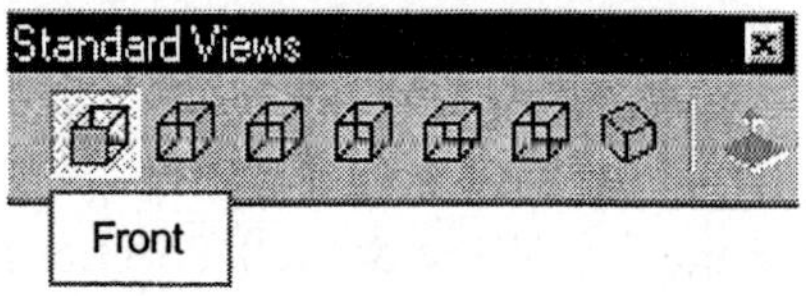

6) Position the **mouse pointer** on an individual toolbar icon to receive a ToolTip.

Enter commands from the Toolbar icons, Main Pull down menu and or Short Cut Keys. Examples:

Toolbar icon	Save  icon
Main Pull down menu	File, Save
Short Cut Key	Ctrl + S

On-line Help contains step-by-step instructions for a command.

Some commands contain a short AVI file. The Help icon appears in the Feature toolbars.

7) Click **Help** from the Main menu. Select **SolidWorks 2001Plus Help Topics**. Click **Index**. Enter **rectangle**. Double-click on **rectangle**. The description appears in the right window. Click the **Show Me** button to play the Rectangle .AVI file. Close Help. Click **Close** .

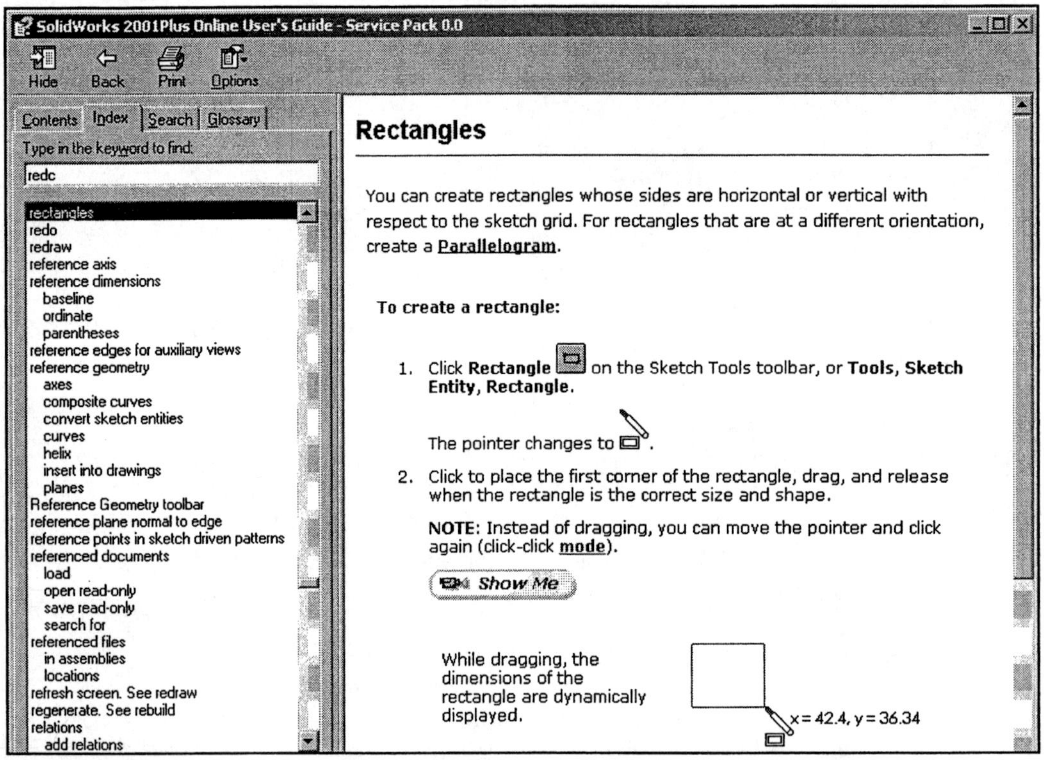

The Welcome to SolidWorks interactive tutorial and Getting Started Manual contains additional information on using SolidWorks.

To simplify the commands in this book, the default Toolbar icons and mouse Right-click Pop-up menus are used throughout the text. The Standard Main Pull down menu is referred to as the Main menu throughout the remainder of the text.

Create the Part Document Template

A part document template contains user-defined parameters such as:

- Dimensioning Standard
- Units
- Part geometry

Use a part document template to conserve setup time.

Dimensions for the FLASHLIGHT ASSEMBLY are provided both in English and Metric units. The Primary units are in inches. Three decimal places are displayed to the right of the decimal point. The Secondary units are in millimeters. Secondary units are displayed in brackets [x]. Two decimal places are displayed to the right of the decimal point. The PARTENGLISH TEMPLATE contains System Options and Document Properties settings for the parts contained in the FLASHLIGHT ASSEMBLY. Substitute the PARTMETRIC TEMPLATE to create the same parts in millimeters.

Set the Document Properties.

8) Click **Tools**, **Options**, from the Main menu. The System Options - General dialog box is displayed.

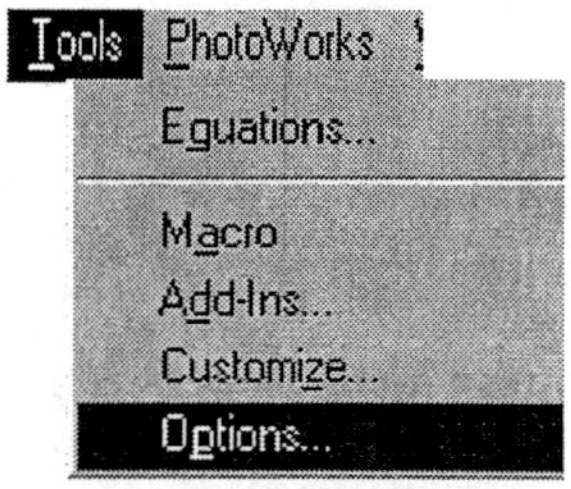

Set the Dimension Standard to ANSI.

9) Click the **Document Properties** tab. Select **ANSI** from the Dimensioning standard drop down list.

Set the Units.

10) Click the **Units** option. Enter **inches, [millimeters]** from the Linear units list box. Click the **Decimal** button. Enter **3**, **[2]** in the Decimal places spin box.

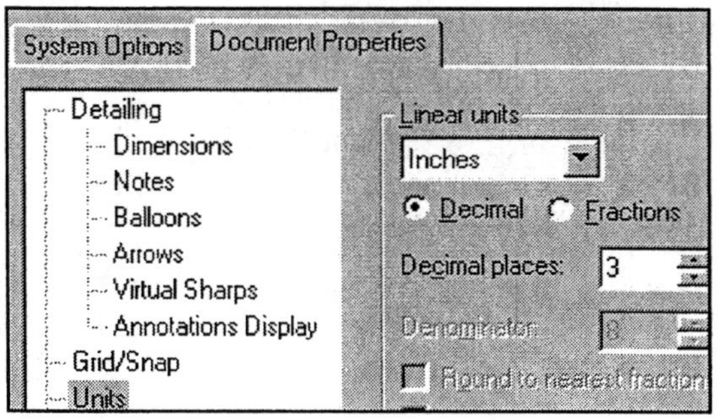

Set the Grid/Snap.

11) Select the **Grid/Snap** option. Uncheck all **Grid** and **Snap check boxes**.

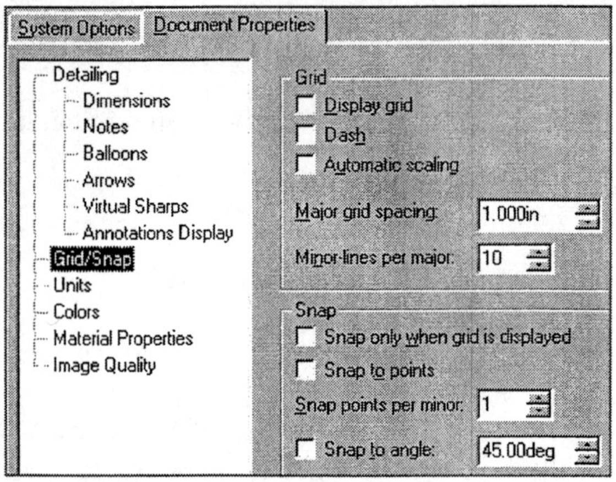

Change Grid/Snap settings at any time through Tools, Options or through the Sketch toolbar.

Document properties are stored in the current document. System Options are stored in the registry of the computer.

Set the Length increment.

12) Click the **System Options** tab. Click the **Spin Box Increments** option. Click the English units **text box**. Enter **.100**. Click the Metric units **text box**. Enter **2.5**.

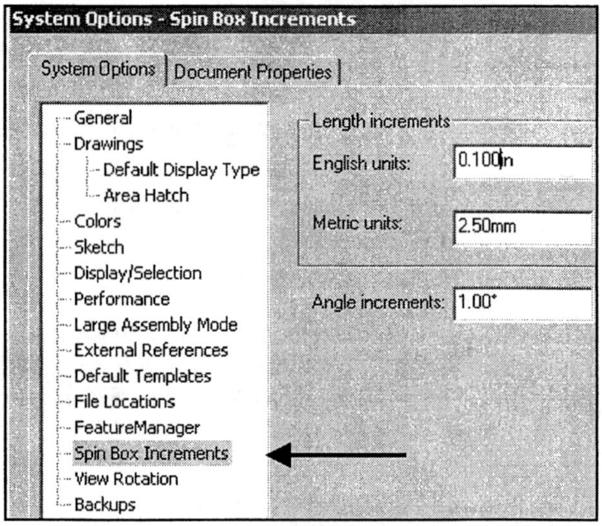

Verify the General options.

13) Click the **General option**. Verify the default options. Check the **Input dimension value**. Check **Use shaded face highlighting**. Check **Auto-show Property Manager**. Update the current part document. Click **OK**.

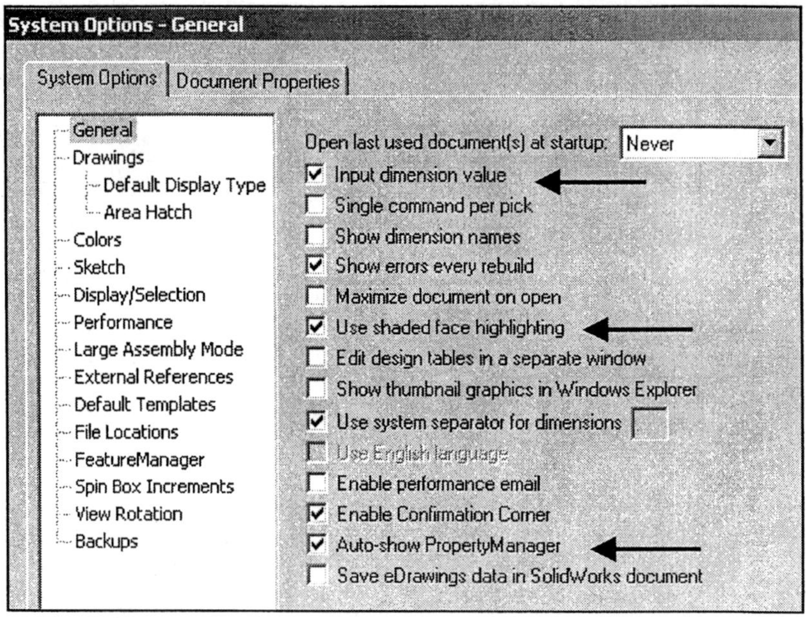

ASME Y14.5M defines the types of decimal dimension display for inches and millimeters. The Primary units are in inches. The Secondary units are in millimeters. Secondary units are displayed in brackets [x].

The precision is set to 3 decimal places for inches. Example: 2.700 is displayed. If you enter 2.7, the value 2.700 is displayed. The precision is set to 2 decimal places for millimeters. Example: [68.58] is displayed. For consistency, the inch part dimension values for the text include the number of decimal places required. The drawings utilizes the decimal dimension display as follows:

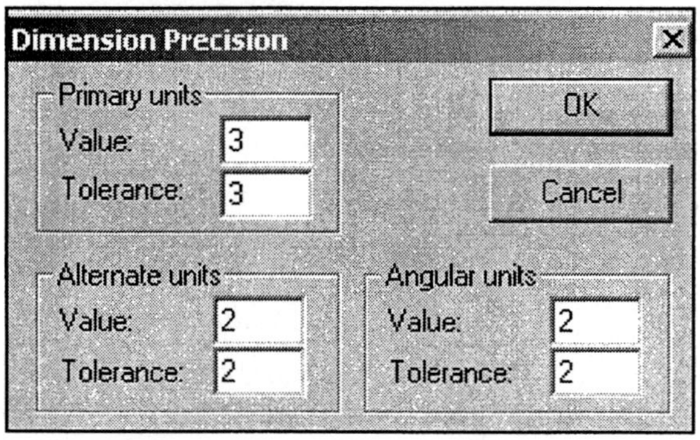

TYPES of DECIMAL DIMENSIONS (ASME Y14.5M)			
Description	**Example MM**	**Description**	**Example INCH**
Dimension is less than 1mm. Zero precedes the decimal point.	0.9 0.95	Dimension is less than 1 inch. Zero is not used before the decimal point.	.5 .56
Dimension is a whole number. No decimal point. Display no zero after decimal point.	19	Express dimension to the same number of decimal places as its tolerance. Add zeros to the right of the decimal point. If the tolerance is expressed to 3 places, the dimension contains 3 places to the right of the decimal point.	1.750
Dimension exceeds a whole number by a decimal fraction of a millimeter. Display no zero to the right of the decimal.	11.5 11.51		

Save the part document template.

14) Click **File** from the Main menu. Click **Save As**. Click ***.prtdot** from the Save As type list box. The default Templates file folder is displayed. Enter **PARTENGLISH TEMPLATE, [PARTMETRIC TEMPLATE]** in the File name text box. Click **Save**.

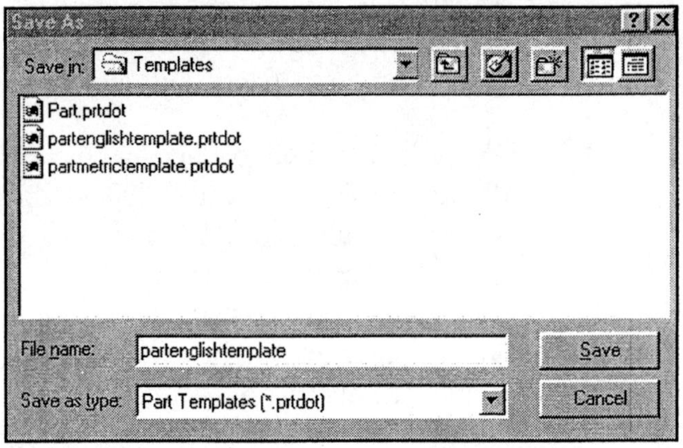

15) Verify the new template. Click **New**. The new template, PARTENGLISH TEMPLATE, [PARTMETRIC TEMPLATE] is added to the New SolidWorks Document Template dialog box. Click **OK**. Part2 is displayed.

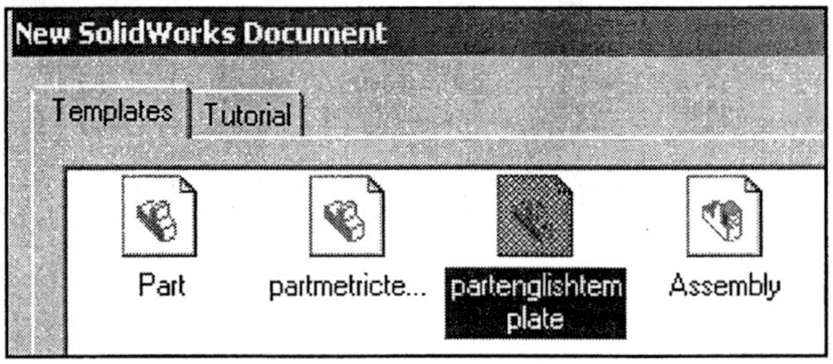

Note: Use the PARTENGLISH TEMPLATE, [PARTMETRIC TEMPLATE] when creating parts for the FLASHLIGHT.

Save the BATTERY

During a SolidWorks session, the first system default part filename is named: Part1.sldprt. The system attaches the .sldprt suffix to all created parts.

The second created part in the same session, increments to Part2.sldprt.

There are numerous ways to manage a part. Companies use Project Data Management (PDM) systems to control, manage and document file names and drawing revisions. Use appropriate filenames that describe the part.

Save the part.

16) Click **Save**. Select **My Documents** folder in the Save In text box. Enter **BATTERY** for the File name. Click **Save**. The part, BATTERY is displayed in the Graphics window.

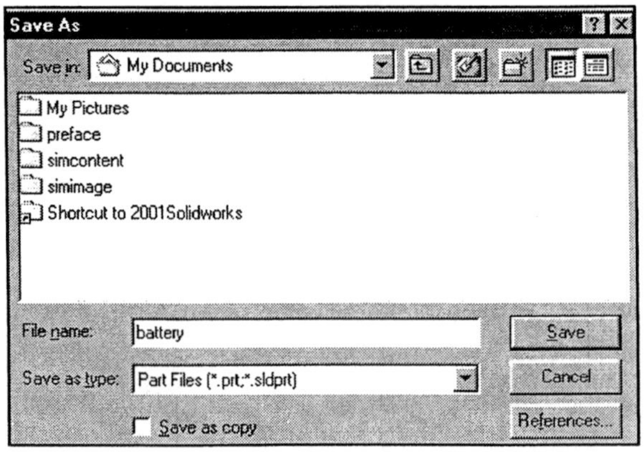

Extruded Base Feature

What is a Base feature? The Base feature is the first feature that is created. The Base feature is the foundation of the part. Note: Keep the Base feature <u>simple!</u>

The Base feature geometry for the BATTERY is an extrusion. The extrusion is named Extruded-Base feature.

Create a solid Extruded-Base feature:

- Sketch a rectangular profile on a flat 2D plane, Figure 1.5a.
- Extend the profile perpendicular (⊥) to the Sketch plane, Figure 1.5b.

The Extruded-Base feature is the 3D block, Figure 1.5c.

Sketch 2D profile

Figure 1.5a

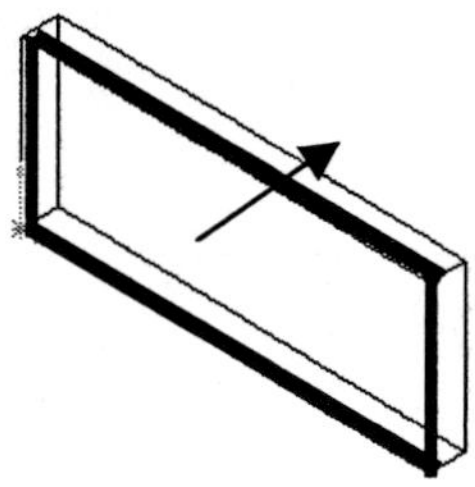

Extrude the Sketch

Figure 1.5b

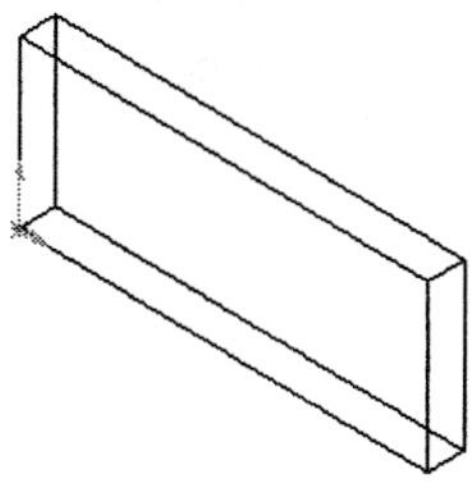

Extruded-Base feature

Figure 1.5c

Reference Planes and Orthographic Projection

The three default ⊥ reference planes, Front, Top and Right represent infinite 2D planes in 3D space, Figure 1.6. Planes have no thickness or mass.

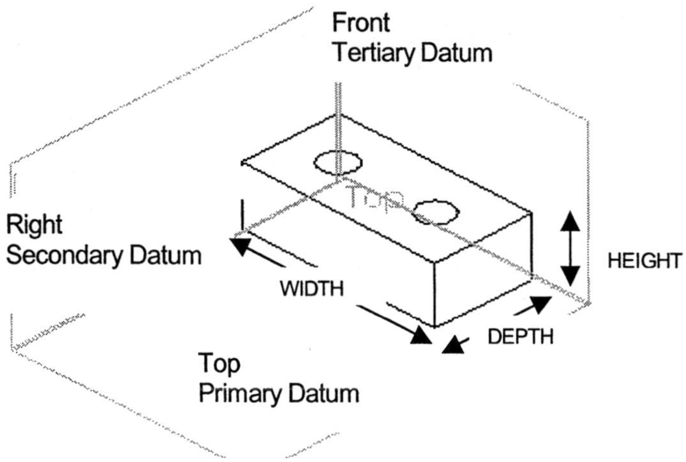

Figure 1.6

Orthographic projection is the process of projecting views onto parallel planes with ⊥ projectors. The default reference planes are the Front, Top and Right side viewing planes.

In geometric tolerancing, the default reference planes are the Primary, Secondary and Tertiary ⊥ datum planes. The following are the planes used in manufacturing:

- Primary datum plane contacts the part at three or more points.
- Secondary datum plane contacts the part at two or more points.
- Tertiary datum plane contacts the part at one or more points.

Orthographic Projection

In third angle Orthographic projection, the standard drawing views are Front, Top, Right and Isometric, Figure 1.7.

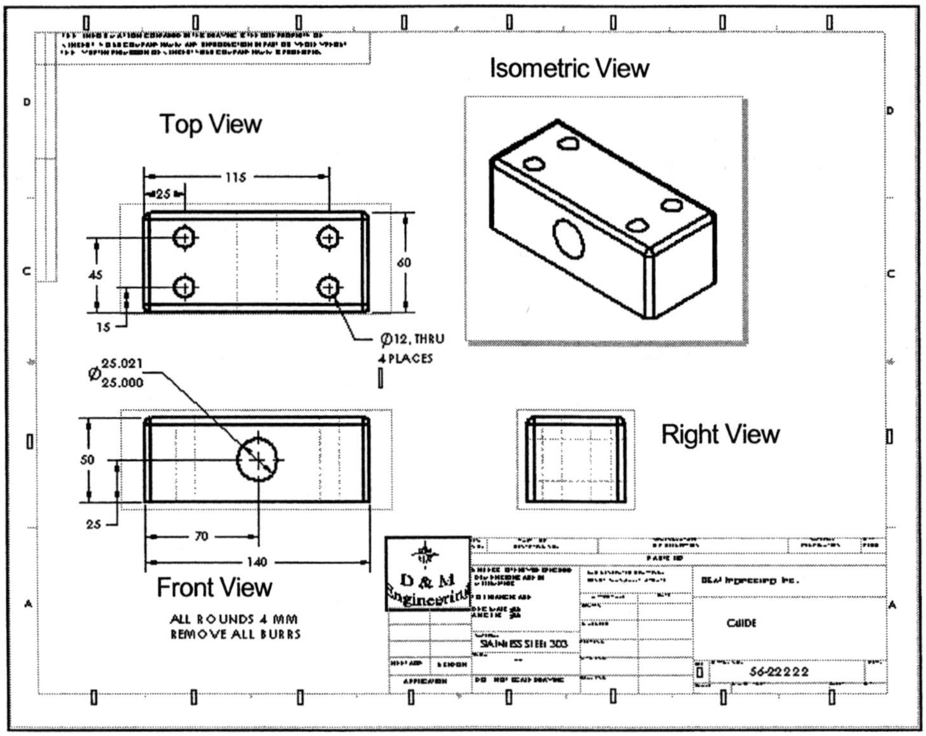

Figure 1.7

There are two Orthographic projection drawing systems. The projection systems are called third angle projection and first angle projection respectively, Figure 1.8. The systems are derived from positioning a 3D object in the third or first quadrant.

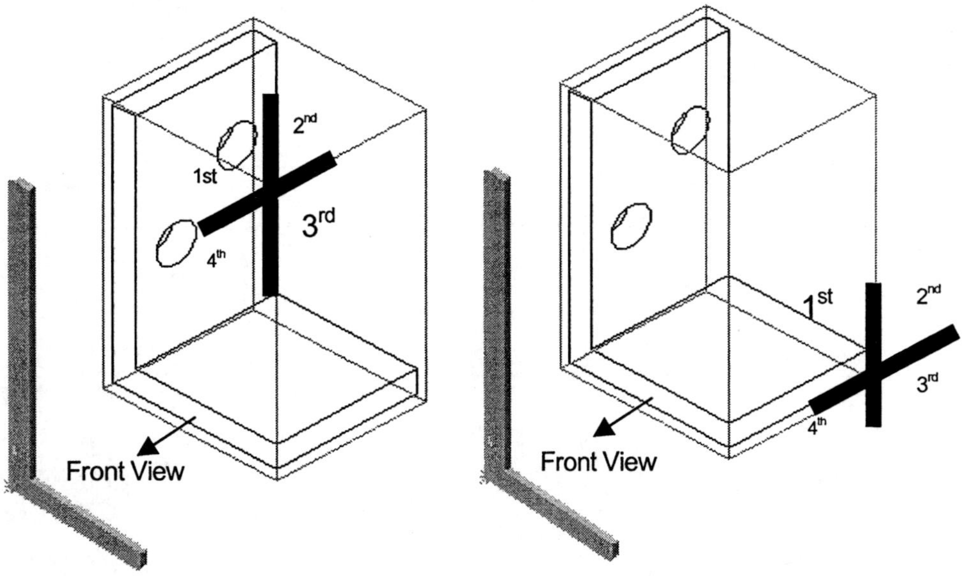

Figure 1.8

In third angle projection, the part is positioned in the third quadrant. The 2D projection planes are located between the viewer and the part. The projected views are placed on a drawing, Figure 1.9.

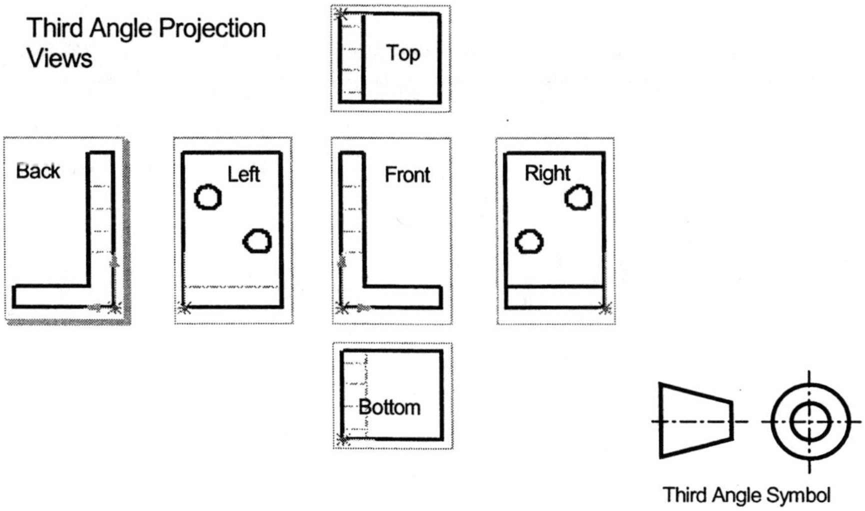

Figure 1.9

In first angle projection, the part is positioned in the first quadrant. Views are projected onto the planes located behind the part. The projected views are placed on a drawing, Figure 1.10.

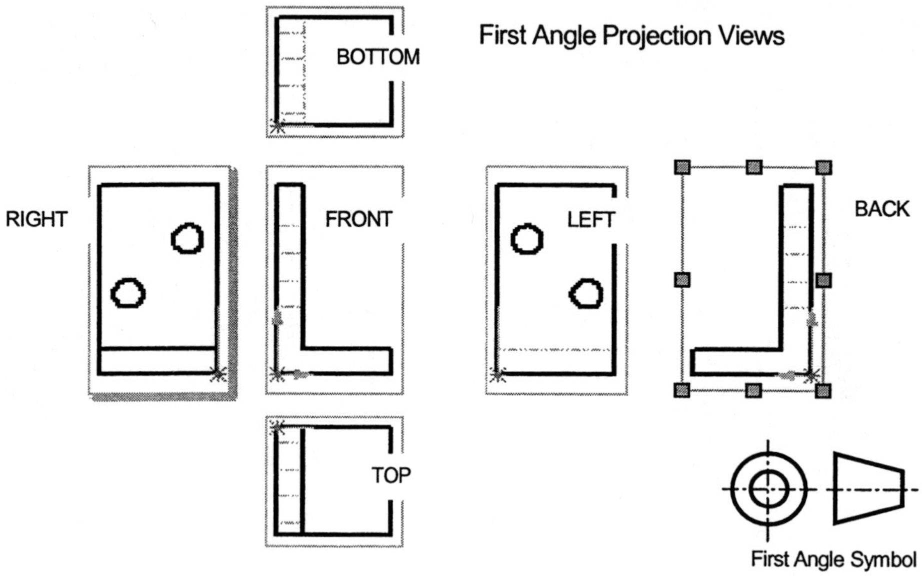

Figure 1.10

First angle projection is primarily used in Europe and Asia. Third angle projection is primarily used in the U.S. and is the ANSI standard. Designers should have knowledge and understanding of both systems. There are numerous multi national companies. Example: A part is designed in the U.S., manufactured in Japan and destined for a European market.

Note: Third angle projection is used in this text. A truncated cone symbol appears on the drawing to indicate the projection system:

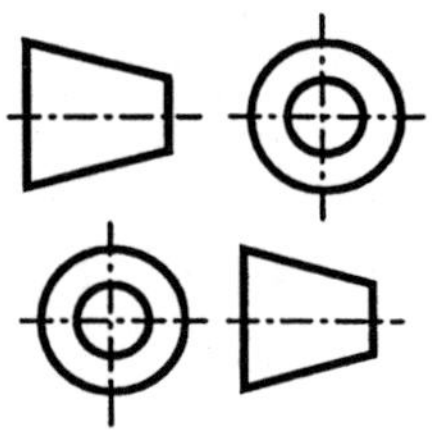

Third Angle Projection Symbol

First Angle Projection Symbol

The selected Sketch plane defines the orientation of the part in 3D space, Figure 1.11.

Create three examples by selecting the Front, Top and Right planes as the Sketch planes for the Extruded Base feature. The overall dimensions are physically the same, however the orientation differs.

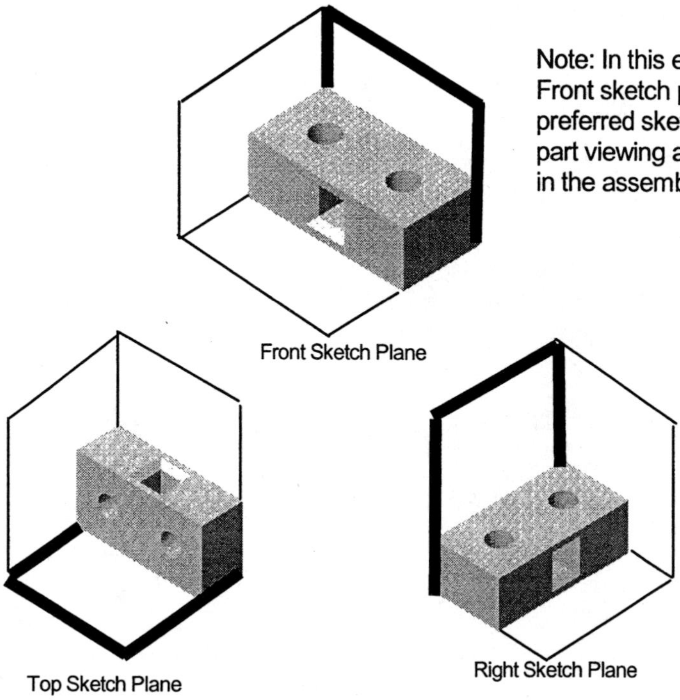

Note: In this example, the Front sketch plane is the preferred sketch plane for part viewing and orientation in the assembly.

Figure 1.11

Before incorporating your design ideas into the Sketch plane, ask yourself a question:

How will the part be oriented in the assembly? Answer: A part should be oriented or aligned to assist in the final assembly.

Default Planes

Review the three default planes, Front, Top and Right. The Front view is the default view. The Front plane is displayed as a green rectangle. In the Front view, the Top plane is displayed as a green horizontal line. The Right plane is displayed as a green vertical line. Each plane is displayed as a green rectangle in an Isometric view.

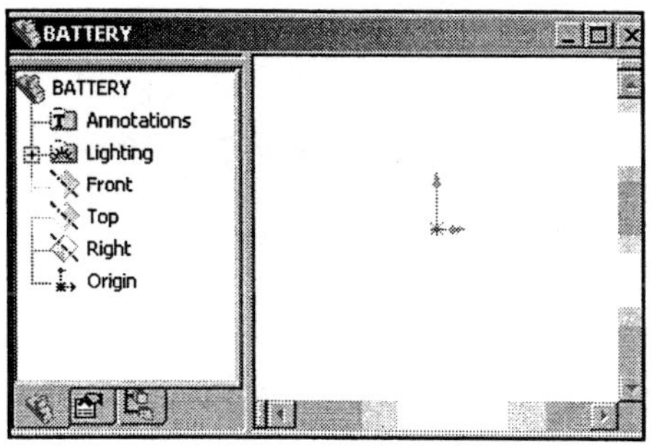

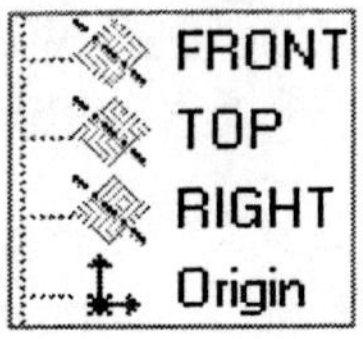

Display the planes in the Front view.

17) Display the Front plane. Click **Front** from the FeatureManager. Display the Top plane. Click **Top**. Display the Right plane. Click **Right**.

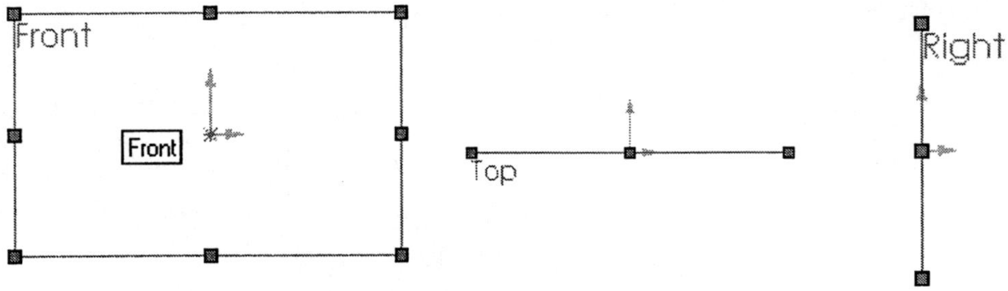

18) Display the planes in an Isometric view. Click **Isometric** from the Standard Views toolbar. Click **Front** from the FeatureManager. Display the Top plane. Click **Top**. Display the Right plane. Click **Right**.

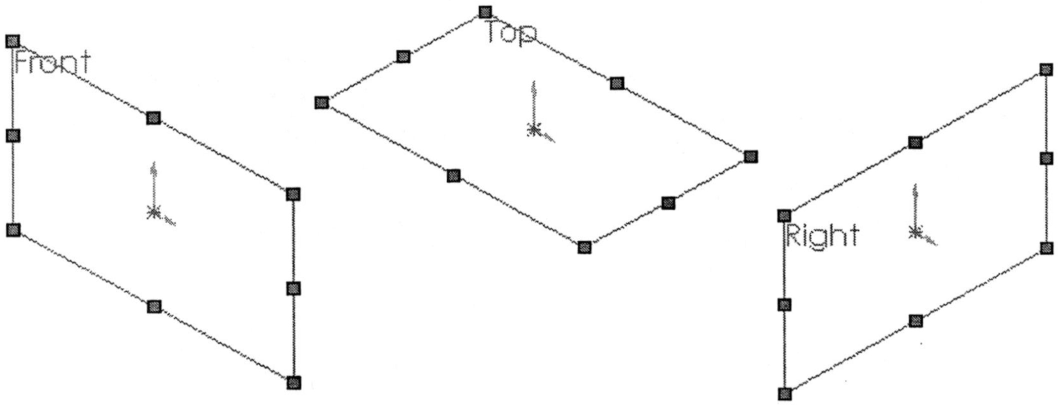

Create the BATTERY – Extruded Base Feature

The BATTERY utilizes an Extruded Base feature. The Sketch requires a Sketch plane and a 2D profile. The Sketch uses the Top plane. The 2D profile is a rectangle. The rectangle is centered about the Origin. The rectangle is extruded perpendicular to the Sketch plane.

Select the Top plane for the Sketch plane.

19) Select the Sketch plane. Click **Top** plane from the Feature Manager.

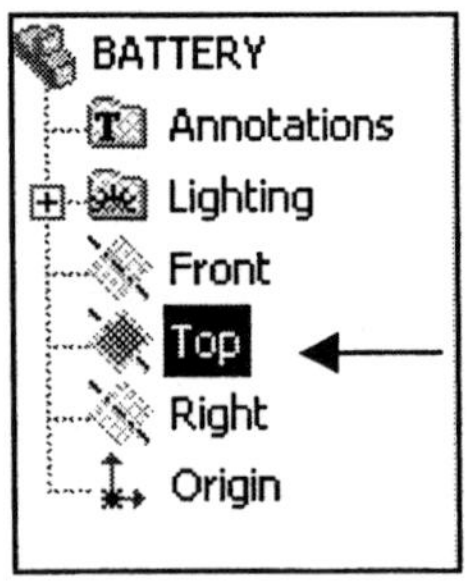

Create a new Sketch.

20) Click **Sketch** from the Sketch toolbar . The Sketch opens on the Top plane. The Origin represents the intersection of the Front, Top and Right planes. The Top plane is displayed as a horizontal line through the Origin.

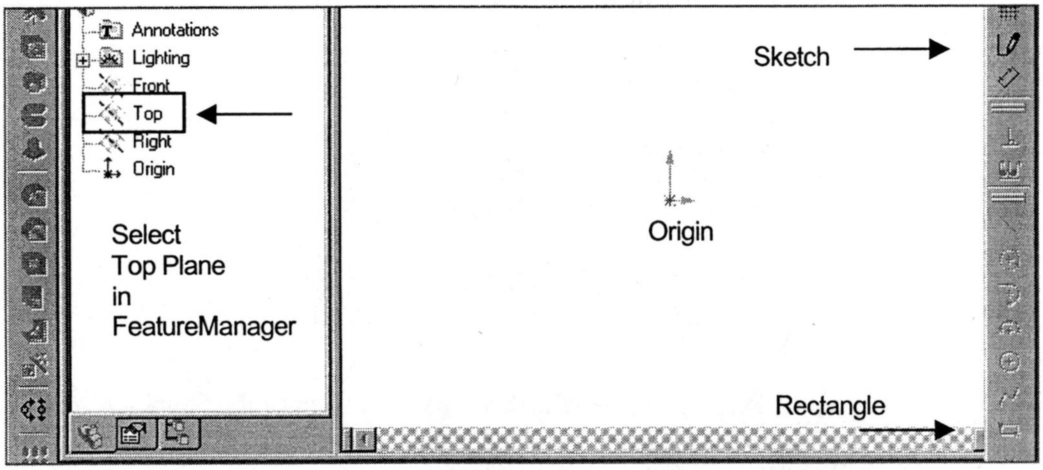

21) Display the Top view. Click **Top** from the Standards View toolbar. The Top plane is displayed.

Sketch a Profile.

22) Sketch the rectangle profile. Click **Rectangle** from the Sketch Tools toolbar. The mouse pointer is displayed as a pencil with a rectangle. Click the **first point** in the lower left quadrant. Click the **second point** in the upper right quadrant. The Origin is approximately in the middle of the Rectangle.

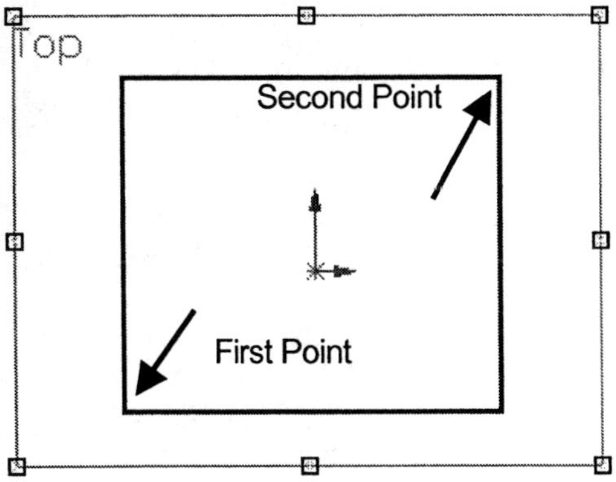

23) Sketch the Centerline.

Click **Centerline** from the Sketch Tools toolbar. When the mouse pointer is located on the Origin, the pointer displays a pencil with an orange point, . Sketch a diagonal centerline from the **upper left corner** to the **lower right corner**. The endpoints of the centerline are coincident with the corner points of the Rectangle.

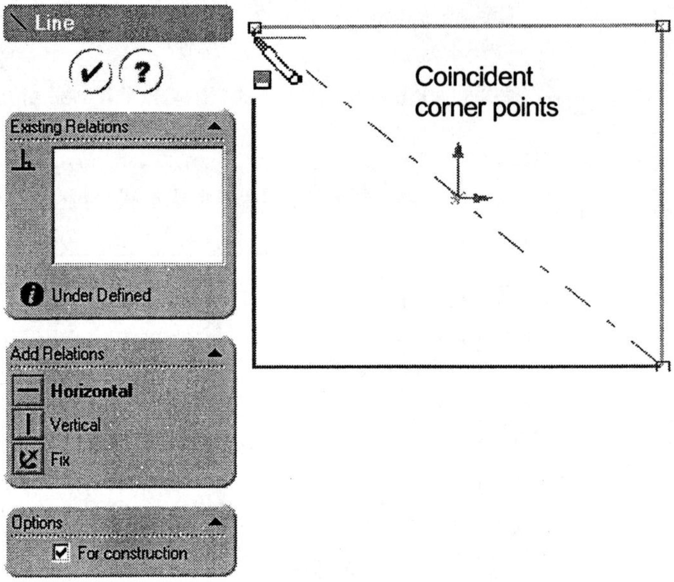

The Line PropertyManager is displayed to the left of the Graphics window.

Dimensions add location and size information. All models require dimensions for manufacturing. Dimensions are not required for feature creation in SolidWorks.

Add dimensions.

24) Add a dimension. Click **Dimension** from the Sketch toolbar. The mouse pointer changes shape to the dimension symbol, . Select the **top horizontal line.** Drag the **mouse pointer** off the Sketch. Position the dimension text. Click the **text location** above the horizontal line. Enter **2.700**, **[68.58]** for width. Click the **green Check mark** .

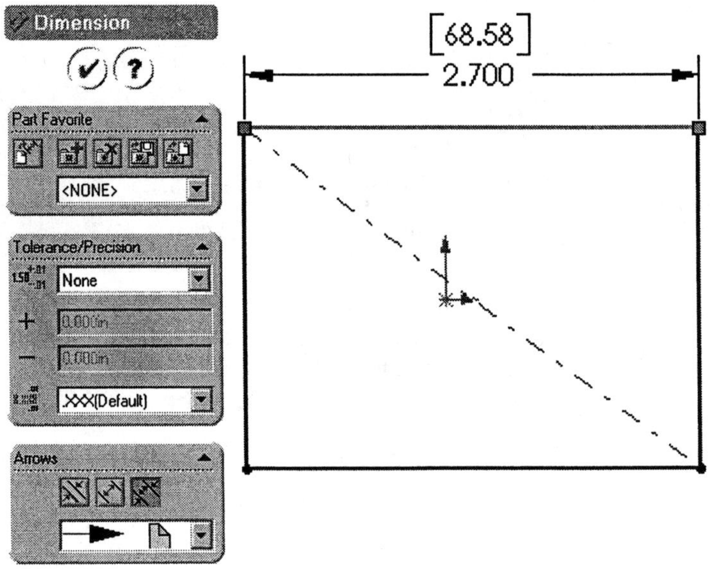

The Dimension PropertyManager is displayed to the left of the Graphics window. Geometric relations are dimensional relationships between one or more creations. Caution should be used. All four sides may appear equal and the Origin may appear centered, but they are not!

25) Add Geometric Relations. Click **Select** . Add a midpoint relation. Hold down the **Ctrl** key. Click the diagonal centerline, **Line5**. Click the **Origin**. Release the **Ctrl** key. Click the **Midpoint** button. Click **OK.**

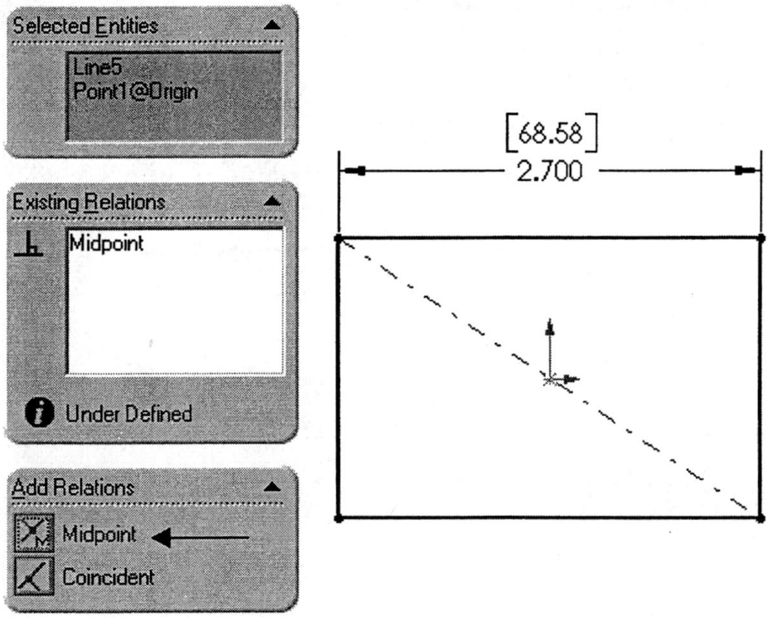

Note: The Line# may be different than the numbers above. The Line# is dependent on the Line# order creation.

26) Add an equal relation. Hold down the **Ctrl** key. Click the top horizontal profile line, **Line1**. Click the left vertical profile line, **Line2**. Click the **Equal** button. Release the **Ctrl** key.

Click **OK**. The black Sketch is fully defined.

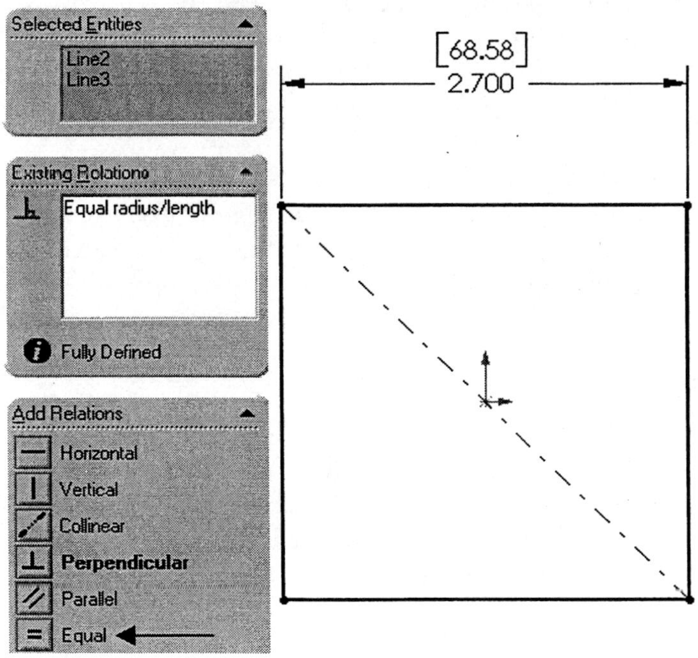

27) Display the sketch relations. Click **Display/Delete Relations** from the Sketch Relations toolbar. The Distance relation is created from a dimension. The Vertical and Horizontal relations are created from the Rectangle Sketch tool. Click **OK**.

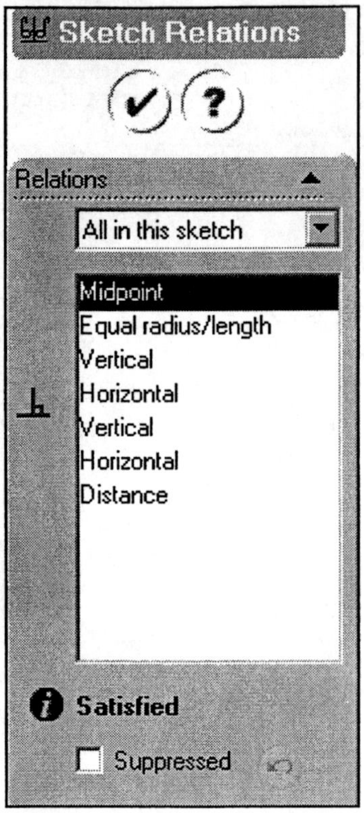

28) Click **Select**. Click a **vertical line.** Individual geometric relations are displayed in the Existing Relations text box.

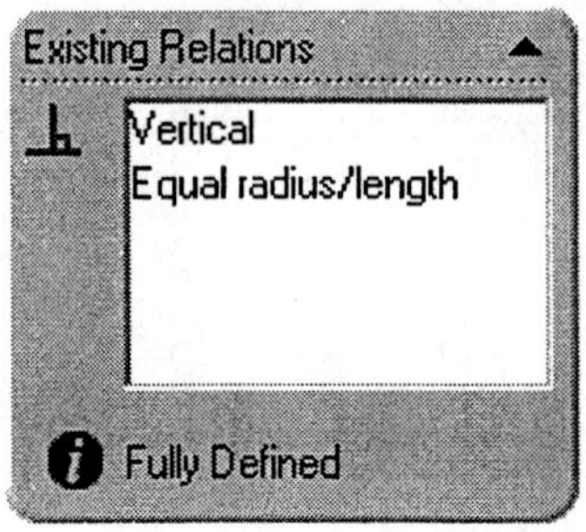

Color and cursor feedback is used in SolidWorks to aid in the sketching process.

In a fully defined Sketch, all entities are displayed in black. A fully defined Sketch has a defined position, dimensions and or relationships.

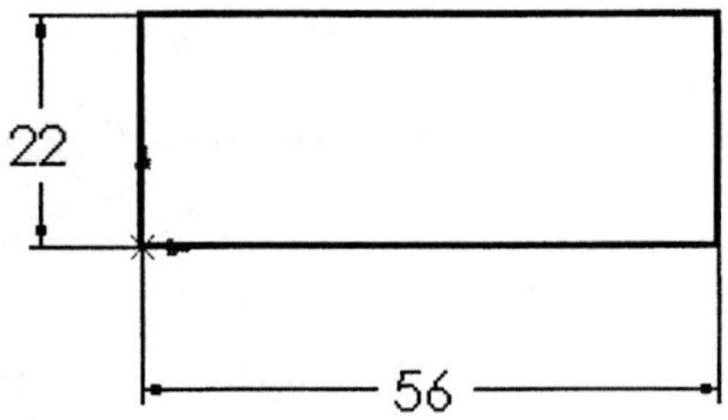

In a under defined Sketch, the entities that require position, dimensions or relationships are displayed in blue. Click and drag under defined geometry to modify the Sketch.

In an over defined Sketch, there is geometry conflict with the dimensions and or relationships. In an over defined Sketch, entities are displayed in red.

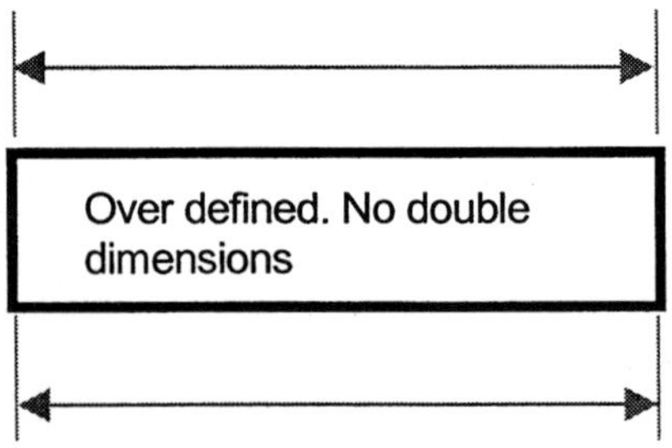

Extrude the Sketch.

29) Extruding the Sketch adds depth to the rectangle. Click **Extruded Boss/Base** in the Features toolbar. The Extrude PropertyManager is displayed. The extruded Sketch is previewed in an Isometric view. The preview displays the direction of the feature creation. Specify the depth of the extrusion. Blind is the default Type option. Drag the **handle view arrow** in the extruded direction. View the mouse pointer feedback for depth. The depth of the extrusion increases respectively with the depth spin box value. Enter **4.700, [119.38]** for Depth.

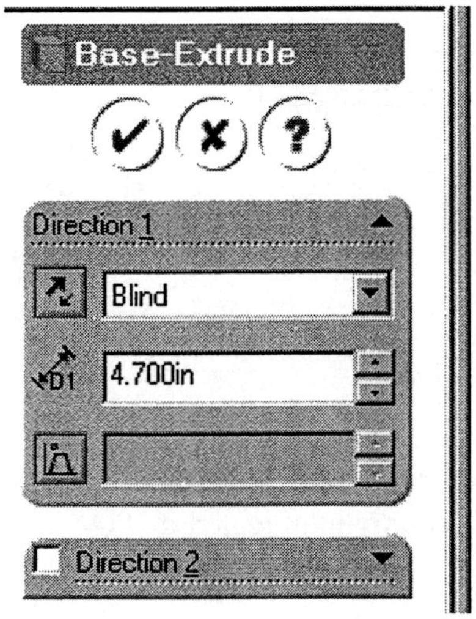

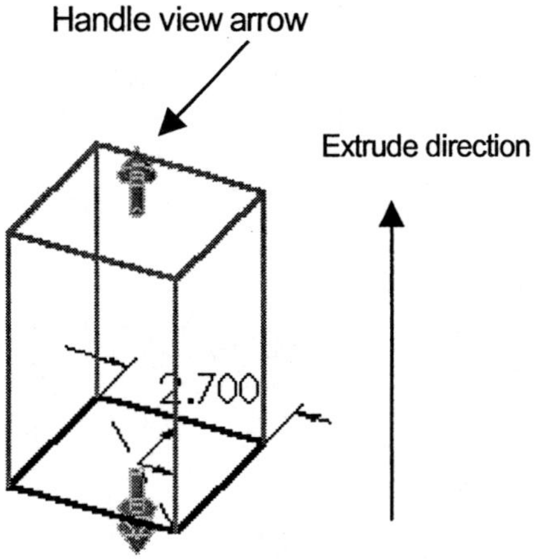

30) Display the Base-Extrude feature. Click **OK**.

Note: If the mouse pointer is positioned inside the Graphics window, then select the view direction arrow before entering a Depth.

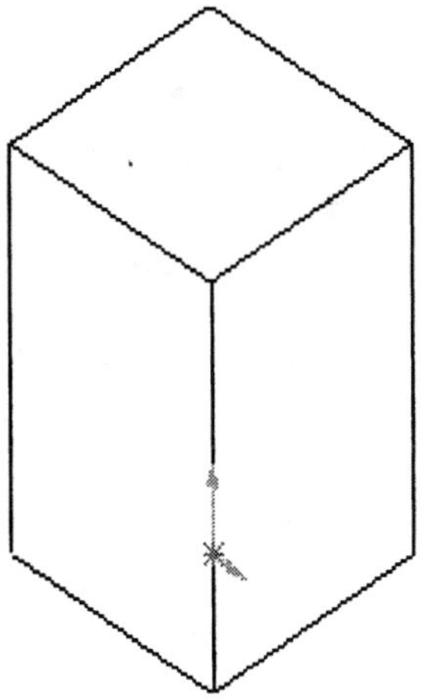

31) Fit the part to the Graphics window. Click **Zoom to Fit**.

32) Save the BATTERY. Click **Save**.

Create the BATTERY - Modify Dimensions

The BATTERY dimensions are not correct. The Extruded height is 4.700, [119.39]. Modify the Extruded height to 4.100, [104.14].

Modify the BATTERY.

33) Double-click on the **Base-Extrude** feature. The Dimension PropertyManager is displayed. Change the height dimension. Double-click **4.700**, [**119.38**]. Enter **4.100**, [**104.14**] in the Modify dialog box. Click the **Check mark** √ from the Modify dialog box.

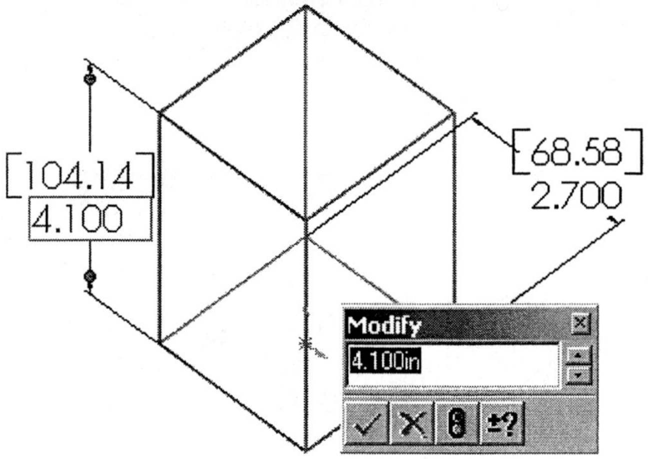

The Check saves the current value and exits the Modify dialog box. Note:

- Restore restores the original value and exits.
- Rebuild rebuilds the model with the current value.
- Reset modifies the spin box increment.

34) Update the dimension. Click **Rebuild** from the Standard toolbar.

35) Fit the model to the screen. Press the **f** key.

An Extruded-Base feature is named Base-Extrude in the FeatureManager design tree.

The Plus Sign icon indicates that additional feature information is available.

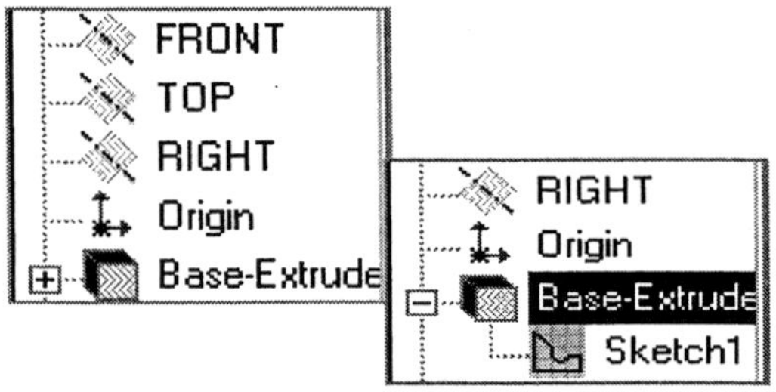

36) Display the FeatureManager. Click **FeatureManager**. Click the **Plus Sign** of the Base-Extrude feature. Sketch1 is the name of the Sketch used to create the Base feature. The Minus Sign indicates that feature information is expanded. Collapse the feature. Click the **Minus Sign**.

Display Modes and View Modes

Display modes create models in a variety of views. Display the GUIDE. Click the following icons from the View Display toolbar to assist in model visualization.

View different display modes.

37) Click **Wireframe** .

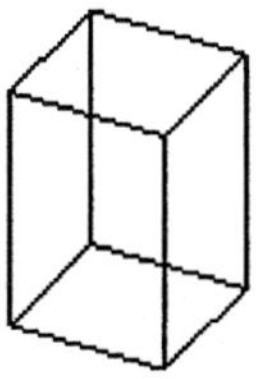

38) Click **Hidden In Gray** .

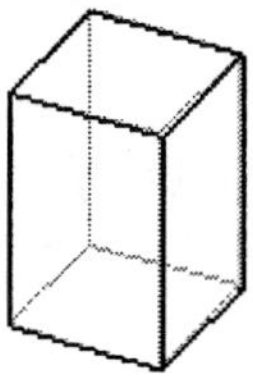

39) Click **Hidden Lines Removed** .

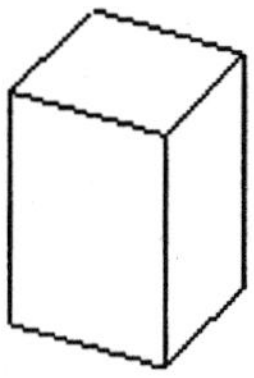

40) Click **Shaded** .

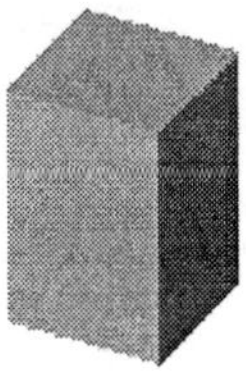

Parts take longer to display as file size increases. Use Fast HLR/HLG  to display models quickly.

Fast HLR/HLG  toggles between coarse and fine model display for the Hidden Lines Removed (HLR) and Hidden Lines Gray (HLG) mode.

Depending on your computer configuration and the part, the Hidden Lines Removed option will take longer to display than the Shaded option. The text examples use different display modes to illustrate features.

The view modes manipulate the model in the Graphics windows.

Display different view modes.

41) Click **Hidden In Gray**. Click the following icons from the View toolbar to assist in view visualization.

42) Click **Zoom to Fit**. Display the full size of the part in the current window.

43) Click **Zoom to Area**. Select two **opposite corners** of a rectangle to define the boundary of the view. The defined view fits to the current window.

44) Click **Zoom In/Out**. Drag the **mouse pointer** upward to zoom in. Drag the **mouse pointer** downward to zoom out. Press the lower case **z** key to zoom out. Press the upper case **Z** key to zoom in.

45) Click **Zoom to Selection**. Click the **front edge**. The selected geometry fills the current window. Note: Select on a Vertex, an Edge or a feature.

46) Click **Rotate**. Drag the **mouse pointer** to rotate about the screen center. Note: Use the arrow keys to rotate in 15-degree increments.

47) Click **Pan**. Drag the **mouse pointer** up, down, left or right. The model scrolls in the direction of the mouse.

48) Right-click in the **Graphics window** area to display the zoom options.

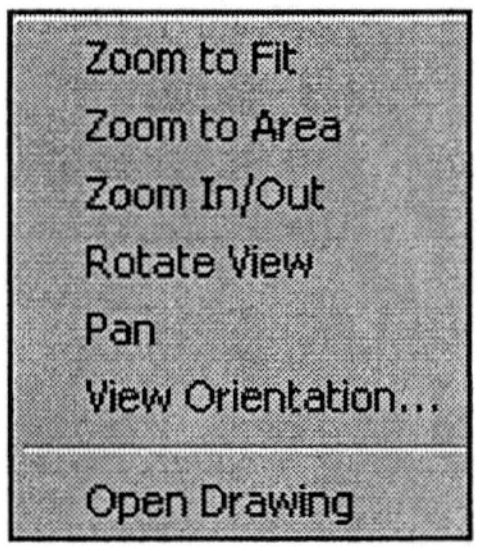

49) Return to the standard Isometric view. Click **Zoom to Fit**.

50) Click **Isometric** from the Standards View toolbar.

View Orientation

The view orientation defines the preset position of the model in the Graphics window. The Standard View toolbar displays eight view options: Front, Back, Left, Right, Top, Bottom, Isometric and Normal To.

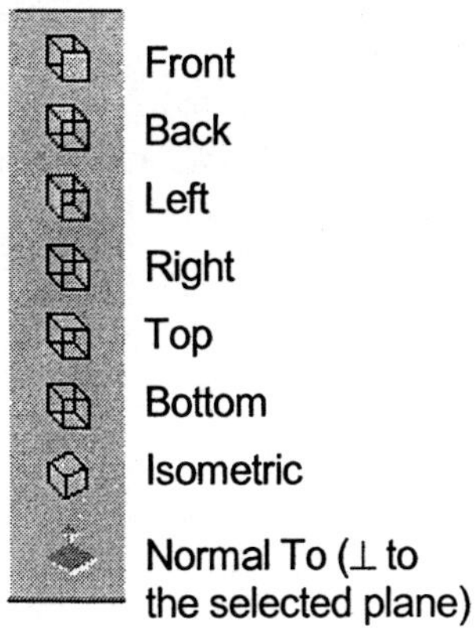

The Isometric view displays the part in 3D with two equal projection angles. The Normal To view displays the part ⊥ to the selected plane.

Different View orientation is selected for sketches, parts and assemblies.

Selecting Geometry.

51) Position the mouse pointer on the **front face** of the PLATE. Drag the **mouse pointer** over the part geometry. The pointer displays various items depending upon geometry: Line, Face, Point or Vertex. Example:

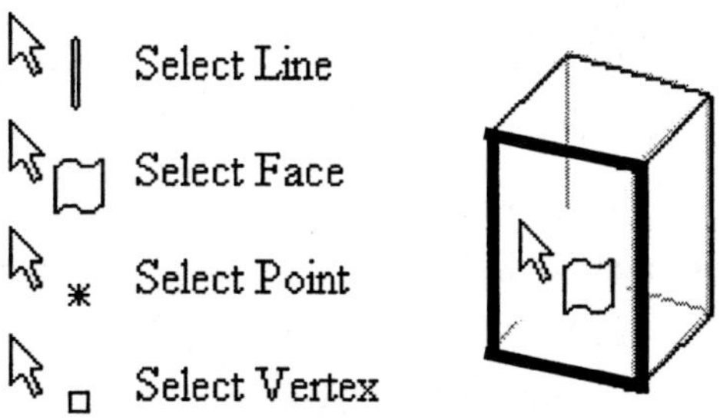

If the Sketch is too large in the Graphics window, you may not be able to view the dimension. Press the f key to fit the Sketch to the screen.

Create the BATTERY – Fillet Feature

Sharp edges and square corners can create part stress. The Fillet feature removes sharp edges, strengthens corners and or cosmetically improves appearance. Fillets blend inside and outside surfaces.

On castings and plastic modeled parts, implement Fillets and Rounds into the initial design.

If you are uncertain of the exact radius value, input a small test radius of 0.05inch or 1mm. It will take less time for a supplier to modify an existing dimension than to create one.

Minimize the number of Fillet Radius created in the FeatureManager. Combine Fillets and Rounds that have a common radius. Select all Fillet/Round edges. Add them to the Items to Fillet list in the Fillet/Round feature dialog box.

The vertical sides on the BATTERY are rounded. Use the Fillet feature to round the 4 side edges.

Create the Fillet feature.

52) Display the part's hidden edges in gray. Click **Hidden In Gray** from the View toolbar.

53) Create a Fillet feature. Click **Fillet** from the Feature toolbar. Click the **4 vertical edges**. Enter **.500**, **[12.7]** for Radius. Display the Fillet feature. Click **OK**.

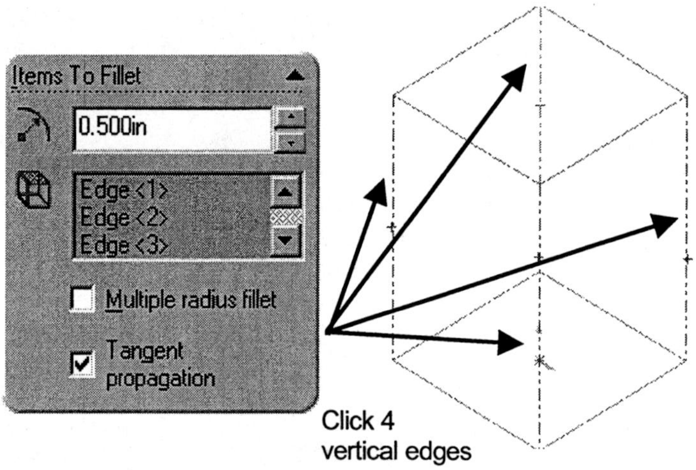

54) Rename **Fillet1** to **Side-Fillets** in the Feature Manager.

55) Save the BATTERY. Click **Save**.

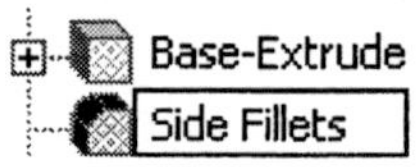

Create the BATTERY - Extruded Cut Feature

The Extruded Cut feature removes material. An Offset Edge takes existing geometry, extracts it from an edge or face and locates it on the current sketch plane. Offset the existing Top face. Create a Cut feature.

Create the Extruded Cut feature.

56) Select the Sketch plane. Click the **Top** face.

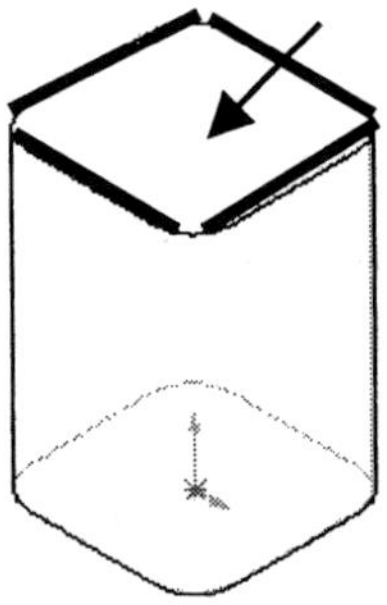

57) Display the face. Click **Top** from the Standards View toolbar.

58) Sketch the profile. Click **Sketch**.

59) Offset the existing geometry from the boundary of the Sketch plane. Click **Offset** from the Sketch Tools toolbar. Enter **.150**, **[3.81]** for the Offset distance. Click the **Reverse** check box. The new Offset profile displays inside the original profile. Click **OK** from the Offset Entities dialog box.

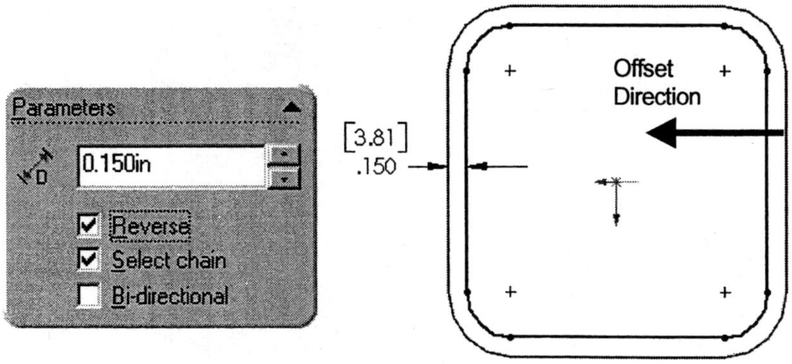

Note: A leading zero is displayed in the spin box. For inch dimensions less than 1, the leading zero is not displayed in the part dimension.

60) Display the profile. Click **Isometric** from the Standards View toolbar.

61) Extrude the Offset profile. Click **Extruded Cut** from the Feature toolbar. Enter **0.200**, **[5.08]** for Depth of Cut. Display Cut-Extrude1. Click **OK**.

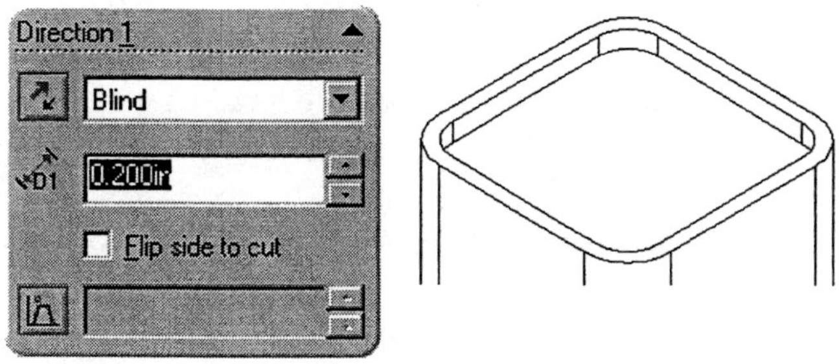

62) Rename **Cut-Extrude1** to **Top-Cut**.

63) Save the BATTERY. Click **Save**.

Create the BATTERY - Fillet Feature on the Top Face

Create fillets for the top outside edges. Use a face to create a constant radius Fillet feature. The top narrow face is too small to select. Use the Face Selection Filter to select faces. Turn the filters off to select geometry.

Create the Fillet feature on the top face.

64) Display the Selection Filter toolbar. Click **View** from the Main menu. Click **Tools**, **Selection Filter**.

65) Create the Fillet. Click **Face Filter** from the Selection Filter toolbar.

66) Click the **top thin face**. Select **Fillet** from the Feature toolbar. Face<1> is displayed in the Edge fillet items box. Click **Constant Radius** for Fillet Type. Enter **0.050**, [**1.27**] for Fillet Radius.

67) Display the Fillet on the inside and outside top edges. Click **OK**.

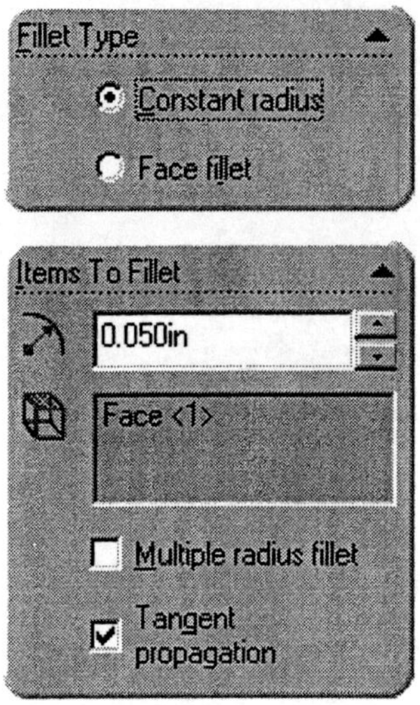

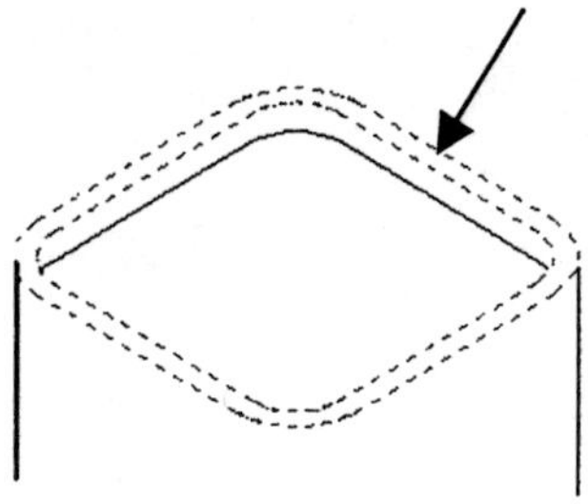

68) Turn the Face Filter off. Click **Face Filter**.

69) Rename **Fillet2** to **Top Face Fillet**.

70) Save the BATTERY. Click **Save**.

Note: Do not select a Fillet radius which is larger that the surrounding geometry.

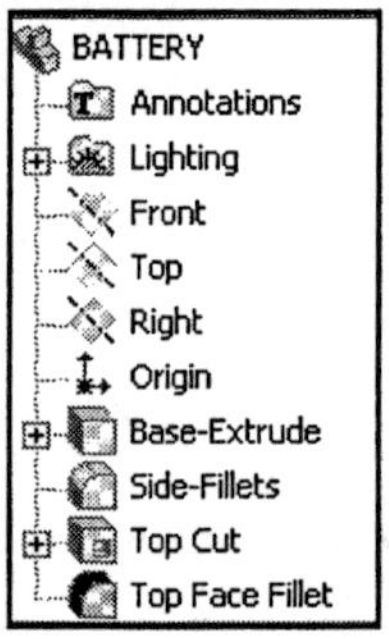

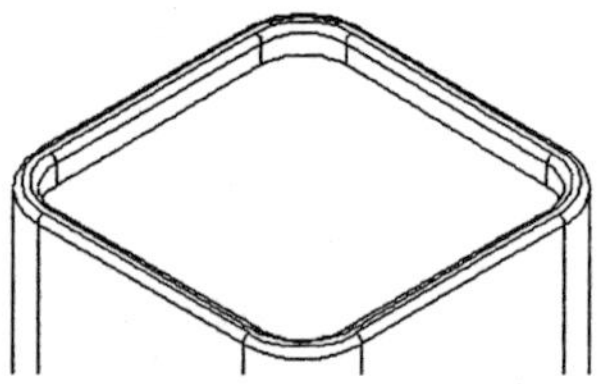

Example: The top edge face width is .150, [3.81]. The Fillet is created on both sides of the face. A common error is to enter a Fillet too large for the existing geometry. A minimum face width of .200, [5.08] is required for a Fillet radius of .100, [2.54].

The following error occurs went the Fillet radius is too large for the existing geometry:

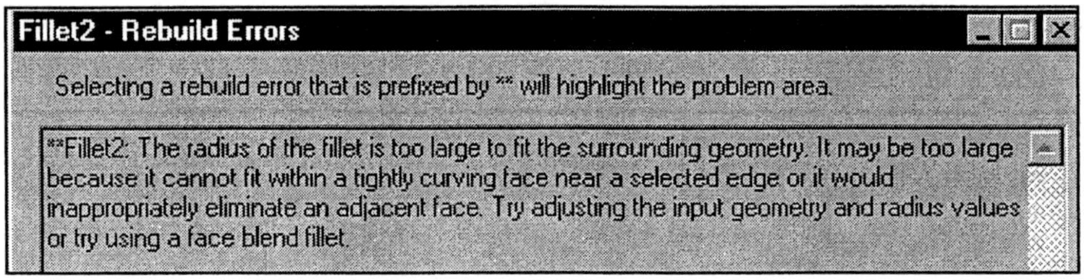

Avoid the Fillet Rebuild error. To avoid this error, reduce the Fillet size or increase the face width.

Create the BATTERY - Extruded Boss Feature

Two Battery Terminals are required. Conserve design time, represent the terminals as cylindrical Extruded Boss feature. An Extruded Boss feature adds material.

Create the Extruded Boss feature.

71) Select the Sketch plane. Click the **face** of the Top-Cut feature.

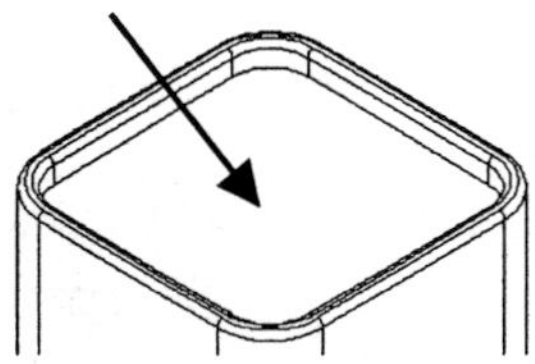

72) Create the Sketch. Click **Sketch**.

73) Display the Sketch plane. Click **Top** from the Standards View toolbar.

74) Sketch the Profile. Click **Circle** from the Sketch Tools toolbar. Create the first point. Click the **center point** of the circle coincident to the Origin. Create the second point. Drag the **mouse pointer** to the right. Release the **left mouse button**.

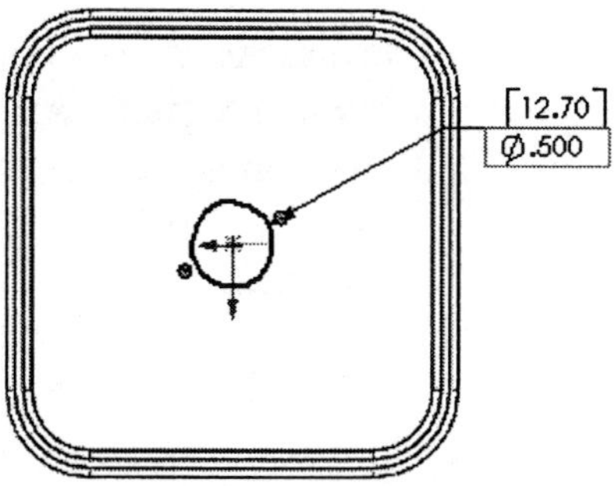

75) Click **Dimension**. Select the **circumference** of the circle. Click the **text location**. Enter **.500, [12.7]**.

76) Copy the sketched circle. Click **Select**. Hold the **Ctrl** key down. Click the **circumference** of the circle. Drag the **circle** to the upper left quadrant. Create the second circle. Release the **mouse button**. Release the **Ctrl** key.

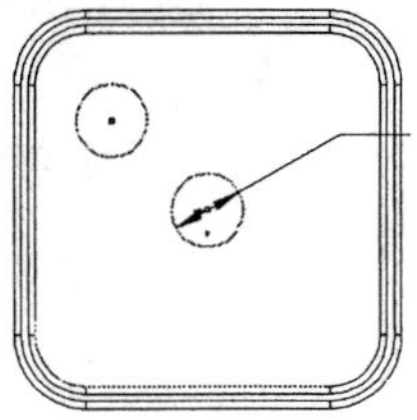

77) Add an equal relation. Click **Select**. Hold down the **Ctrl** key. Click the **circumference of the first circle**. Both circles are selected. Click **Equal**. Release the **Ctrl** key. Click **OK**.

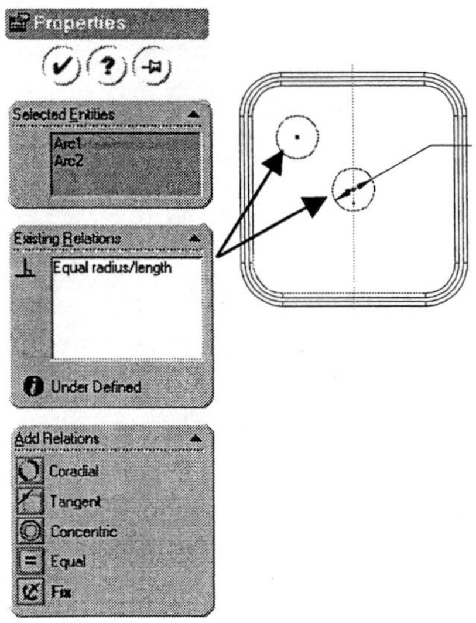

By default, the PropertyManager is displayed when the Sketch is active. The FeatureManager is displayed when the Part is active.

The dimension between the center points is critical. Dimension the distance between the two center points with an aligned dimension.

78) The Right plane is the dimension reference. Right-click the **Right** plane from the FeatureManager. View the plane. Click **Show**.

79) Add a dimension. Click **Dimension**. Click the **two center points** of the circles. Drag the **dimension text** off the profile. Release the **mouse button**. Enter **1.000, [25.4]** for the aligned dimension.

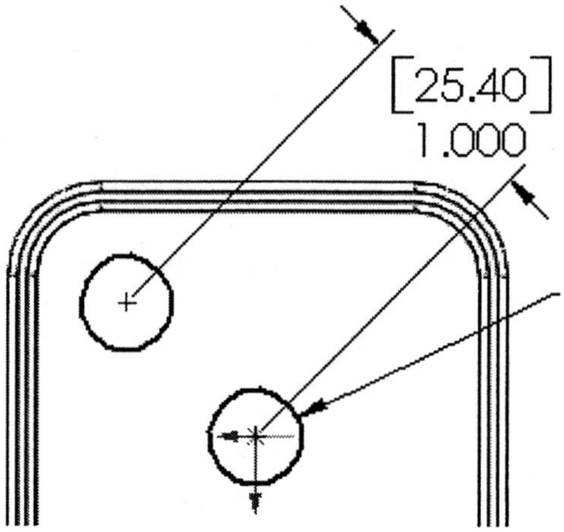

The dimension text toggles between linear and aligned. An aligned dimension is created when the dimension is positioned between the two circles.

80) Create an angular dimension. Click **Centerline**. Sketch a centerline between the **two circle center points**.

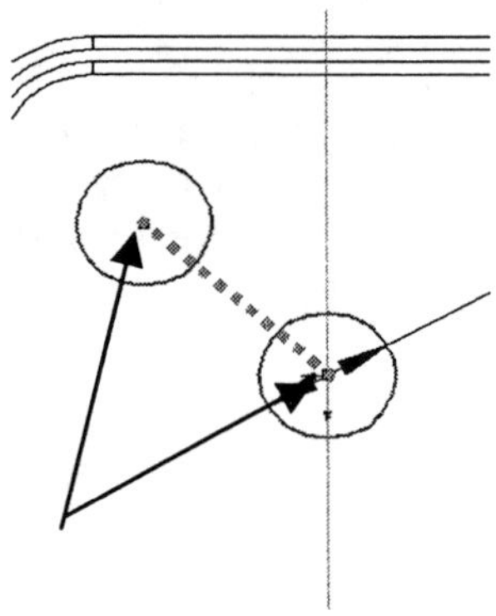

81) Create an acute angular dimension. Click **Dimension** . Click the **centerline** between the two circles. Click the **Right** plane. Drag the **dimension text** between the centerline and the Right plane off the profile. Release the **mouse button**. Enter **45**.

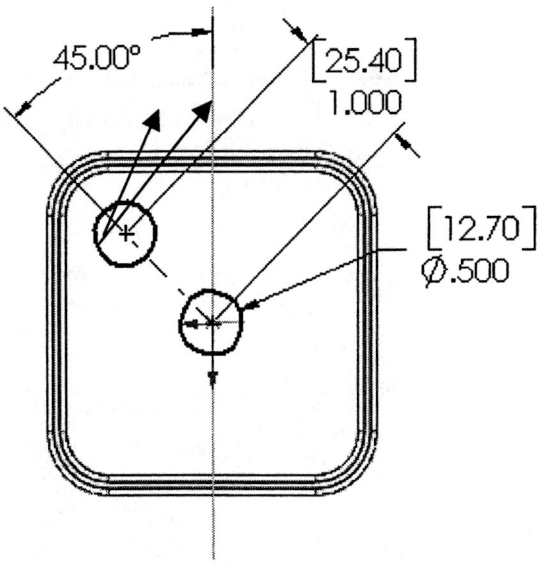

Note: Acute angles are less than 90°. Acute angles are the preferred dimension standard.

The overall battery height is a critical dimension. The battery height is 4.500 inch, [114.30mm]. Calculate the depth of the extrusion:

For Inches: 4.500in. – (4.100in. Base-Extrude height – .200in. Offset cut depth) = .600in. The depth of the extrusion is .600in.

For Millimeters: 114.3mm – (104.14mm Base-Extrude height – 5.08mm Offset cut depth) = 15.24mm. The depth of the extrusion is 15.24mm.

82) Extrude the Sketch. Click **Extruded Boss/Base** from the Feature toolbar. Blind is the default Type option. Enter **.600, [15.24]** for Depth. Create a truncated cone shape for the battery terminals. Click the **Draft ON/OFF** button. A draft angle is a taper. Enter **5** in the Draft Angle text box.

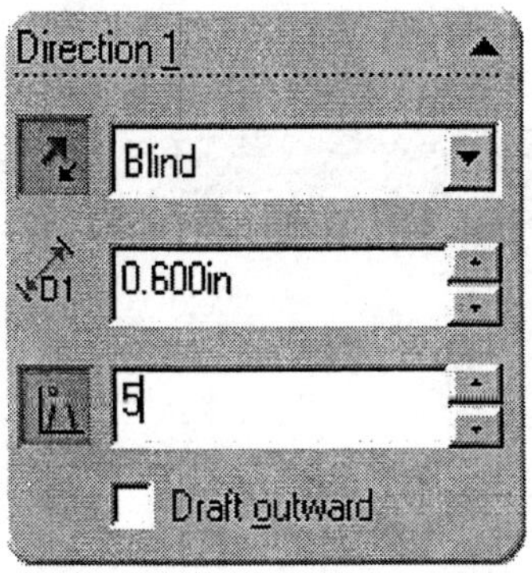

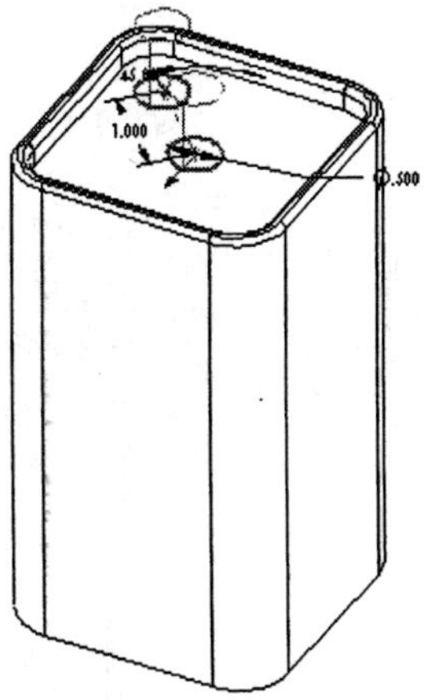

83) Display the Boss-Extrude1 feature. Click **OK**.

84) Rename **Boss-Extrude1** to **Terminals**. Rename **Sketch3** to **Sketch-TERMINALS**.

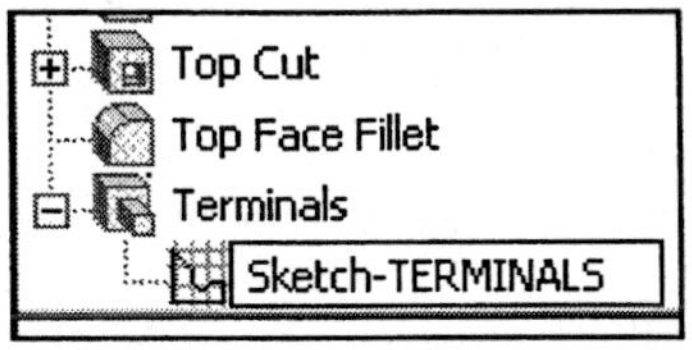

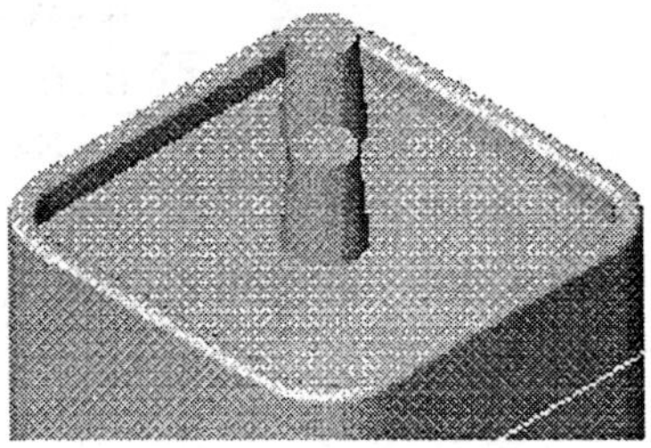

Measure the overall height.

85) Verify the overall height. Click **Tools, Measure** from the Main menu. Click **Right** from the Standard Views toolbar. Click the **top edge** of the battery terminal. Click the **bottom edge** of the battery. The overall height, Y is 4.500inch, [114.3mm]. Click **Close.**

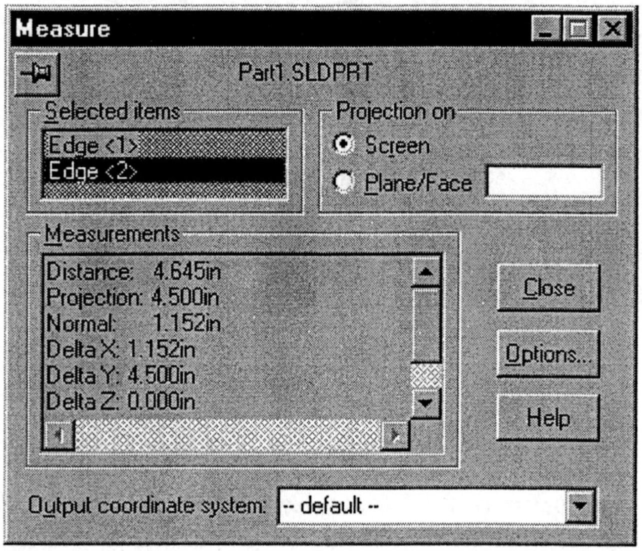

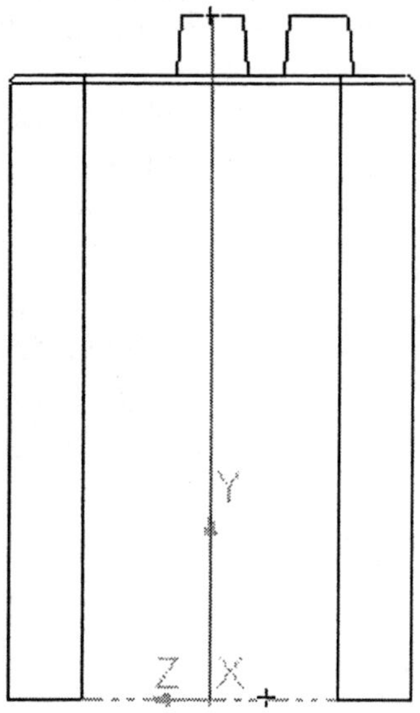

86) Hide all planes. Click **View** from the Main menu. Click **Planes.**

87) Display the Trimetric view. Click **View Orientation**. Double-click **Trimetric.**

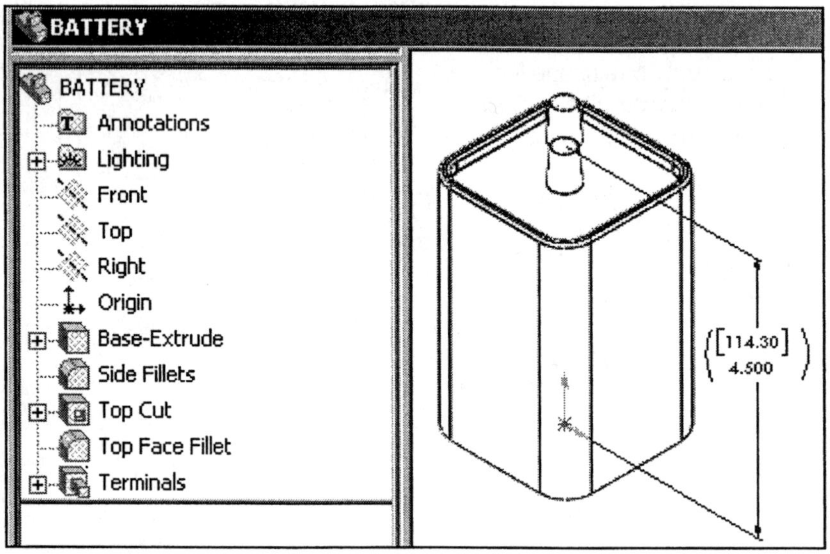

88) Save the BATTERY. Click **Save**.

BATTERY PLATE

The BATTERY PLATE has a variety of design functions. The BATTERY PLATE:

- Aligns the LENS.
- Creates an electrical connection between the SWITCH assembly, BATTERY and LENS.

Design the BATTERY PLATE, Figure 1.12.

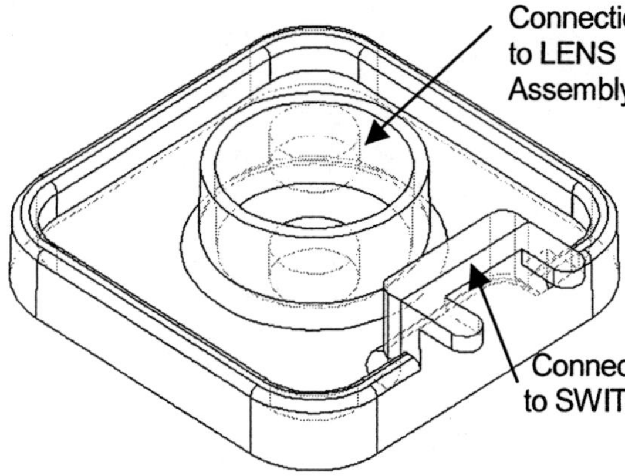

Figure 1.12

Utilize features from the BATTERY to develop the BATTERY PLATE.

BATTERY PLATE Feature Overview

Create the BATTERY PLATE. Modify the BATTERY features. Create two holes from the original sketched circles. Use the Extruded Cut feature, Figure 1.13.

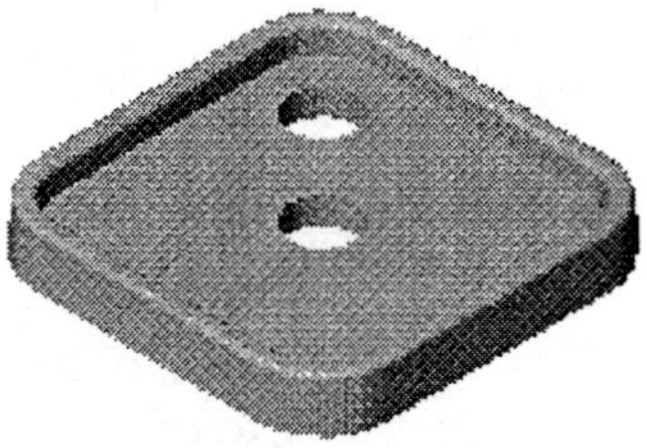

Figure 1.13

Modify the dimensions of the Base feature. Add a 1-degree draft angle.

Note: A sand pail contains a draft angle. The draft angle assists the sand to leave the pail when the pail is flipped upside down.

Create a new Extruded Boss feature. Offset the center circular sketch, Figure 1.14.

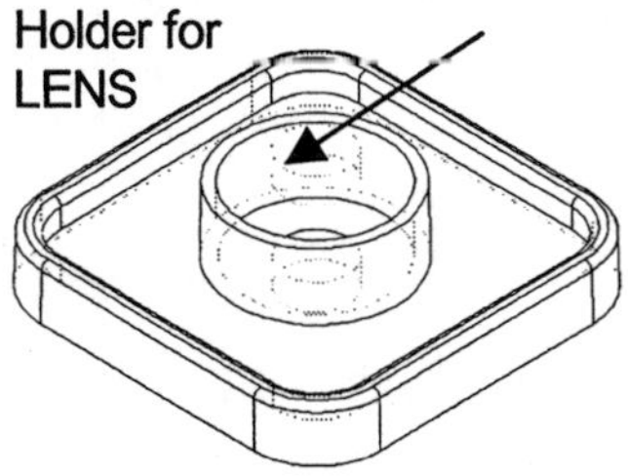

Figure 1.14

The Extruded Boss feature contains the LENS. Create an inside draft angle. The draft angle assists the LENS into the Holder.

Create the first Extruded Boss feature using two depth directions, Figure 1.15. Create the second Extruded Boss feature using sketched mirror geometry, Figure 1.16.

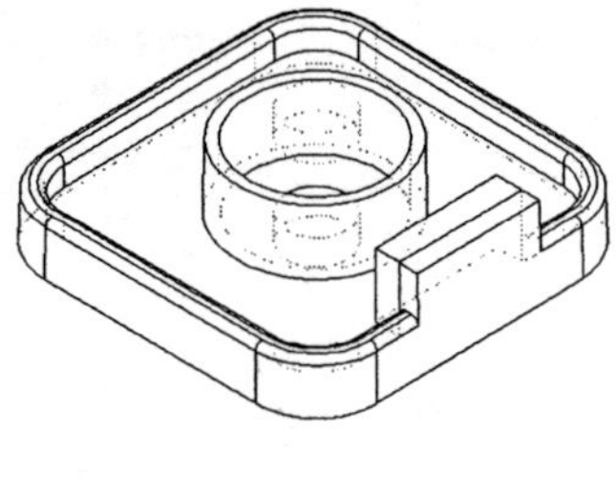

Figure 1.15

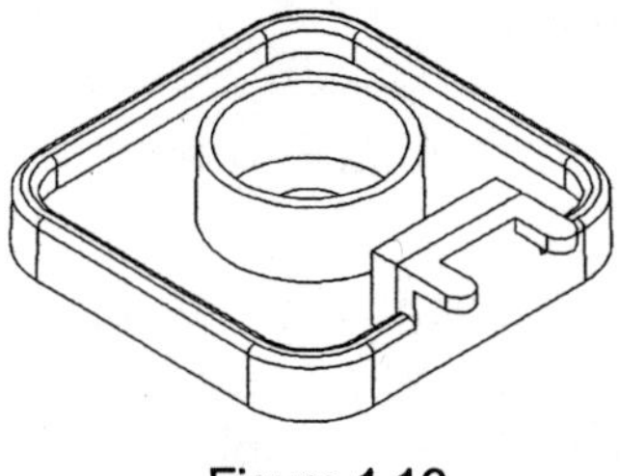

Figure 1.16

Create Face and Edge Fillet features to remove sharp edges, Figure 1.17.

Let's create the BATTERYPLATE.

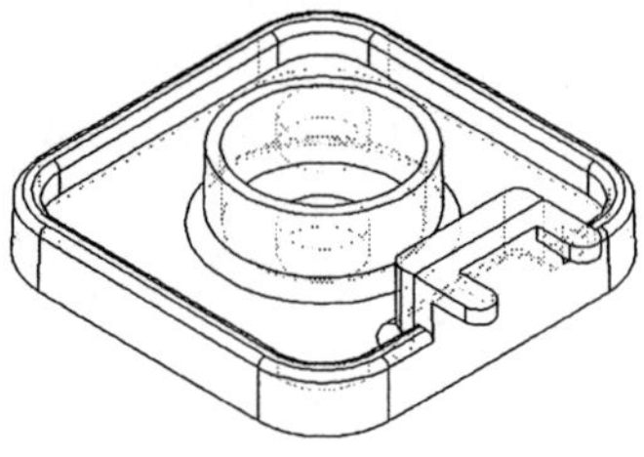

Figure 1.17

Create the BATTERYPLATE

Create the BATTERYPLATE from the BATTERY.

Create a new part from an existing part.

89) Create the BATTERYPLATE from the BATTERY. Click **File**, **SaveAs**. Select **My Documents** folder in the Save In text box. Enter the name of the part. Enter **BATTERYPLATE**. Click **Save**.

The BATTERYPLATE part icon is displayed at the top of the FeatureManager

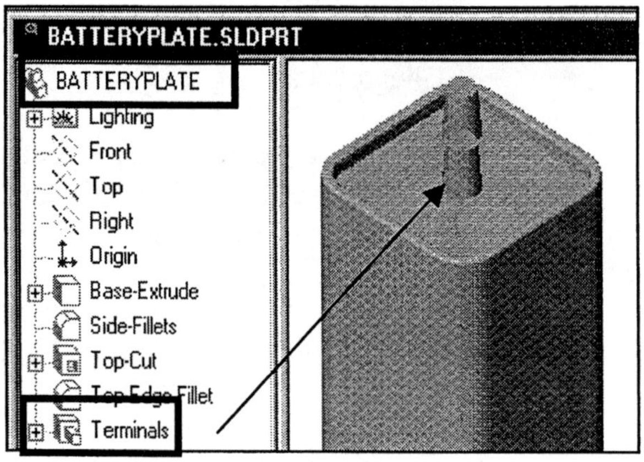

Create the BATTERYPLATE - Delete and Edit Features

Create two holes. Delete the Terminals feature and reuse the circle sketch.

Delete and Edit Features.

90) Remove the Terminals (Extruded Boss) feature. Click **Edit** from the Main menu. Click **Delete**. Click **Yes** from the Confirm Delete dialog box. Do not delete the two-circle sketch, Sketch-TERMINALS.

91) Create an Extruded Cut feature from the two circles. Click **Sketch-TERMINALS** from the FeatureManager. Click **Extruded-Cut**. Click **Through All** for the Depth. Create the cut holes. Click **OK**.

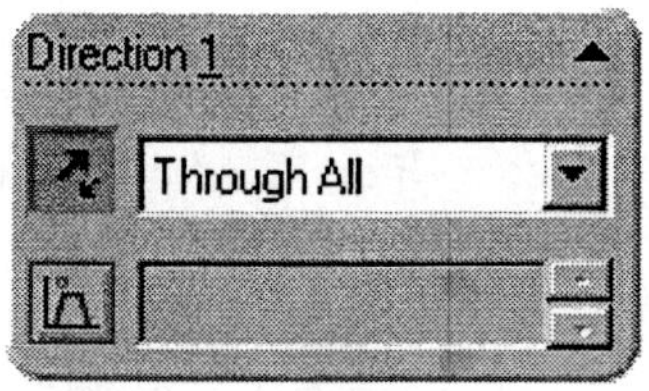

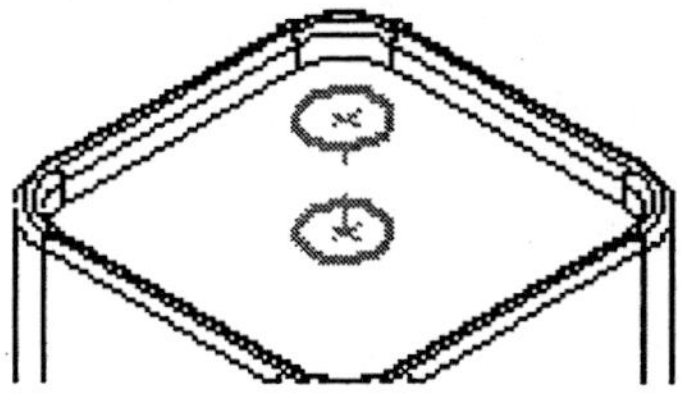

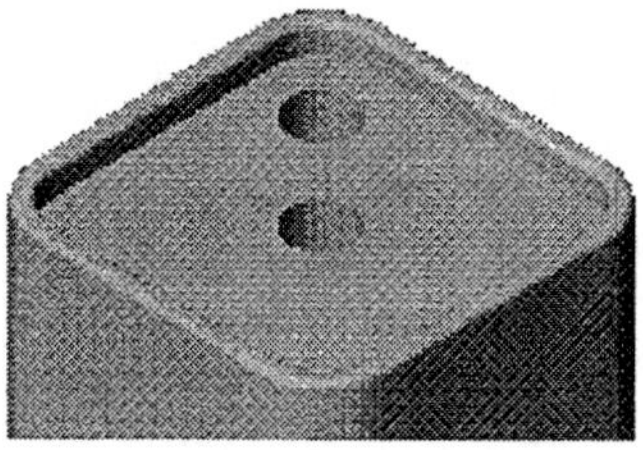

92) Rename **Cut-Extrude** to **Holes**.

93) Save the BATTERYPLATE. Click **Save**.

94) Edit the Base-Extrude feature. Right-click the **Base-Extrude** feature. Click **Edit Definition** from the Pop-up menu. Change the overall Depth to **.400**, [**10.16**]. Click the **Draft ON/OFF** button. Enter **1.00** in the Angle text box.

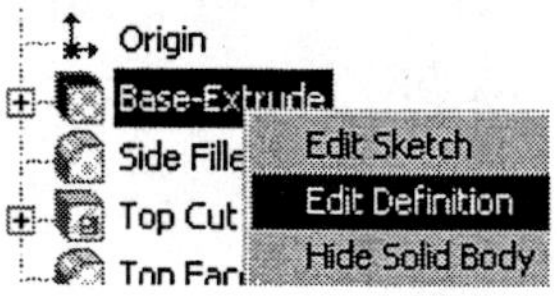

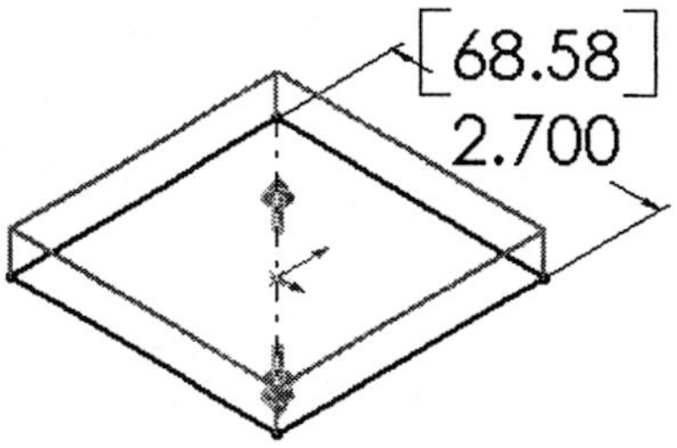

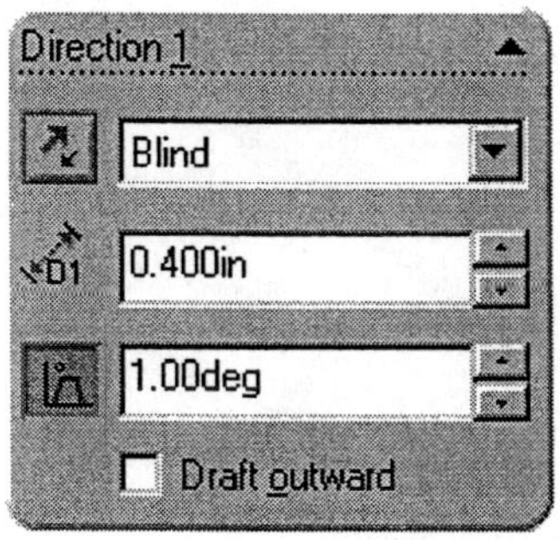

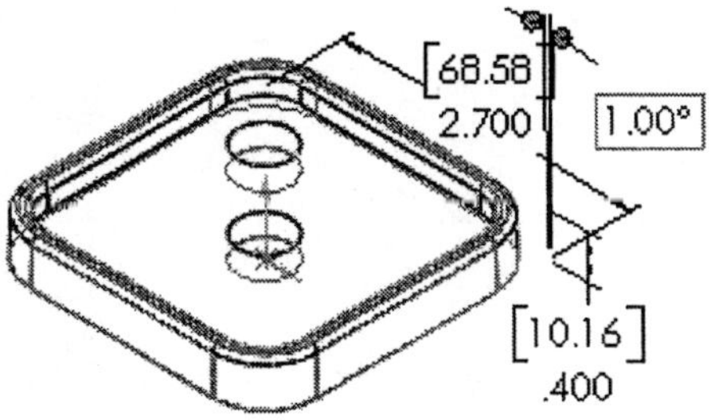

95) Display the modified Base feature. Click **OK**.

96) Save the BATTERYPLATE. Click **Save**.

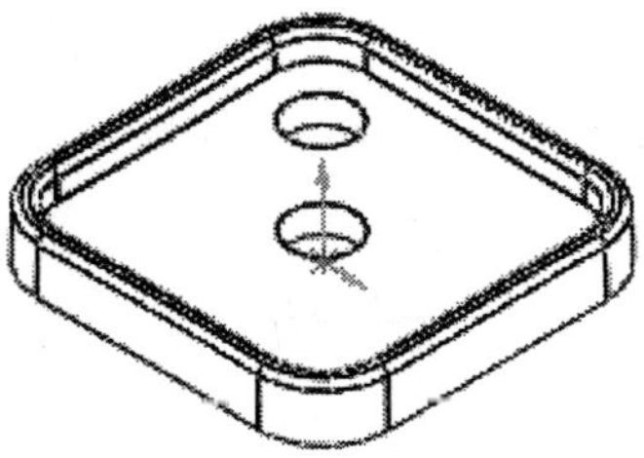

Create the BATTERYPLATE - Extruded Boss Feature

The Holder is created with a circular Extruded Boss feature.

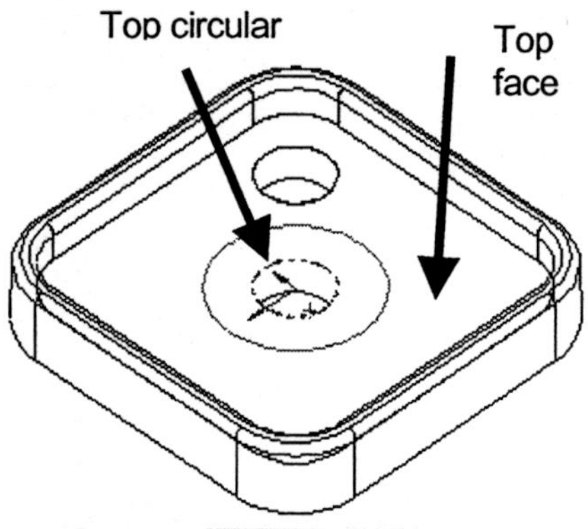

Create the Extruded Boss feature.

97) Select the Sketch plane. Click the **top face**.

98) Create the Sketch. Click **Sketch**. Offset the center circular edge. Click the **top circular edge** of the center Hole feature.

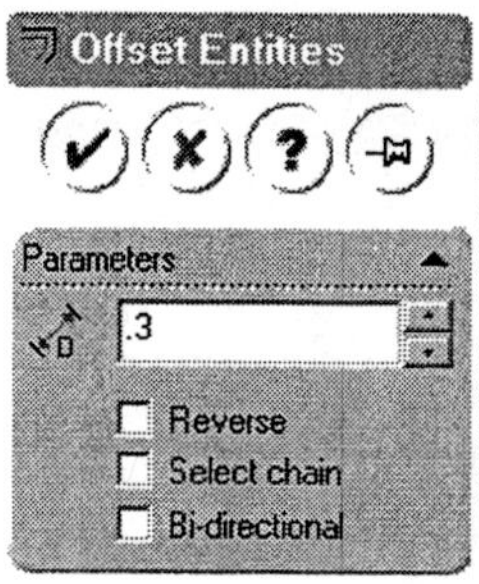

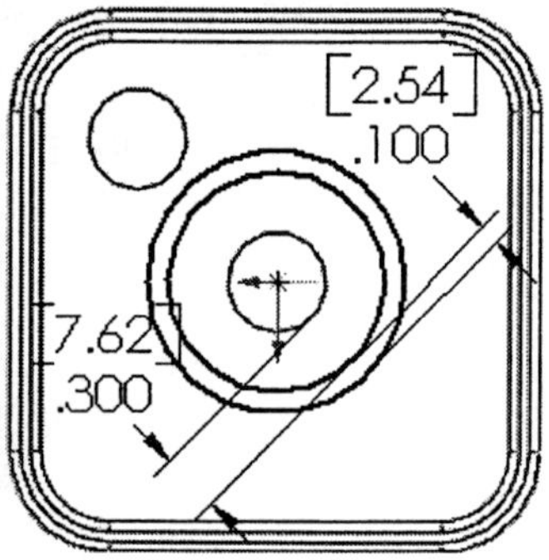

99) Click **Offset Entities**. Enter **.300**, [**7.62**]. Click **OK**. Create the second offset circle. Select the first **offset circle**.

Click **Offset Entities**. Enter **.100**, [**2.54**]. Click **OK**.

100) Extrude the Sketch. Click **Extruded Boss/Base**. Blind is the default Type option. Enter **.400**, [**10.16**] for Depth. Click the **Draft ON/OFF** button. Enter **1** in the Angle text box. Display the Extrude Boss feature. Click **OK**.

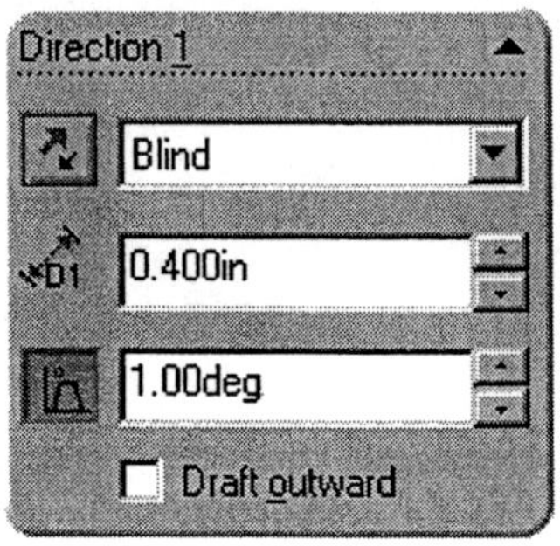

101) Rename **Boss-Extrude** to **Holder**.

102) Save the BATTERYPLATE. Click **Save**.

The outside face tapers inward and the inside face tapers outward when applying the Draft Angle to the two concentric circles.

Draft Angle displayed at 5 degrees

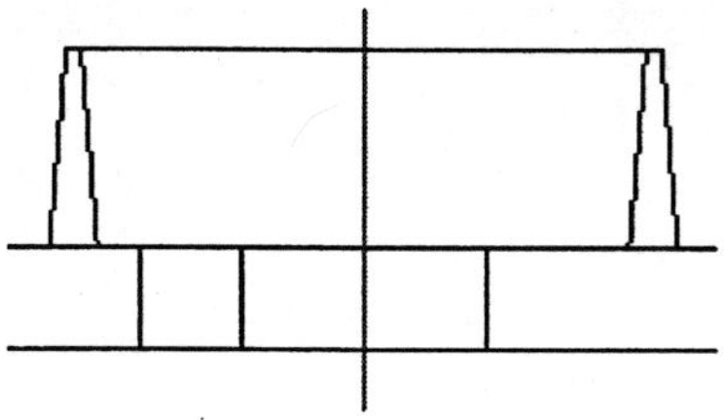

Create the BATTERYPLATE - Extruded Boss Feature

The next two Extruded Boss features are used to connect the BATTERY to the SWITCH. The first Sketch is extruded in two directions. The second Sketch is extruded in one direction. Both sketches utilize symmetry with the Origin and the Mirror Sketch Tool. The sketches utilize smaller dimensions than the current Grid Snap settings. Turn off the Snap to Points setting before you sketch the profiles.

Create the first Extruded Boss feature.

103)Zoom and **Rotate** the view to clearly display the inside right face.

Note: Press the arrow keys to rotate in 15-degree increments.

104) Select the Sketch plane. Click the **inside right face**.

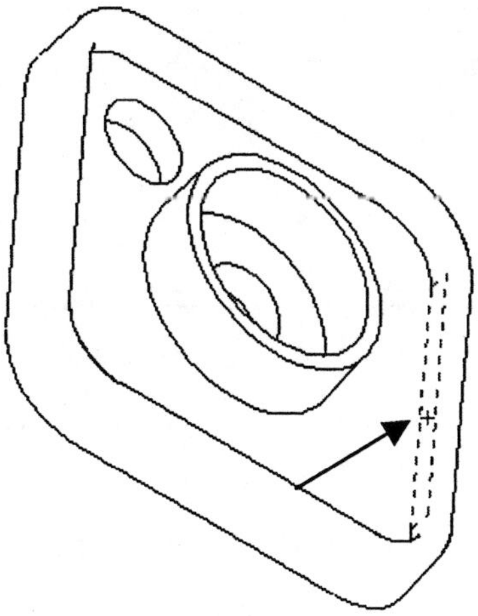

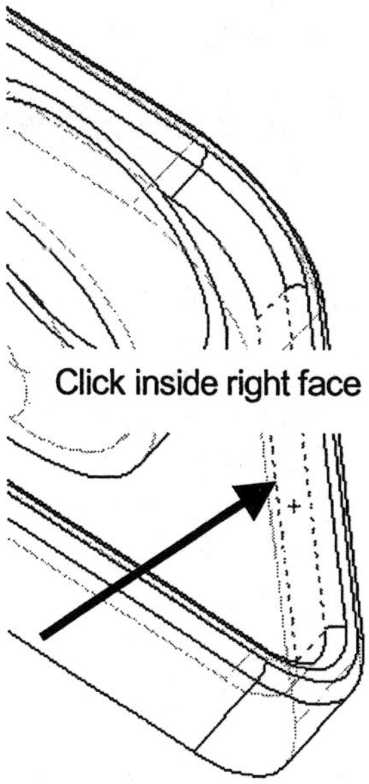

105) Create the Sketch. Click **Sketch** .

106) Display the Right view. Click **Right** . Display **Hidden in Gray** .

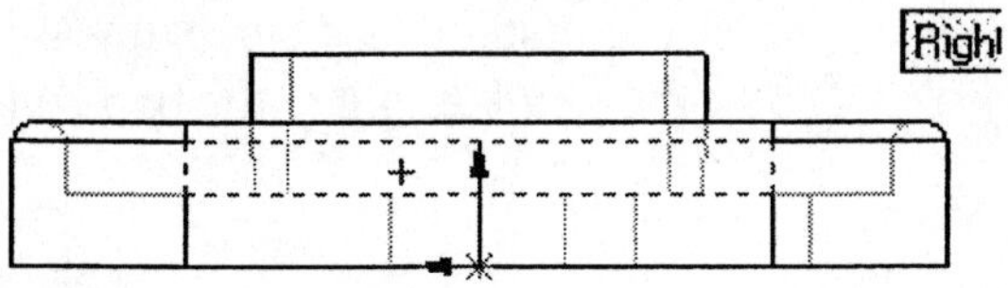

107) Sketch the profile. Click **Rectangle** . Sketch a **rectangle** coincident with the bottom and top edges. The Origin is approximately in the middle of the bottom line.

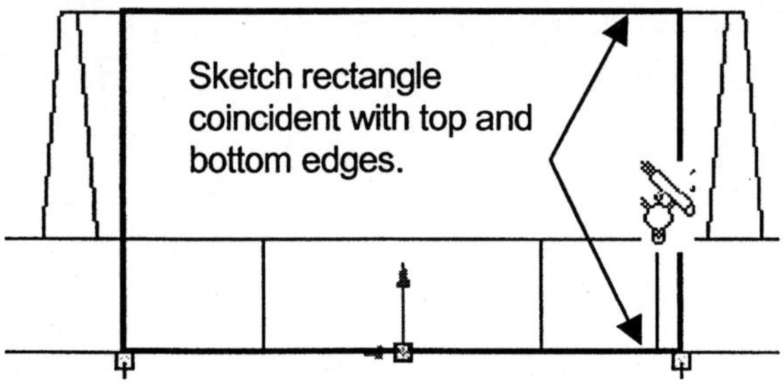

Geometric relationships are captured as you sketch.

The mouse pointer icon displays the following relationships: Horizontal , vertical , coincident , midpoint , intersection , tangent and perpendicular .

Note: If Automatic Relations are not displayed, Click Tools from the Main menu. Click Options, General, Automatic Relations in the Sketch box.

108) Add geometric relations. Click **Select** . Click **Origin** from the FeatureManager. Hold down the **Ctrl** key. Click the **bottom line** of the rectangle. Click **Midpoint** from the Relations dialog box. Release the **Ctrl** key. Click **Apply**.

The sketch is symmetric about the Origin.

109) Dimension the Sketch.

Click **Dimension** . Click the **bottom horizontal line**. Click the **text location**. Enter **1.000, [2.54]**.

The black Sketch is fully defined.

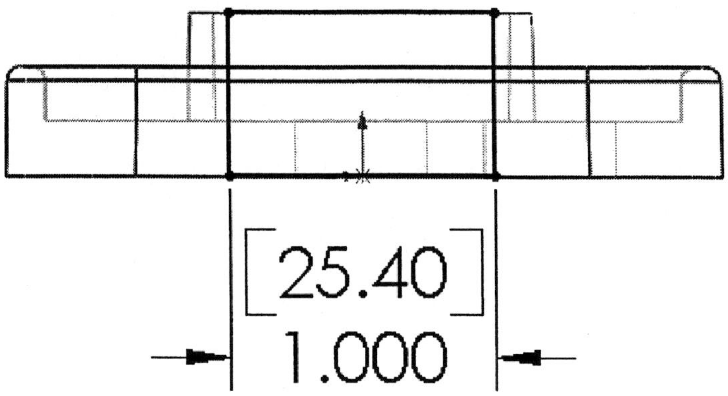

110) Display the Isometric view. Click **Isometric** .

111) Extrude the Sketch. Click **Extruded Boss/Base** . Create the first depth direction, Direction 1. Blind is the Type option. Enter **.400**, [**10.16**] for Depth. Click the **Draft IN/OUT** button. Enter **1.00** for Draft Angle. The sketch is extruded towards the Holes.

112) Create the second depth direction. Click the **Direction 2** check box. Select **Up to Surface** for Type. Select the outside **right face** for the second extruded depth. The Selected Items text box displays Face<1>.

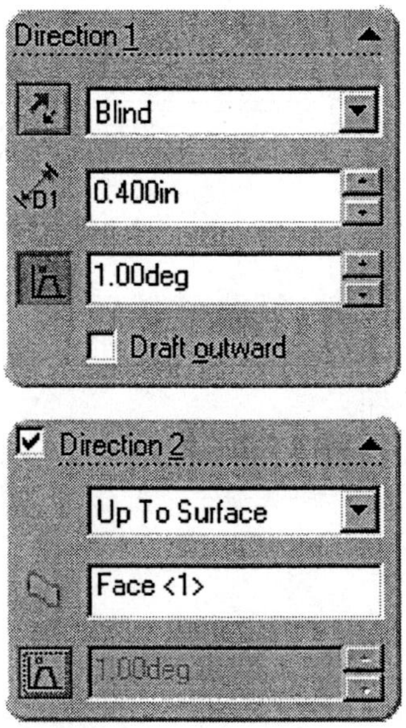

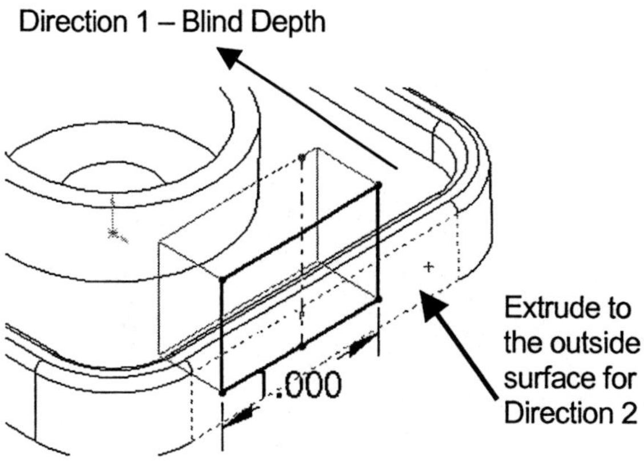

113) Display the Boss-Extrude2 feature. Click **OK**.

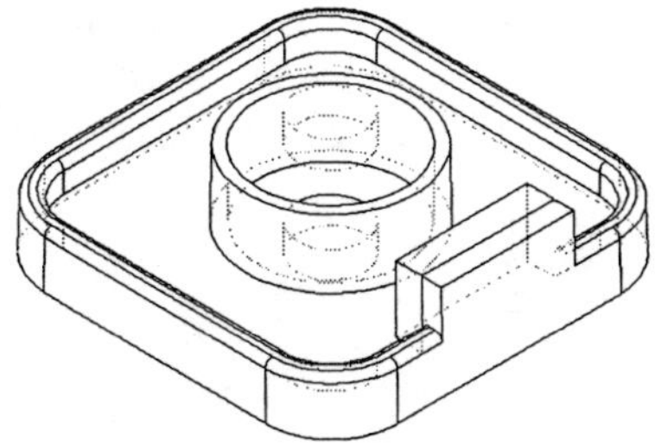

114) Rename **Boss-Extrude2** to **Connector Base**.

115) Show the Connector Base sketch. Click **Plus** to expand Connector Base in the FeatureManager. Right-click **Sketch5**. Click **Show Sketch**.

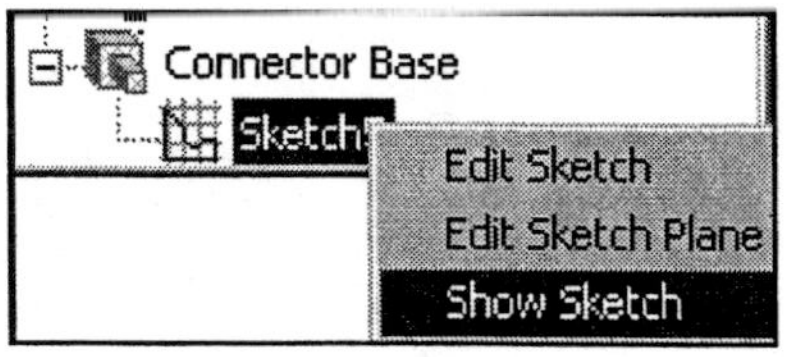

116) Save the BATTERYPLATE. Click **Save** .

Create the second Extruded Boss feature.

117) Select the Sketch plane. Click the **top narrow face** of the first Extruded Boss feature.

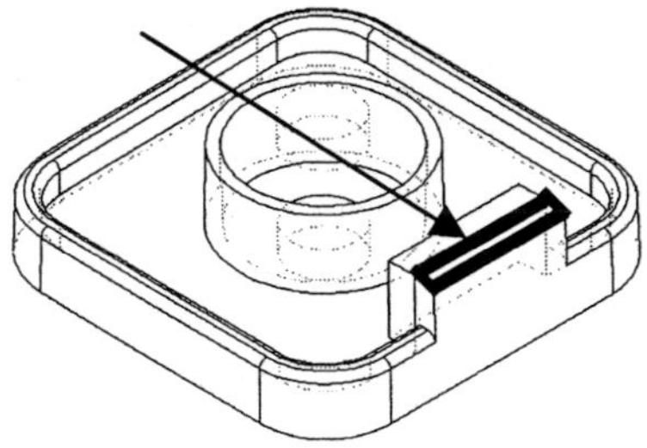

118) Create the Sketch. Click **Sketch** . Display the Top view. Click **Top** .

Convert a line segment from the Connector Base sketch to the current sketch plane.

119) Click the **vertical line** of the Connector Base Sketch. Click **Convert Entities**.

120) Sketch the centerline. Click **Centerline**. Sketch a **horizontal centerline** with the first point coincident to the Origin. Create the mirrored centerline. Click **Mirror**. The centerline displays two parallel mirror marks.

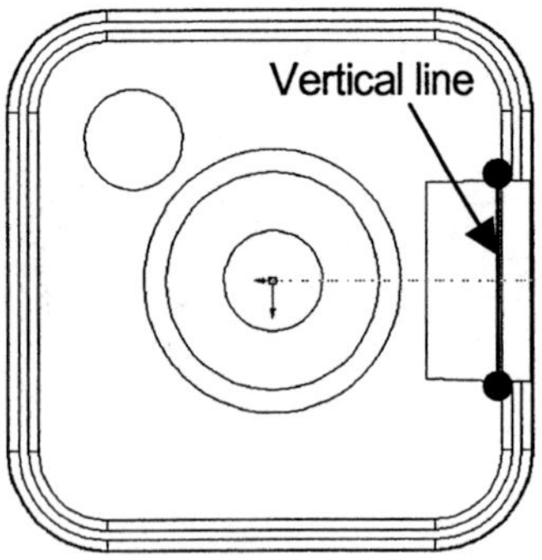

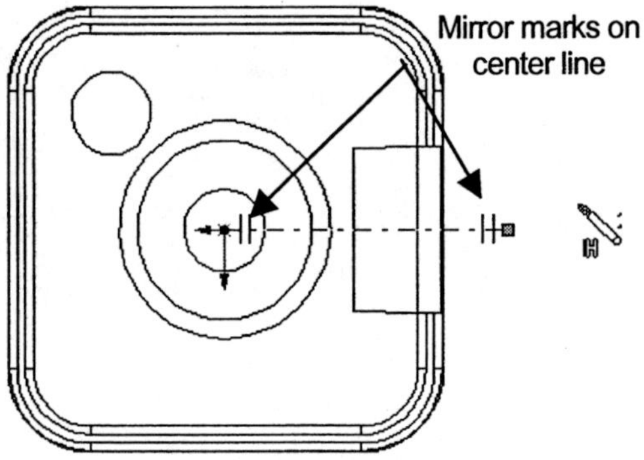

121) Sketch the profile. Create the Sketch on one side of the mirror centerline. Click **Line**. Create a **horizontal line** coincident with the endpoint of the converted Connector Base sketch. The line is automatically mirrored.

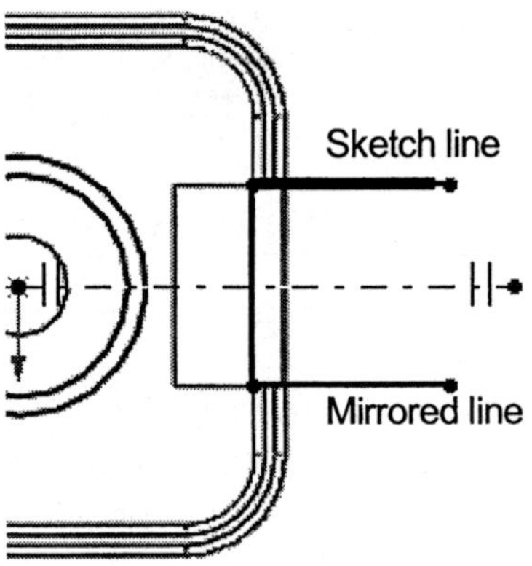

122) Create a Tangent Arc. Click **Tangent Arc**. Create the first arc point. Click the **endpoint** of the horizontal line. Create a 180° arc. Drag the **mouse pointer** downward until the start point, center point and end point are vertically aligned. Release the **mouse button**.

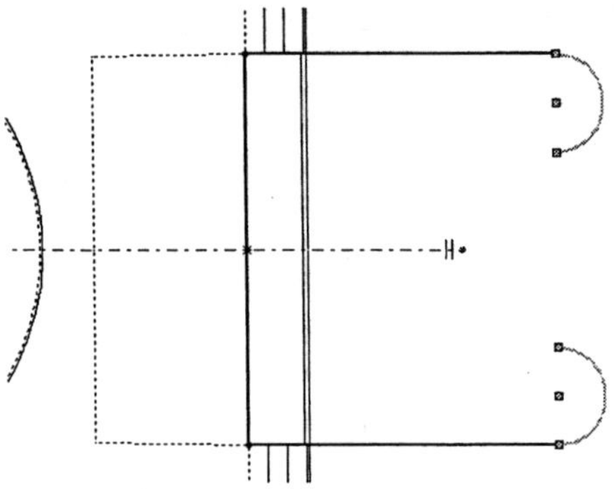

123) Complete the Sketch. Click **Line** . Create a **horizontal line**. Create a **vertical line** coincident with the inside top right edge.

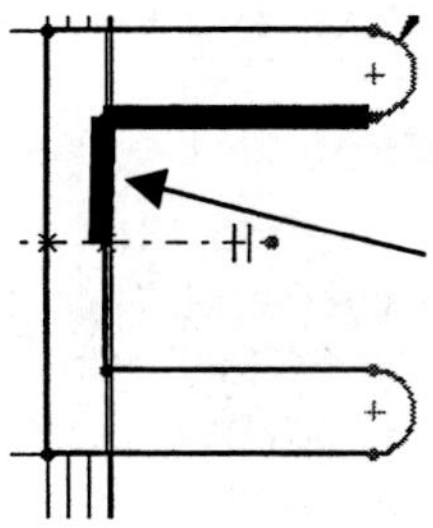

124) Turn the Mirror function off. Click **Mirror** .

125) Dimension the Sketch.
Create a radial dimension.

Click **Dimension** . Click the **arc edge**. Click the **text location**. Enter **.100**, **[2.54]** for the Radius.

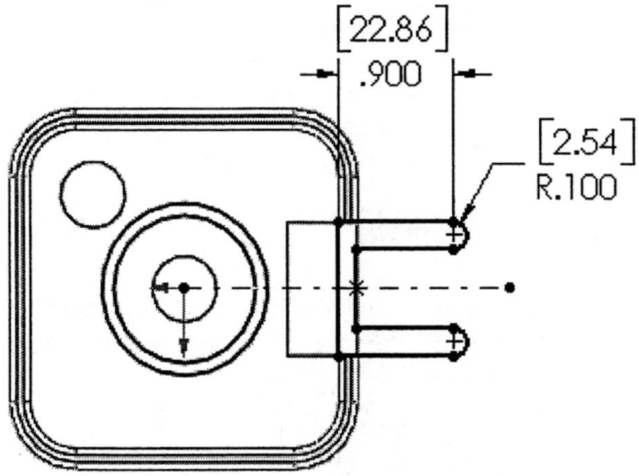

126) Create a linear dimension. Click the left most **vertical line**. Click the **arc edge**. The arc edge displays red. Click the **text location**.

Note: Click the arc edge, not the arc center point to create a max. dimension. The linear dimension uses the arc center point as a reference. Modify the Properties of the dimension. The Maximize option references the outside tangent edge of the arc.

127) Right-click on the **dimension text**. Click **Properties** from the Pop-up menu. Click the **Max** button from the First arc condition option. Enter **1.000**, **[25.4]** in the Value list box. Display the dimension. Click **OK**. The black Sketch is fully defined.

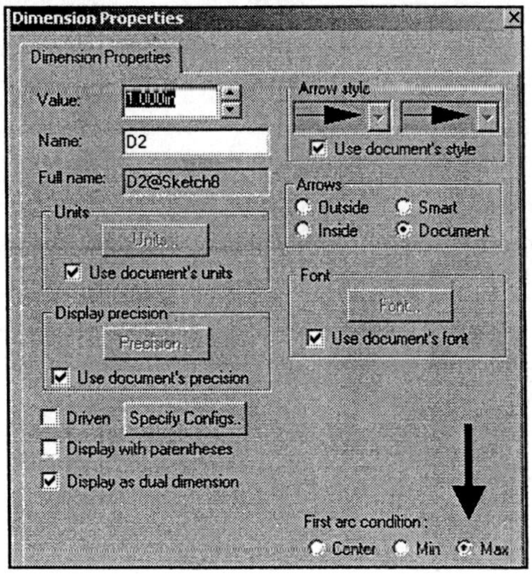

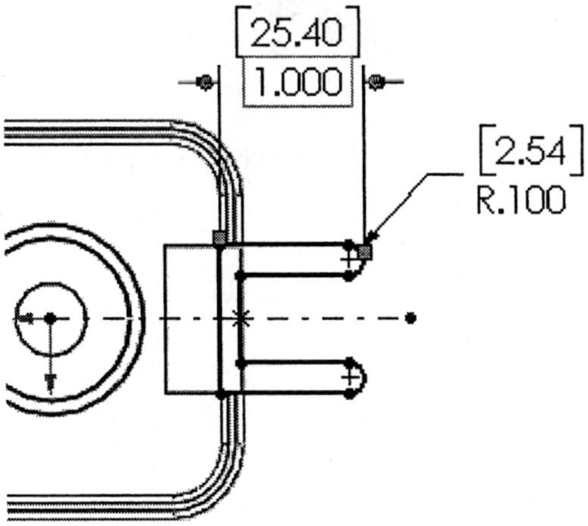

128) Display the Isometric view. Click **Isometric** .

129) Extrude the Sketch. Click **Extruded Boss/Base** . Blind is the default Type option. Enter **.100**, **[2.54]** for Depth. Click the **Reverse** check box. Display the Boss-Extrude3 feature. Click **OK**.

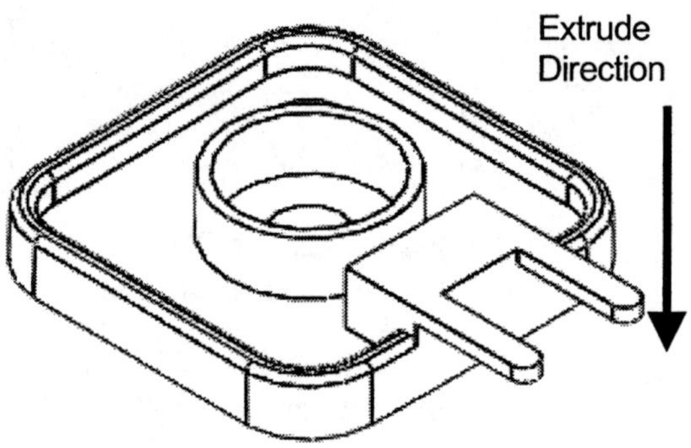

130) Rename **Boss-Extrude3** to **ConnectorSwitch**.

131) Save the BATTERYPLATE. Click **Save** .

Disjoint Geometry

Incorrect selection of edges and faces leads to disjointed bodies. Disjointed bodies occur when the geometry contains gaps.

Example: Create a Sketch from the outside edge. Reverse the extrusion direction to create disjoint geometry.

The feature is not created and a Rebuild error is displayed.

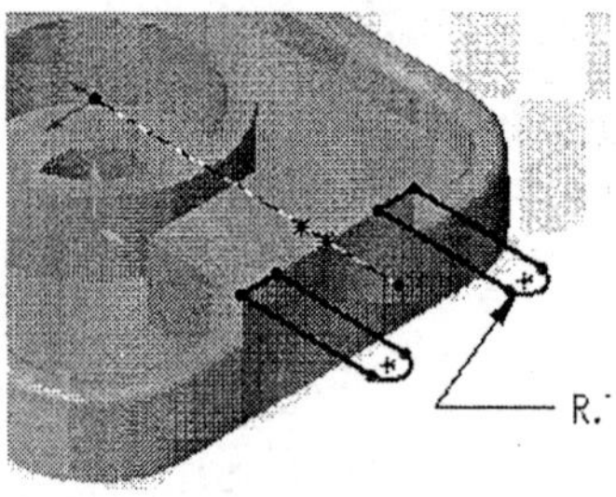

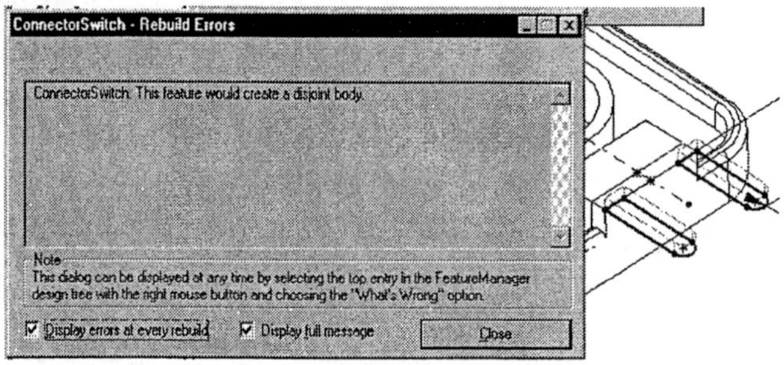

Profiles of disjointed and joined geometry are displayed.

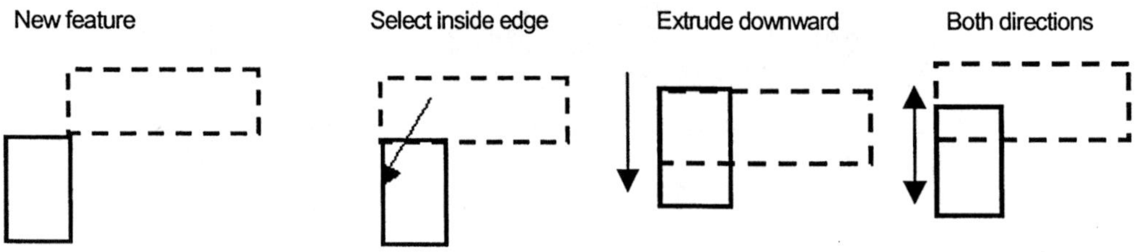

Disjoint profile geometry Joined profile geometry Joined profile geometry

Create the BATTERYPLATE - Edge and Face Fillets

Both edge and face options for Fillet features are used to smooth rough edges.

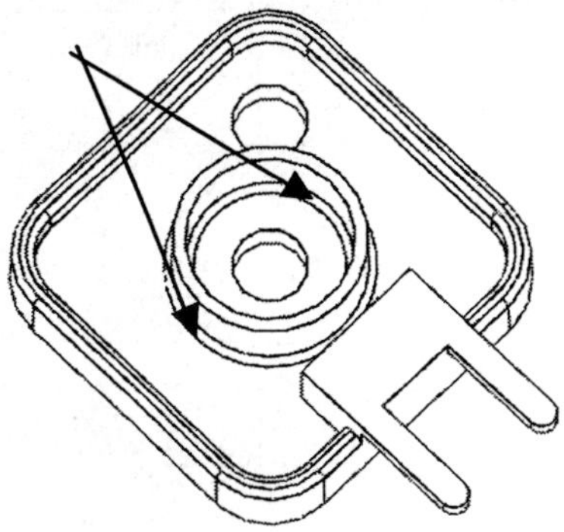

Create a Fillet feature.

132) Create a fillet on the inside and outside edge of the Holder. Create a fillet on all inside tangent edges of the Top-Cut. Click **Fillet**. Enter **.050**, [**1.27**] for Radius. Click the **outside circular edge** of the Holder. Click the **inside circular edge**. Display the Fillet. Click **OK**.

133) Rename **Fillet3** to **HolderFillet**.

Create a Fillet on the outside bottom edge of the Connector. This is a two-step process:

- Create an edge Fillet.
- Create a face Fillet.

Create an edge Fillet on the four vertical edges of the Connector. Create a face blend between two sets of faces.

Create the edge Fillet feature.

134) Click **Fillet**. Click the four **vertical edges**. Enter **.100**, [**2.54**] for Radius. Display the Fillet. Check **Tangent Propagation**. Click **OK**.

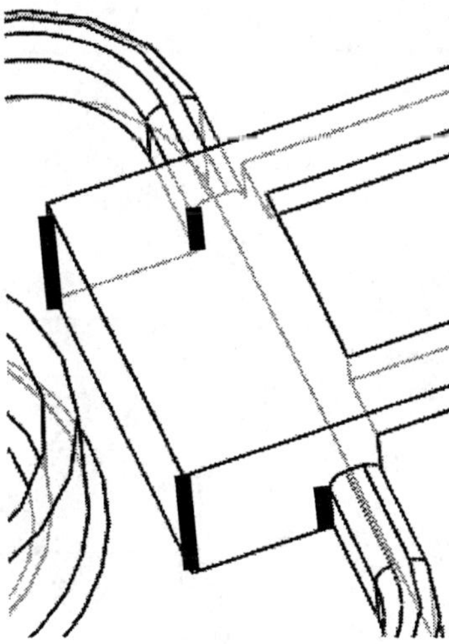

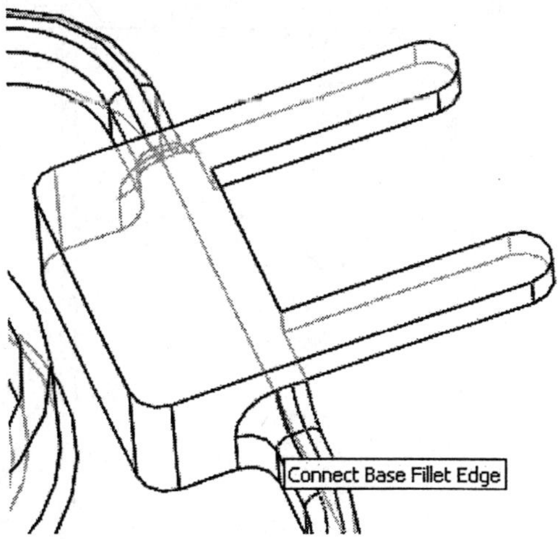

135) Rename **Fillet4** to **Connect Base Fillet Edge**.

Create the face Fillet feature.

136) Click **Fillet** . Click the **Face fillet** button from the Fillet type list box. Enter **.100**, **[2.54]** for Radius. Select the first Face set. Click the **back face**. Select the second Face set. Click **inside** the Face Set 2 list box. Click the **top face** of the Base-Extrude feature. Click **OK**. The Radius is too large. Enter **.050**, **[1.27]** for Radius. Display the Fillet. Click **OK**.

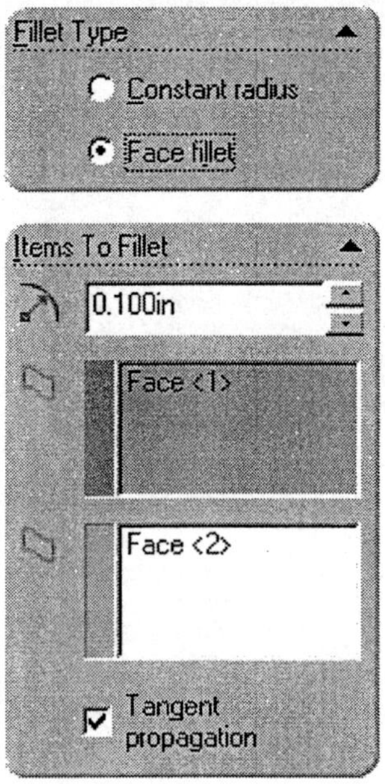

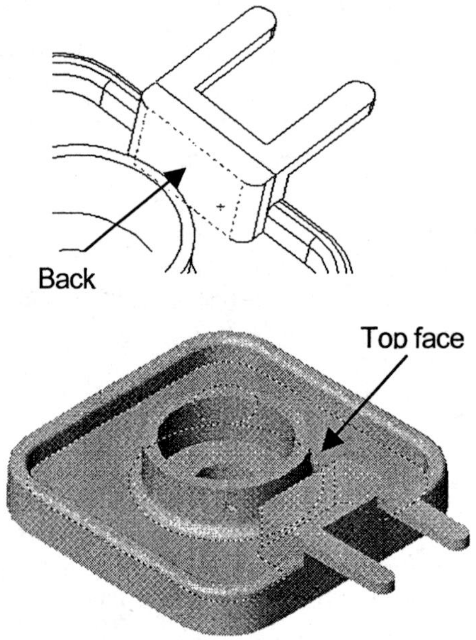

137) Rename **Fillet5** to **Connect Base Fillet Face**.

138) Save the BATTERYPLATE. Click **Save** .

The FeatureManager displays all successful feature name icons in yellow. The rotation of the BATTERYPLATE is completed.

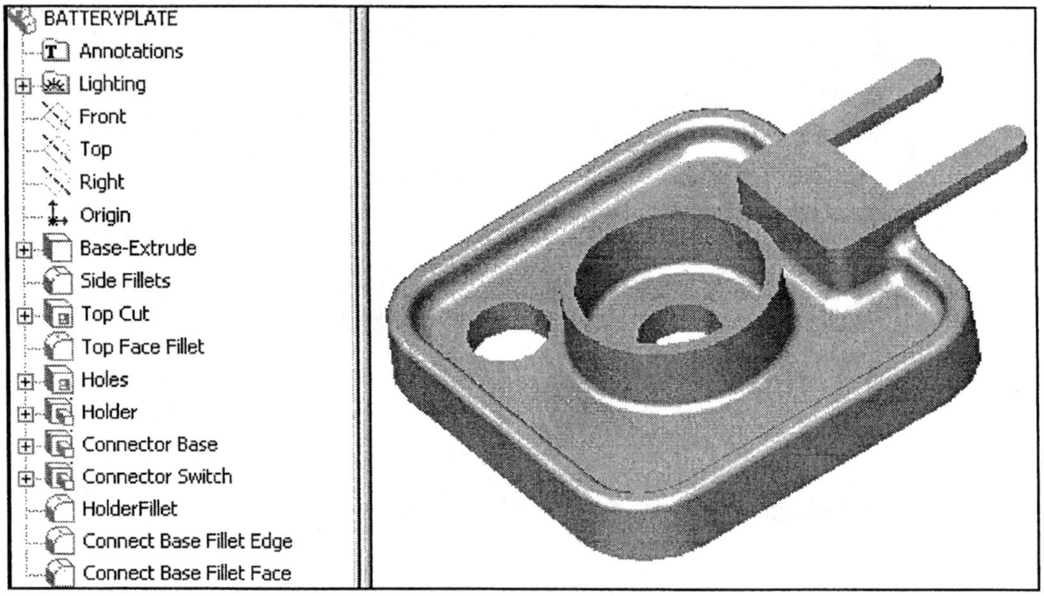

Questions

1. What key design questions should you investigate before starting a design?
2. How do you start a SolidWorks part?
3. How do you create a part document template?
4. Describe the default reference planes.
5. What is the Base feature and why should the Base feature be kept simple?
6. What is the difference between an Extruded Base feature and an Extruded Cut feature?
7. What is a Fillet feature?
8. What command keys are used to Copy sketched geometry?
9. Describe the Edit feature.
10. What type of Geometric Relations can you add to a Sketch?
11. How do you avoid the Fillet Rebuild error?
12. How do you create an angular dimension?
13. What is a draft angle?
14. When do you use a draft angle?
15. When do you use the Mirror command?
16. Describe disjointed geometry.

Exercises

Create the following Extruded Parts:
Exercise 1.1: ROUND-BAR, Hot Rolled Round Bar Steel Stock.
Exercise 1.2: FLAT-PLATE, Cold Finished Flat Steel Stock.
Exercise 1.3: L-BRACKET, Cast Iron.
Exercise 1.4: U-SECTION Aluminum.

Exercise 1.1

Exercise1.2

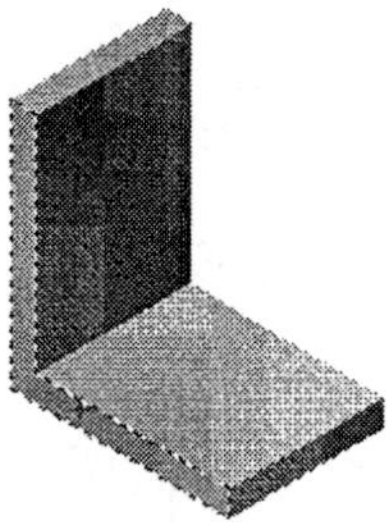
Exercise 1.3

Exercise 1.4

Exercise 1.5: T-SECTION, Cast Iron.
Exercise 1.6: V-SECTION, Cast Iron.
Exercise 1.7: BLOCK, Cast Iron.

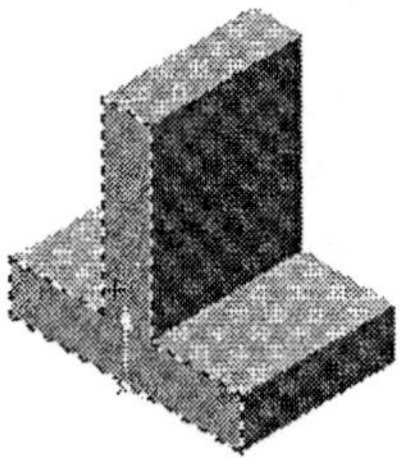
Exercise 1.5

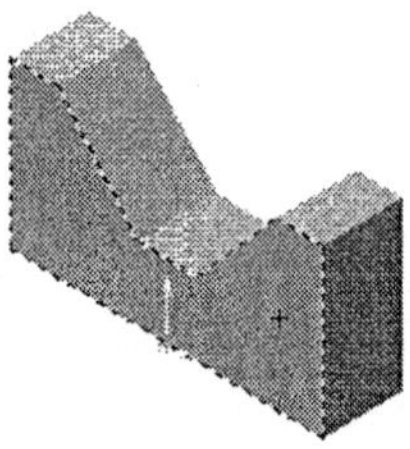
Exercise 1.6

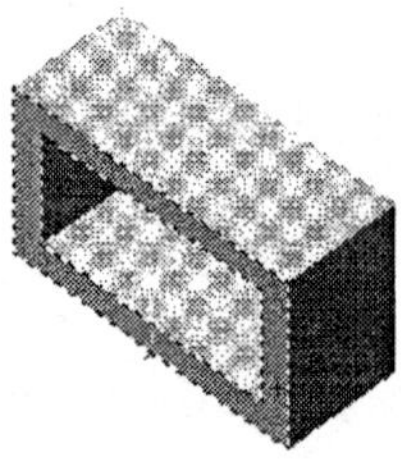
Exercise 1.7

Parts courtesy of the Reid Tool Supply Co. Muskegon, MI USA

Parts manufactured from machine stock can save design time and money.
Machine stock is purchased from material supply companies.
Example: Reid Tool Supply Company (www.reidtool.com).

Create the Extruded Parts:
State the Sketch Plane for each part based upon the orientation displayed.

Exercise 1.8: LOCATING PIN Sketch Plane ______________.
Exercise 1.9: BUSHING Sketch Plane___________.
Exercise 1.10: KEYWAY BUSHING Sketch Plane ____________.

Exercise 1.8 LOCATING PIN

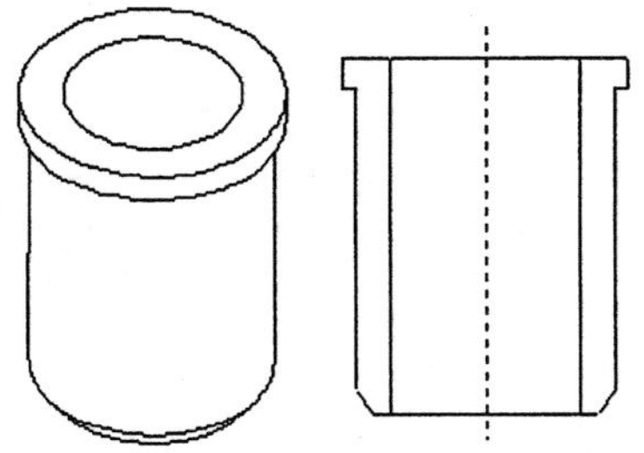

Exercise 1.9 BUSHING

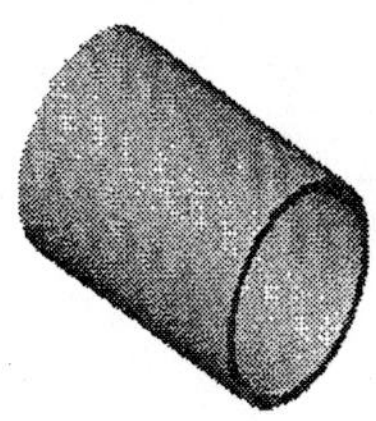

Exercise 1.10
KEYWAY BUSHING

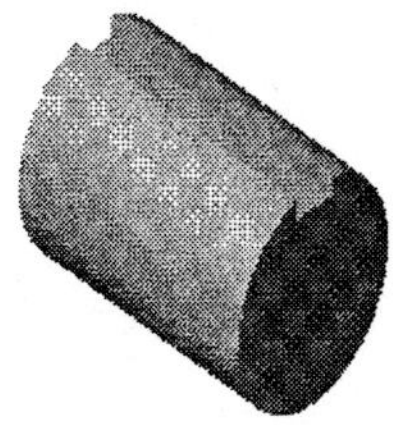

Create the Extruded Parts:
Use symmetry and geometric relations.

Exercise 1.11: JOINING –STRIP.
Exercise 1.12: TEE-JOINING-PLATE.
Exercise 1.13: 90 DEGREE-JOINING-PLATE.

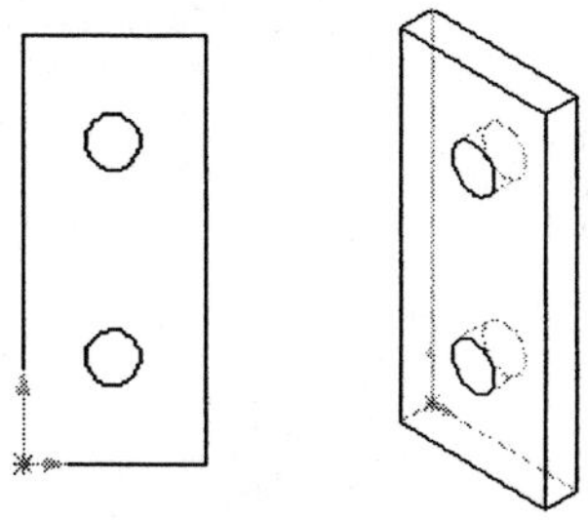

Exercise 1.11

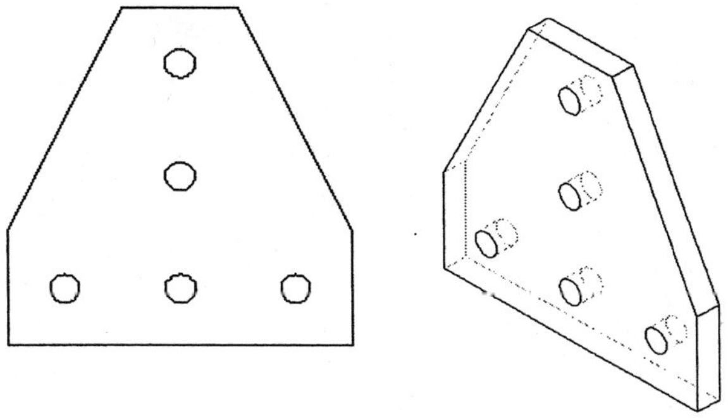

Exercise 1.12

The company, 80/20 Inc., manufactures modular aluminum structural extrusions to reduce fabrication time and cost. 80/20 manufacturers both Metric and English products.

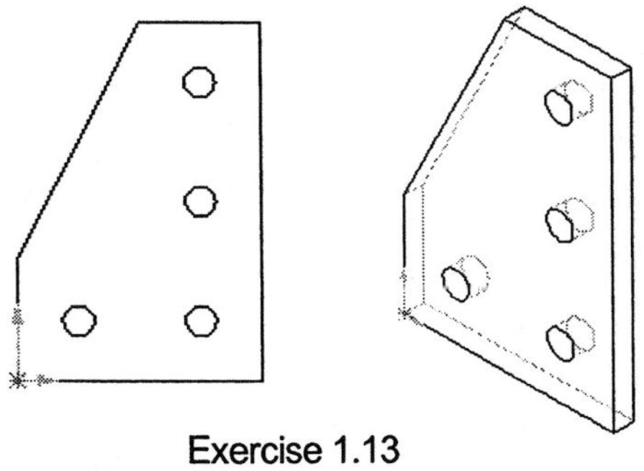

Exercise 1.13

Parts courtesy of 80/20, Inc.
Columbia City, IN USA
(www.8020.net)
and its distributor, Air Inc.
Franklin, MA USA
(www.airinc1.com)

Create the Extruded Parts:
Parts are created from machine stock. No dimensions are provided.

Exercise 1.14: Create one STANDOFF. Create one PLATE. Four STANDOFFs will fit between two PLATEs, (Exercise 2.2).

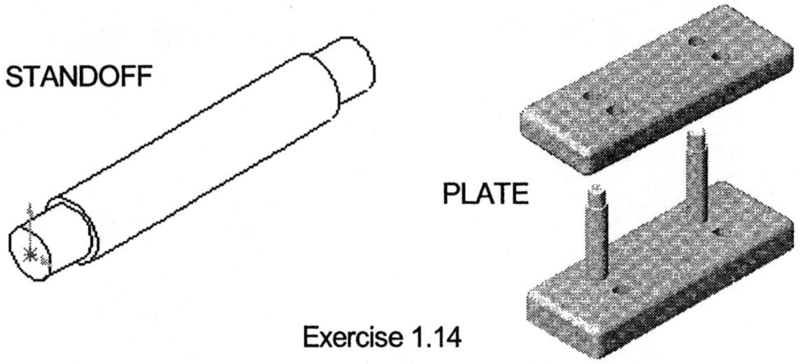

Exercise 1.14

Exercise 1.15: Create a PIN. Create a U-BRACKET.
Exercise 1.16: Create a T-SECTION and MOUNTING PLATE.

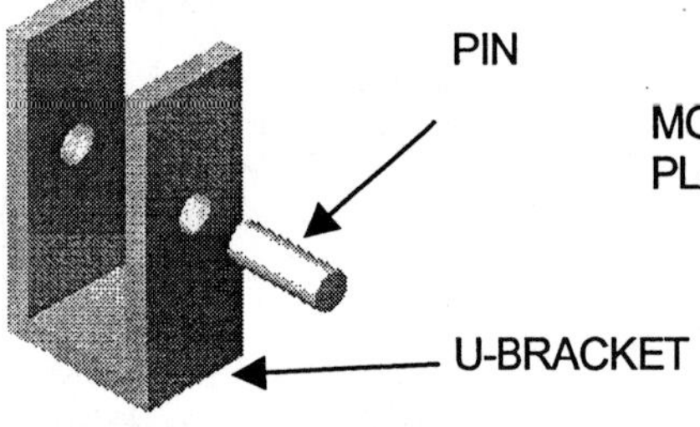

Exercise 1.15

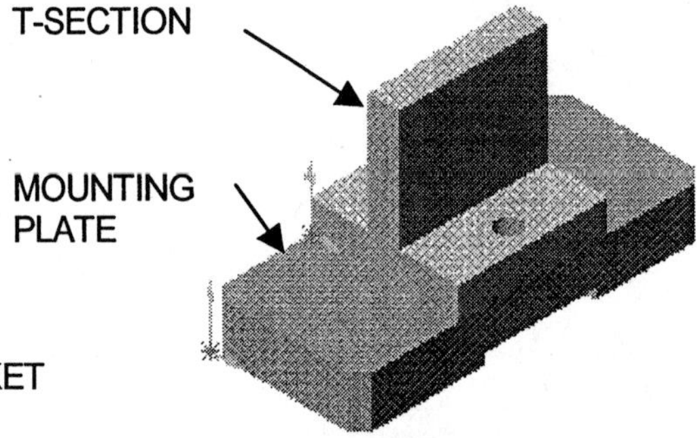

Exercise 1.16

Exercise 1.17: Part document templates:
Create a Metric part document template using an ISO dimension standard.
Create an English part document template using an ISO dimension standard.
Set Drawing/Units to appropriate values for each template.

Exercise 1.18: Create an EMBOSSED-STAMP with your initials.

The initials are created with Extruded Sketched text. How do you create the text? Answer: Explore the command with SolidWorks on-line Help. Click **Help**. Click **Index**. Enter **text**. Click **extruded text on model**. Follow the instructions.

Exercise 1.18

Exercise 1.19: Industry Collaborative Exercise.

On-line intelligent engineering catalogs save time. They help eliminate the requirement for engineers to design, build and test all components from scratch. You now work with a team of engineers on a new project. The senior engineer has a question on the voltage requirements, size requirements and color scheme for a particular connector. You have been given the directive to find an in-line receptacle connector with the following specifications:

Component Type: In-Line Receptacle

Color Scheme: Red Housing-Black Adapter or

Black Housing-Black Adapter

Contact Type: Machine Gold Plated

Circuit Count: 2+2

Key Pattern: 1

Contact Size: 2#16 & 2#22

1) Obtain the Receptacle information from www.aldenproducts.com. Click the On-Line Catalog button. Click the Product Selector, Instrumentation Connectors button. Select the PL700N GP Series.

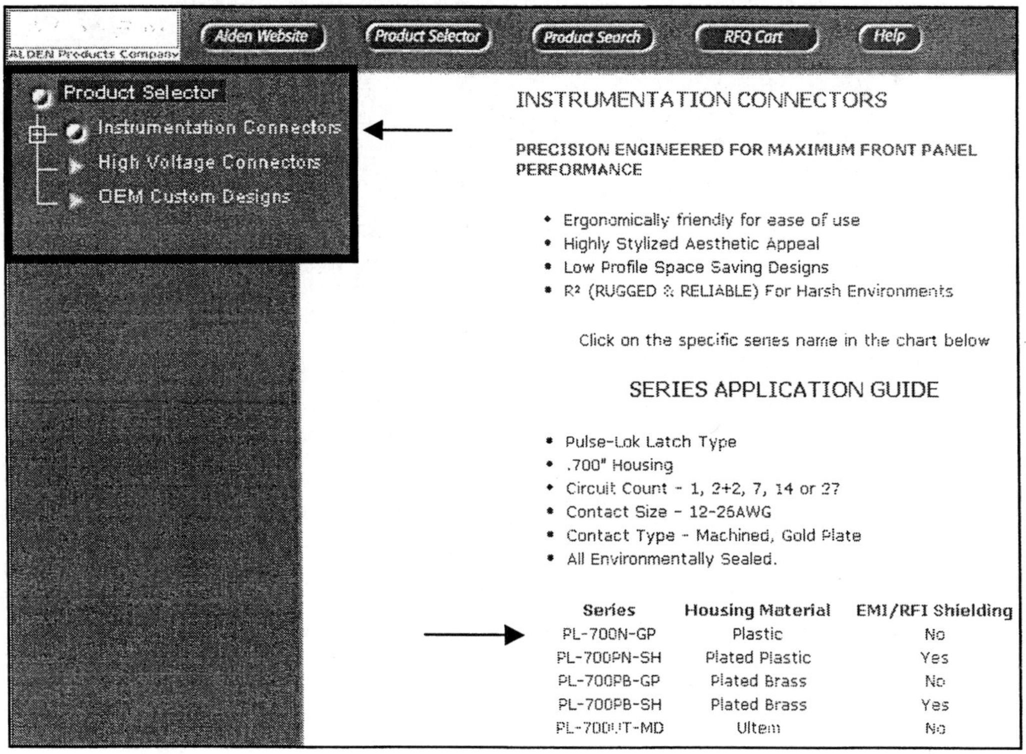

2) Over 370 part numbers are listed. The term "drill down" describes the process of finding the correct component based on your specifications. Select In-Line Receptacle for Component Type. The number of parts to select is reduced to 140. Select Red Housing with Black Adapter. Select Machined, Gold Plated. Select 2+2 for Circuit Count. Select 1 for Key Pattern. Select 2-#16 and 2-#22 for Contact Size. Only one part number, 300250 is listed.

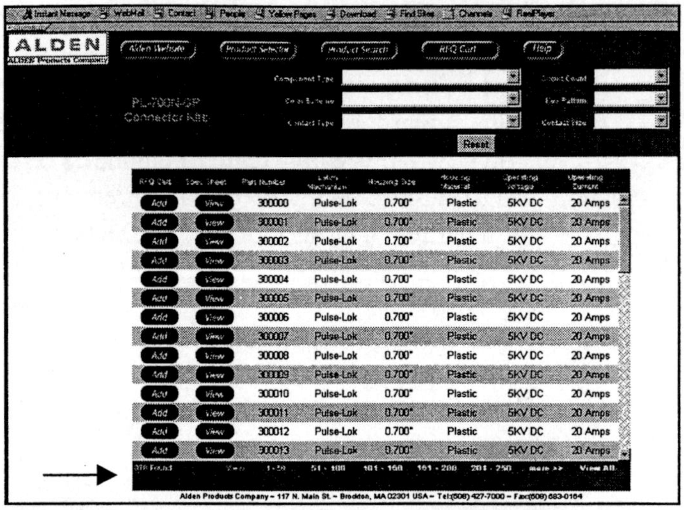

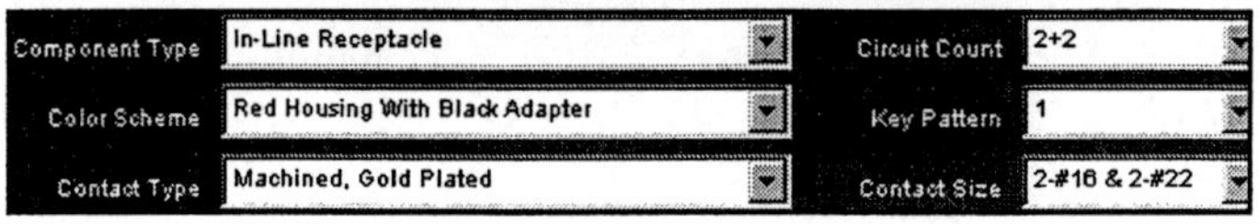

3) Click the View button to display the In-line receptacle.

4) Click Layout Dimensions in the Specification column to obtain the overall size of the In-line receptacle.

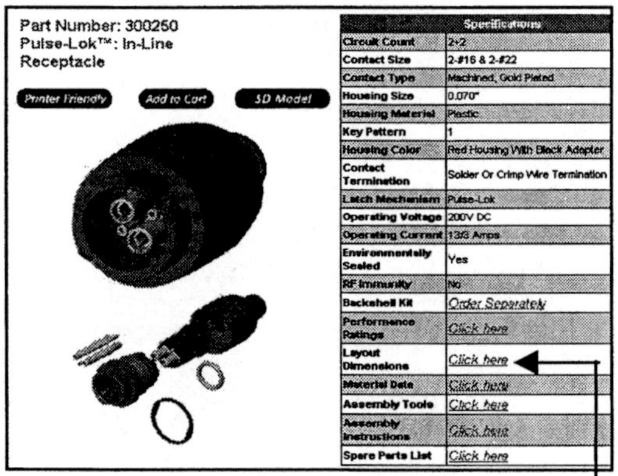

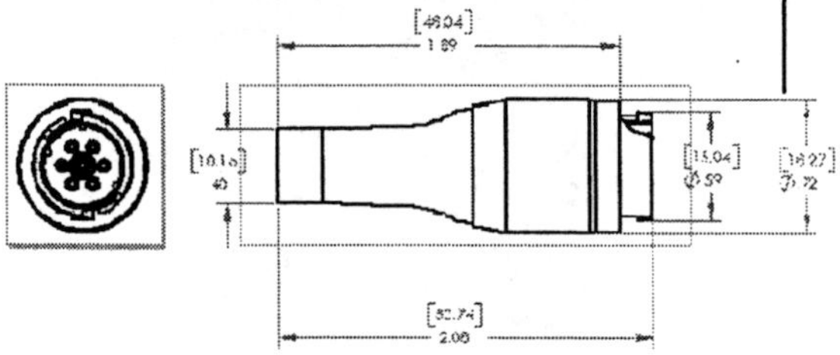

What is the voltage of this component? What is the operating current?

What are the overall dimensions in millimeters?

Is the component available in both color schemes? What are the part numbers for the components?

Components Courtesy of
ALDEN Products. Brockton. MA USA

SmartCAT®, the on line catalog content management (CCM) software utilized on the Alden Products web site was developed by i-MARK®, Hartford, CT (www.imark.com). SmartCAT® enables engineers to apply in-depth research and advanced discovery techniques, such as wizard and performance calculators to specify and purchase exactly what they require. SmartCAT® electronic catalogs guide engineers to the product that meets both geometric and performance characteristics.

The next section required Windows Media Player. Review the video that describes the latch mechanism of the connector.

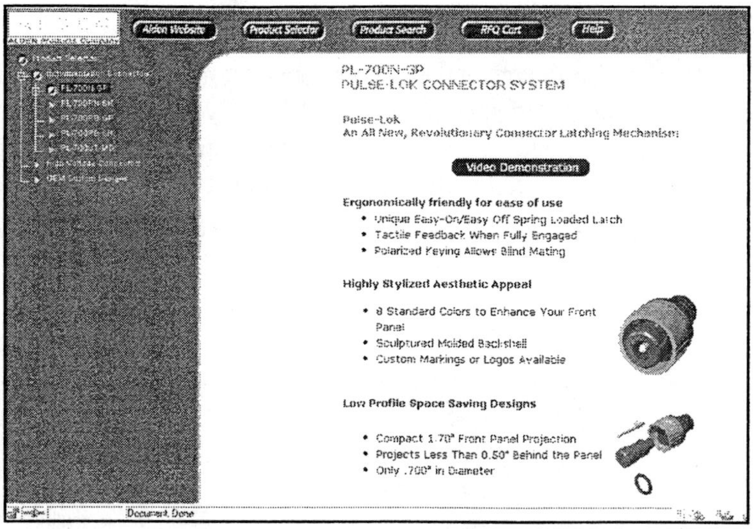

5) Click the Product Selector button.

6) Expand Instrumentation Connectors Click the + button. Select PL700N-GP. Click the Video Demonstration button to view the animation create with SolidWorks documents.

The video file takes a few seconds to download. The Windows Media Player opens. The movie contains animated part files from SolidWorks. Materials and voice have also been added to complete the demonstration.

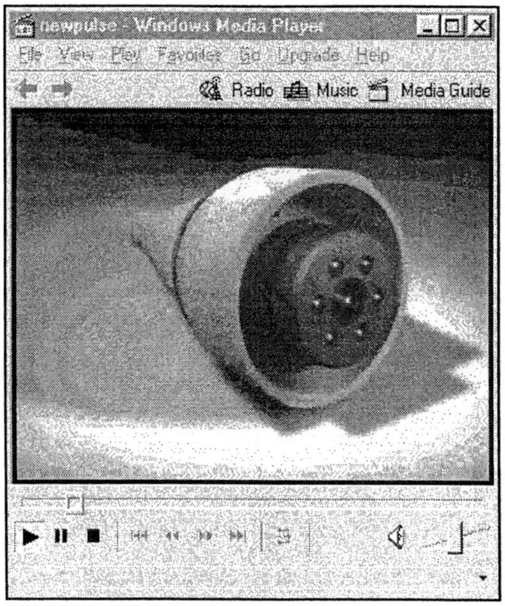

Exercise 1.20: Industry Collaborative Exercise.

Engineers and designers spend a great deal of time searching for product suppliers and part specifications. How do you obtain a supplier for the batteries used in this project? What are the overall dimensions and voltage of a D size battery compared to the current 6-volt battery design? Research suppliers and part information utilizing the URL: http:// www.globalspec.com. Enter Battery. Click the Find button. Select D for battery size.

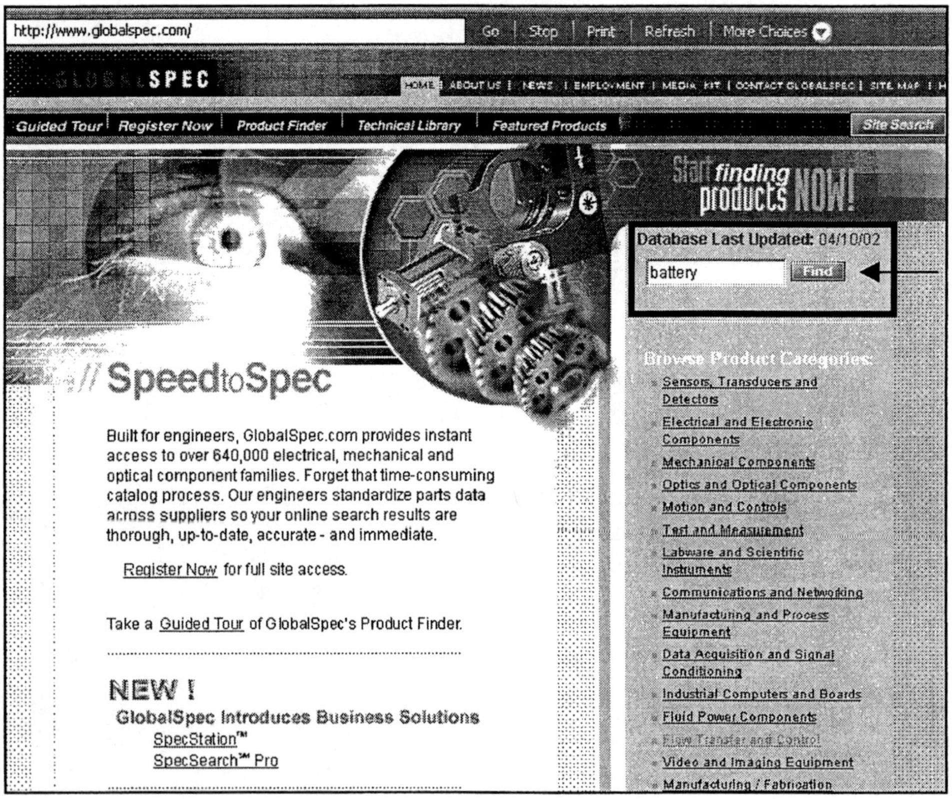

Gold Peak Industries of North America is the supplier. Select Search Results. Record the overall dimensions for the D size battery and voltage requirements.

Add to My Part List | Compare/Print Selected Parts

Select Part	Product Names (☆ indicates an exact match)	Send RFQ	Supplier's Product Info	Print-Friendly Version	Battery Voltage & Cell Size: Battery Size	Battery Voltage & Cell Size: Dimensions	Performance Specifications: Voltage (volts)	Performance Specifications: Capacity (ampere-hour)	Performance Specifications: Internal Resistance (ohms)	Primary Batteries: Standard	Primary Batteries: Lithium
☐	★GP13A Super Alkaline	RFQ	Web	Print	D	34.2mm(D) x 61.5mm (H)	1.50			Manganese	

The second design option for the FLASHLIGHT assembly requires a battery holder and 4-D size batteries. Does a supplier for a 4-D battery holder exist? If so, list the name of the supplier, material and the overall size of the battery holder.

Project 2

Revolve Features

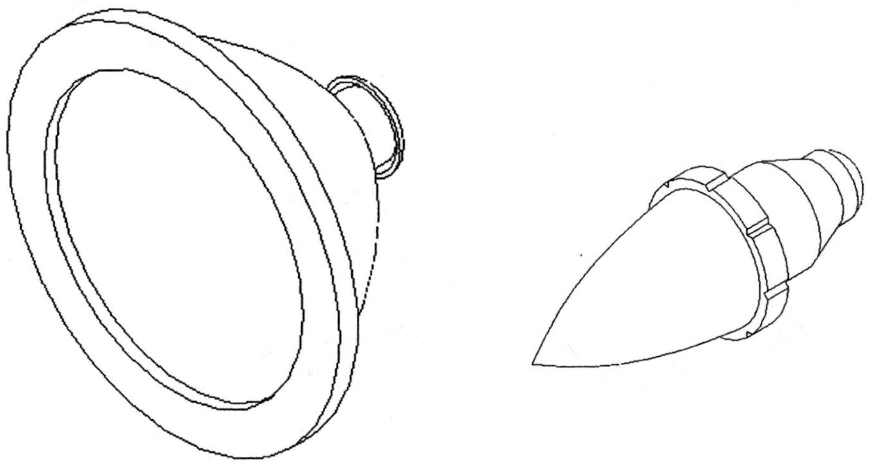

Below are the desired outcomes and usage competencies based upon the completion of Project 2.

Project Desired Outcomes	Usage Competencies
A comprehensive understanding of the customer's design requirements and desires.	To comprehend the fundamental definitions and process of Feature-Based 3D Solid Modeling.
Create two key FLASHLIGHT parts: • LENS • BULB	Specific knowledge of Revolve base features.
	Understanding of the Shell feature, Hole Wizard, Dome feature and Circular Pattern.
	Ability to apply Extrude and Fillet features.

Notes:

Project 2 – Revolve Features

Project Objective

Create two parts for the FLASHLIGHT. Create the LENS and BULB parts.

Project Situation

The LENS is a purchase part utilized in the FLASHLIGHT assembly, Figure 2.1. Obtain dimensional information on the LENS. Review the size, material and construction.

The BULB is a purchased part, Figure 2.2. The BULB is a replacement part and requires a separate part number and order number.

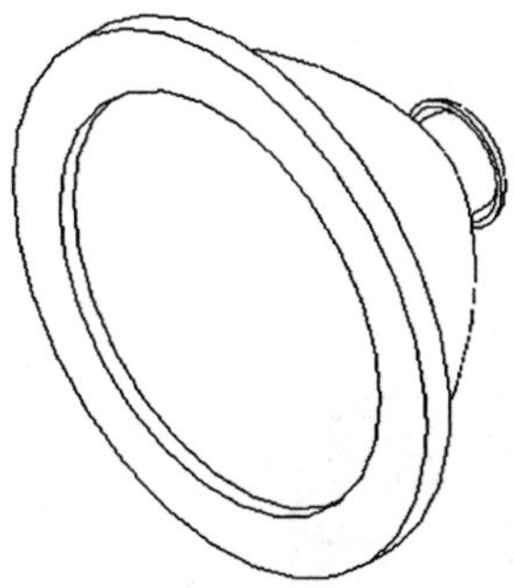

Figure 2.1

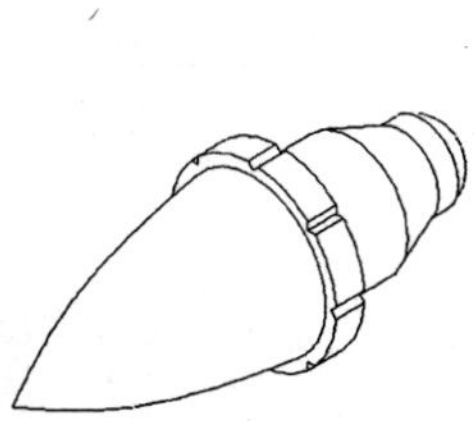

Figure 2.2

Project Overview

Create two parts in this section:

- LENS
- BULB

The LENS and the BULB utilize a Revolve Base feature.

LENS

Determine the key features of the LENS. The Base feature for the LENS is a solid Revolved feature. A solid Revolved feature adds material. The Revolved Base feature is the foundation for the LENS.

A Revolved feature is geometry created by rotating a sketched profile around a centerline, Figure2.3. Close the Sketch profile for a solid Revolved feature, Figure 2.4. Do not cross the centerline.

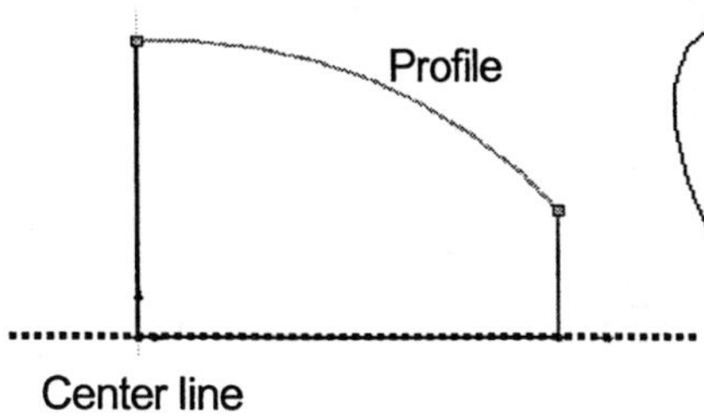

Figure 2.3

Figure 2.4

LENS Feature Overview

Create the LENS. Use the solid Revolved Base feature, Figure 2.5.

Create uniform wall thickness. Create the Shell feature, Figure 2.6.

Create an Extruded-Boss feature from the back of the LENS, Figure 2.7.

Create a Thin-Revolved feature to connect the LENS to the BATTERYPLATE, Figure 2.8.

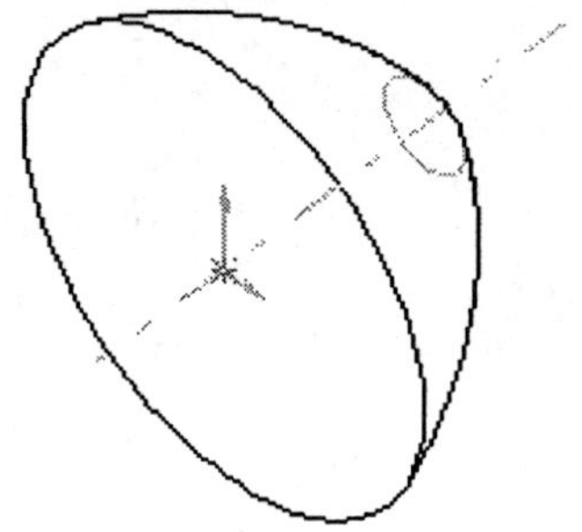

Figure 2.5

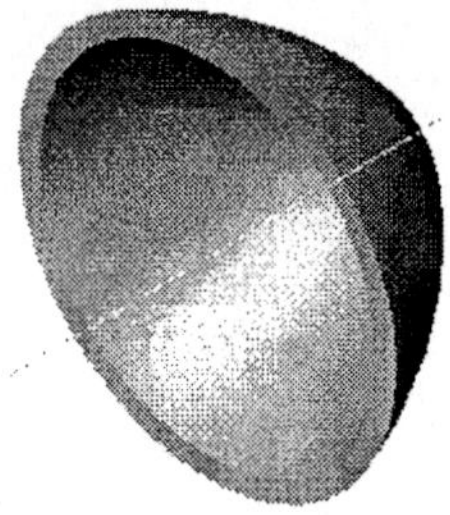

Figure 2.6

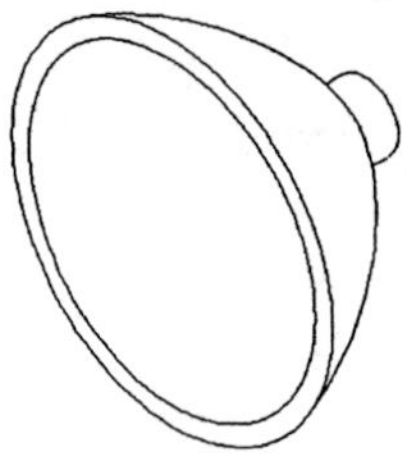

Figure 2.7

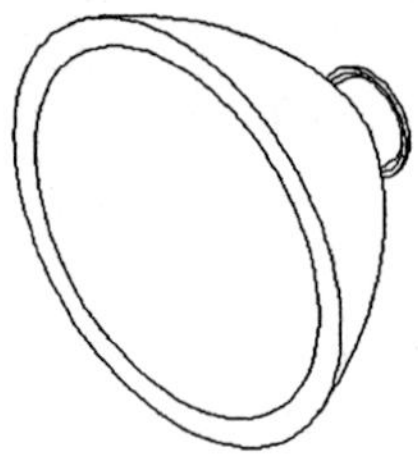

Figure 2.8

Create a Counterbore Hole feature with the HoleWizard, Figure 2.9. The BULB is located inside the Counterbore Hole.

Create the front LensFlange feature. Add a transparent LensShield feature, Figure 2.10.

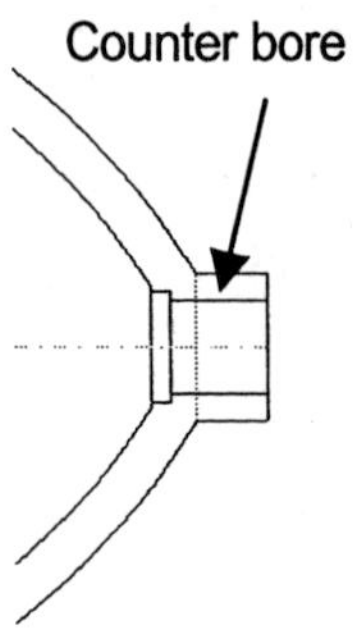

Figure 2.9

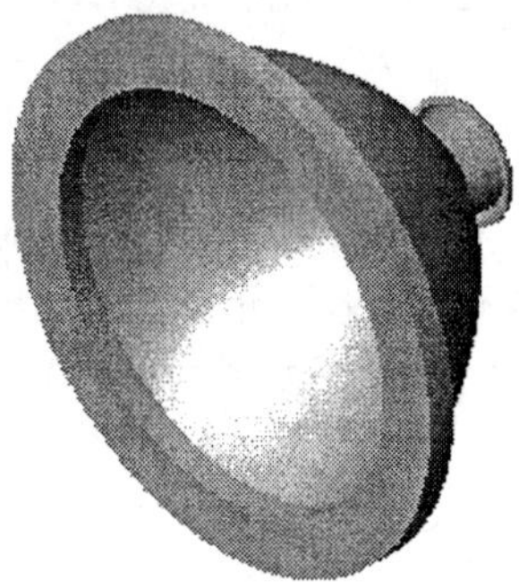

Figure 2.10

Create the LENS

Create the LENS with a Revolved Base feature. The solid Revolved Base feature requires a sketched profile and a centerline. The profile is located on the Right plane with the centerline collinear to the Top plane. The profile lines reference the Top and Front planes. The curve of the LENS is created with a 3-point arc.

Create the LENS.

1) Click **New**. Click **PartEnglishTemplate**, **[PartMetricTemplate]**. Click **OK**. Click **Save**. Select **My Documents** folder in the Save In text box. Enter **LENS**. Click the **Save** button.

2) View the planes. Right click on the **Front** plane in the FeatureManager. Click **Show**. Show the **Top** plane.

3) Select the Sketch plane. Click the **Right** plane.

4) Create the Sketch. Click **Sketch**.

5) Display the view. Click **Right**.

6) Sketch the centerline. Click **Centerline**. Sketch a horizontal **centerline** collinear to the Top plane, through the Origin.

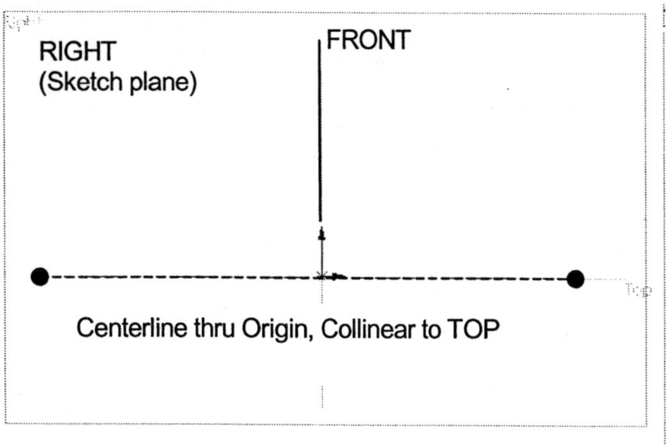

7) Sketch the profile. Create three lines. Click **Line** . Create the first line. Sketch a **vertical line** collinear to the Front plane coincident with the Origin. Create the second line. Sketch a **horizontal line** coincident with the Top plane. Create the third line. Sketch a **vertical line** approximately 1/3 the length of the first line.

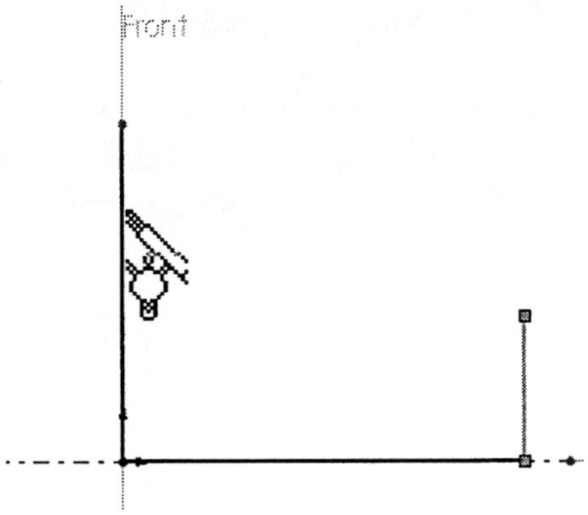

Create an arc. Determine the curvature of the LENS.

A 3 POINT Arc requires a:

- Start point
- End point
- Center point

The arc midpoint is aligned with the center point. The arc position is determined by dragging the arc midpoint or center point above or below the arc.

On-line help contains an animation file to create a 3-point arc. Click **Help**, **Index**, **Arc**, **3Point**. Run the animation. Click the **AVI** icon Show Me . Return to the Graphics window.

8) Create a 3 Point Arc. Click **3Pt Arc** . Create the arc start point. Click the **top point** on the left vertical line. Hold the **left mouse button** down. Drag the mouse pointer to the **top point** on the right vertical line.

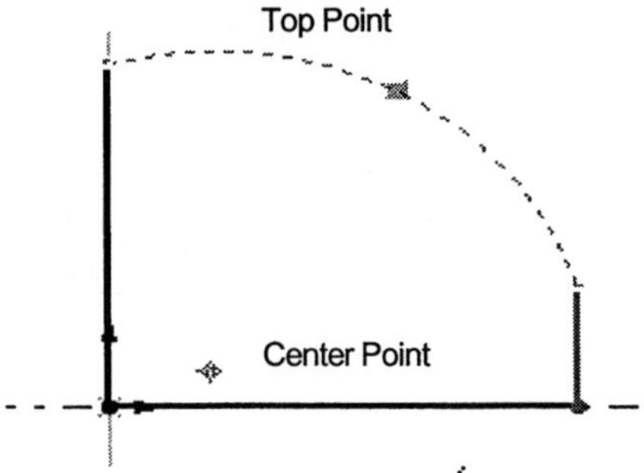

Create the arc end point. Release the **mouse button**.

Click and drag the **arc** until the **center point** is below the Origin. Release the **left mouse button**.

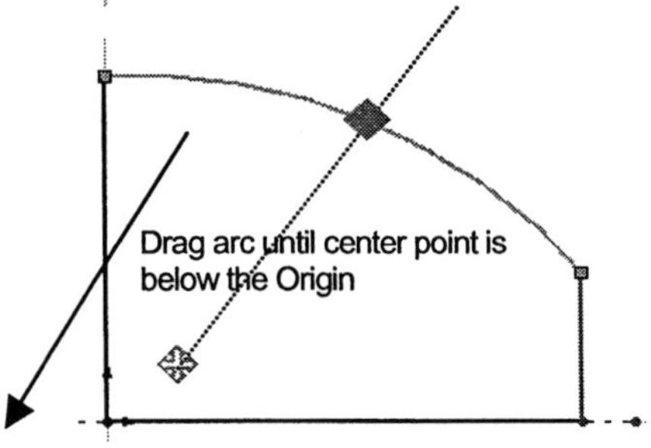

9) Add geometric relationships. The arc is currently selected. Right-click **Select**. The arc is no longer selected. Create an Equal relationship. Hold the **Ctrl** key down. Click the **left vertical line**. Click the **horizontal line**. Click the **Equal** button. Release the **Ctrl** key.

10) Add dimensions. Click **Dimension**. Create a vertical linear dimension for the **left line**. Enter **2.000**, [50.8]. Create a vertical linear dimension for the **right line**. Enter **.400**, [**10.16**]. Create a radial dimension for the **arc**. Enter **4.000**, [**101.6**]. The black Sketch is fully defined.

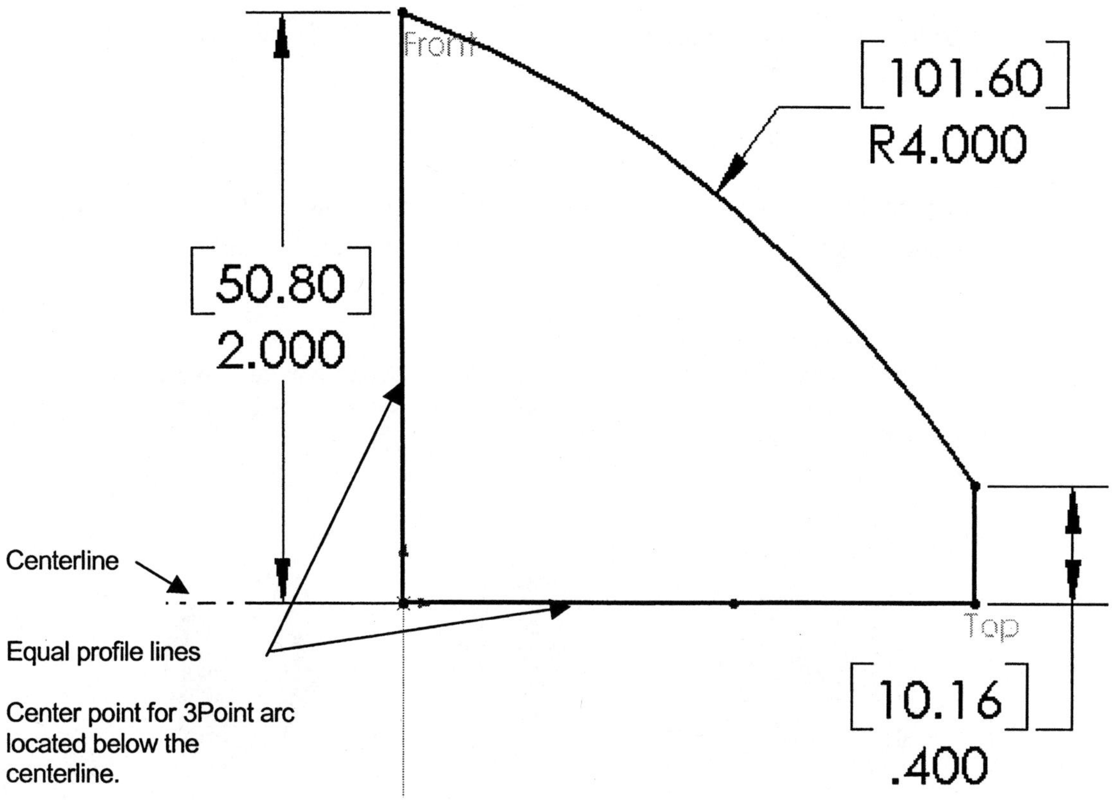

11) Revolve the Sketch. Click **Revolve** from the Feature toolbar. The Revolve Feature dialog box is displayed. Accept the default option values. Create the solid Revolve feature. Click **OK**.

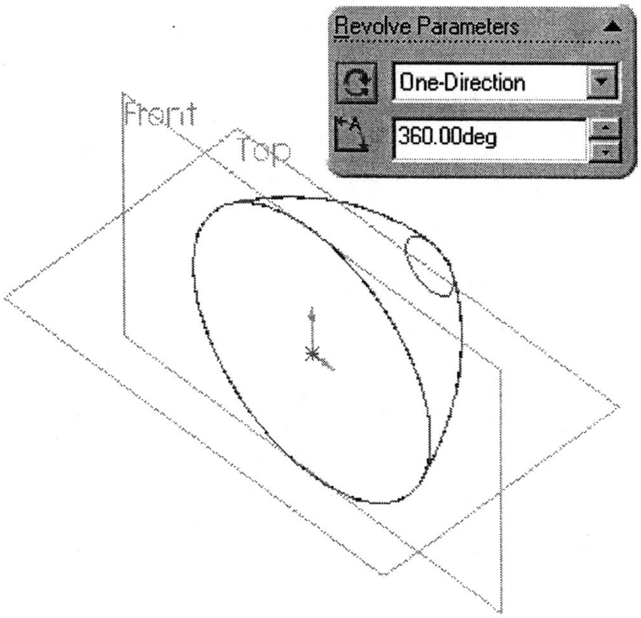

12) Save the LENS. Click **Save**.

Revolve features contain an axis of revolution. The axis is critical to align other features.

13) Display the axis of revolution. Click **View** from the Main menu. Click **Temporary Axis**. A check mark is displayed next to the option. Hide the Temporary axis. Click **Temporary Axis** to remove the check mark.

Solid Revolve features must contain a closed profile. Each revolved profile requires an individual sketched centerline.

Create the LENS - Shell Feature

The Shell feature removes face material from a solid. The Shell feature requires a face and thickness. Use the Shell feature to create thin-walled parts.

Create the Shell feature.

14) Select the face. Click the **front face** of the Base-Revolve feature. Click **Shell**  from the Feature toolbar. Enter **.250**, **[6.35]** in the Thickness text box. Display the Shell feature. Click **OK**.

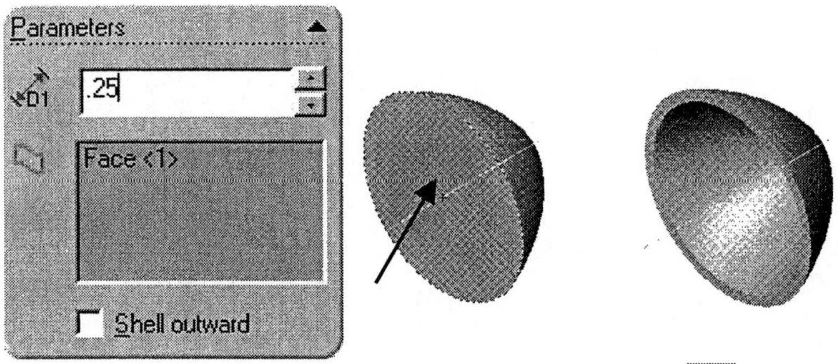

15) Rename **Shell1** to **LensShell**. Save the LENS. Click **Save** 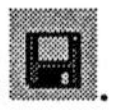.

Create the LENS - Extruded Boss Feature

Create the LensNeck. Use the Extruded-Boss feature. The LensNeck houses the BULB base and is connected to the BATTERY PLATE. The feature extracts the back circular edge from the Base-Revolve feature.

Create the Extruded Boss feature.

16) Select the Sketch plane. Right click near the small hidden **back face**. Click **Select Other**  from the Pop-up menu. Click the **right mouse button** (N) until the back face is displayed. Accept the back face. Click the **left mouse button** (Y).

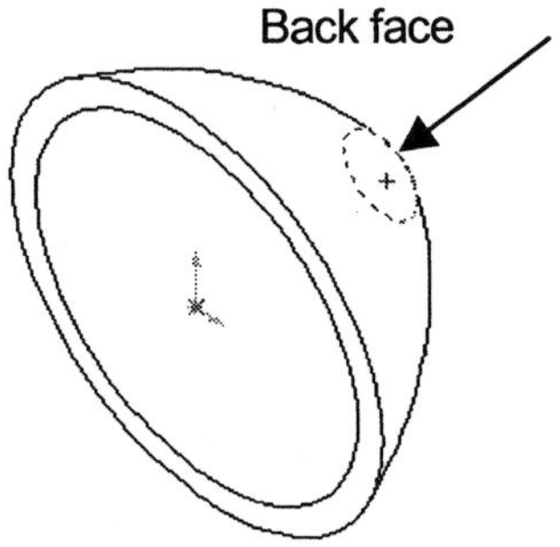

17) **Rotate** the part to view the back face.

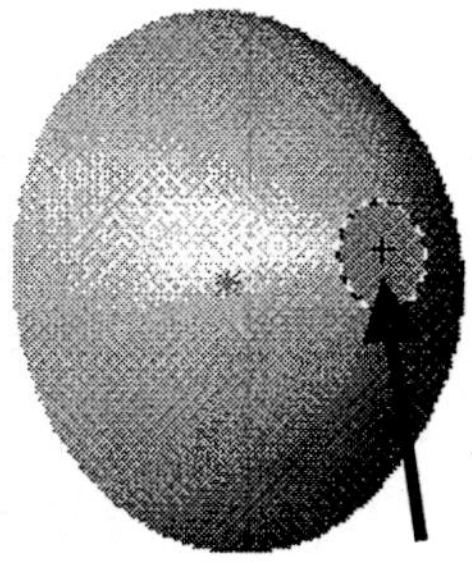

18) Create the profile. Click **Sketch**. Extract the **back face** to the Sketch plane. Click **Convert Entities**.

19) Extrude the Sketch. Click **Extrude Boss/Base**. Enter **.400**, [**10.16**] for Depth. Display the Boss-Extrude1 feature. Click **OK**.

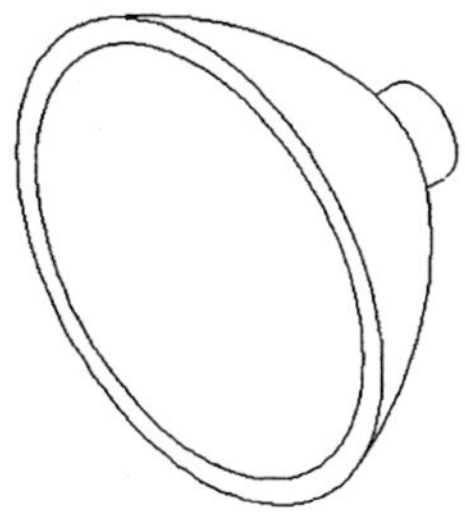

20) Rename **Boss-Extrude1** to **LensNeck**.

21) Save the LENS. Click **Save**.

Create the LENS – Hole Wizard Counterbore Hole Feature

The LENS requires a Counterbore Hole feature. Use the HoleWizard. HoleWizard assists in creating complex and simple Hole features. Specify the user parameters for the custom Counterbore Hole. Dimensions for the Counterbore Hole are provided both in inches and millimeters.

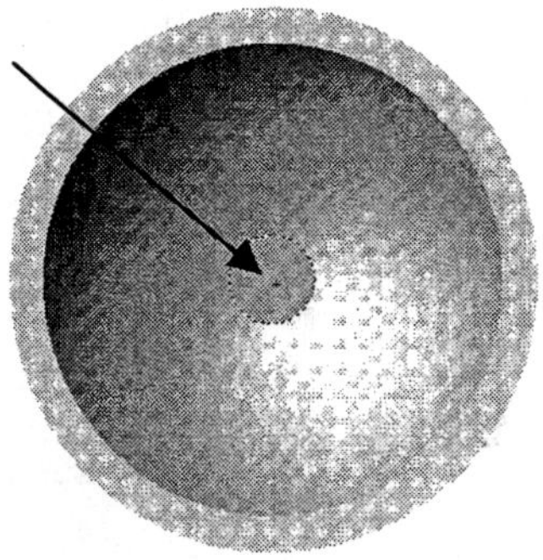

Create the Counterbore Hole.

22) Select the Sketch plane. Click **Front**. Click the small **inside back face** of the Base-Revolve feature.

23) Create the Counterbore Hole. Click **HoleWizard**. The Hole Definition dialog box is displayed. Click the **Counterbore** tab.

24) Define the parameters. Click the Parameter 1 **Binding** in the Screw type property text box. The Parameter 1 and Parameter 2 text boxes are displayed.

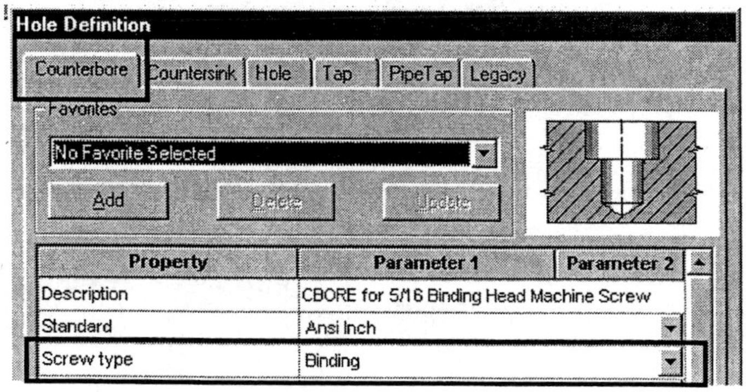

Note: For a metric hole, skip the next step.

For Inch Cbore Hole:

25) Select **Ansi Inch** for Standard. Enter **Hex Bolt** from the drop down list for Screw type. Select ½ from the drop down list for Size. Click **Through All** from the drop down list for End Condition & Depth. Accept the Hole Fit and Diameter value. Click the **C-Bore Diameter** value. Enter **.600**. Click the **C-Bore Depth** value. Enter **.200**.

Property	Parameter 1	Parameter 2
Description	CBORE for 1/2 Hex Head Bolt	
Standard	Ansi Inch	
Screw type	Hex Bolt	
Size	1/2	
End Condition & Depth	Through All	0.394in
Selected Item & Offset		0.000in
Hole Fit & Diameter	Normal	0.5312in
Angle at Bottom	118deg	
C'Bore Diameter & Depth	.6	.2

Note: For an inch hole, skip the next step.

For Millimeter Cbore Hole:

26) Select **Ansi Metric** for Standard. Enter **Hex Bolt** from the drop down list for Screw type. Select **M5** from the drop down list for Size. Click **Through All** from the drop down list for End Condition & Depth. Click the Hole Diameter value. Enter **13.5**. Click the **C-Bore Diameter** value. Enter **15.24**. Click the **C-Bore Depth** value. Enter **5**.

Property	Parameter 1	Parameter 2
Description	CBORE for M5 Hex Head Bolt	
Standard	Ansi Metric	
Screw type	Hex Bolt	
Size	M5	
End Condition & Depth	Through All	16.510mm
Selected Item & Offset		16.510mm
Hole Fit & Diameter	Normal	13.500mm
Angle at Bottom	118deg	
C'Bore Diameter & Depth	15.240mm	5.000mm

27) Add the new hole type to your favorites list. Click the **Add** button. Enter **CBORE FOR BULB**. Click **OK**.

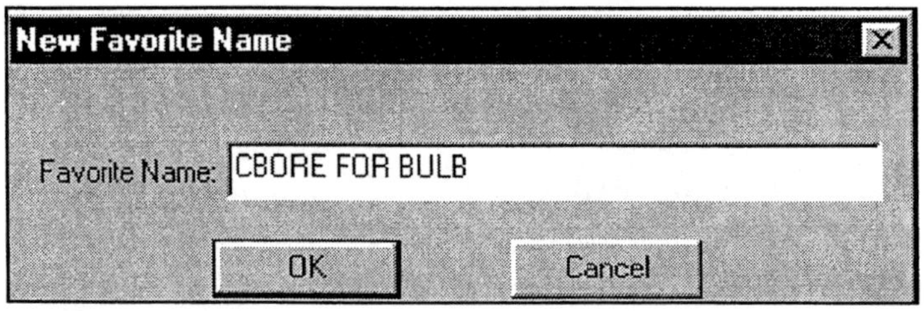

28) Click **Next** from the Hole Definition dialog box. The Hole Placement dialog box is displayed. Position the hole coincident with the Origin. Click **Add Relations** .

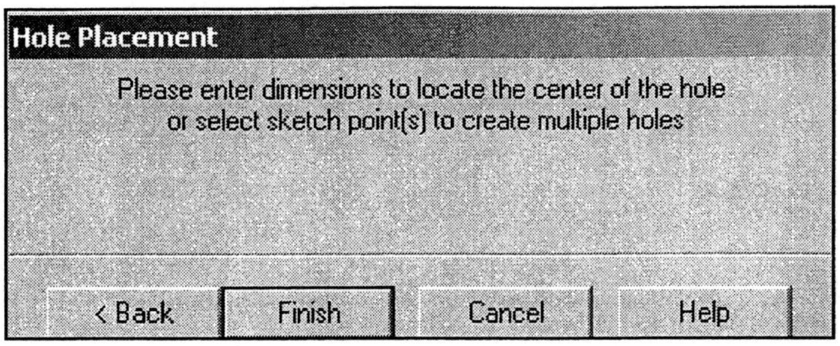

Click the **center point** of the Counterbore hole. Click the **Origin** . Click **Coincident**. Complete the hole. Click **Finish** from the Hole Wizard.

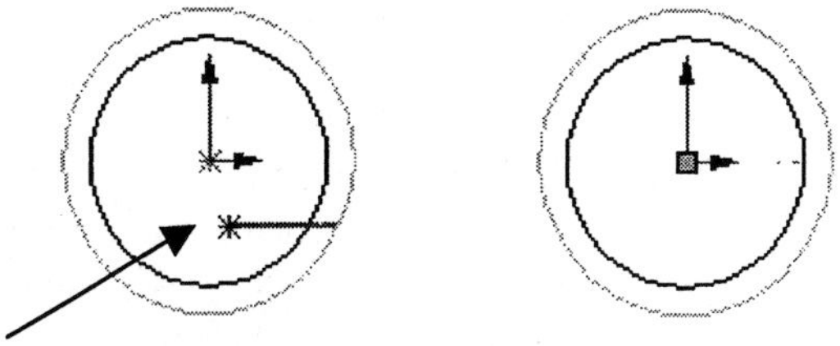

29) Expand the Hole. Click the **Plus Sign** to the left of the Hole feature. Sketch3 and Sketch4 are used to create the Hole feature.

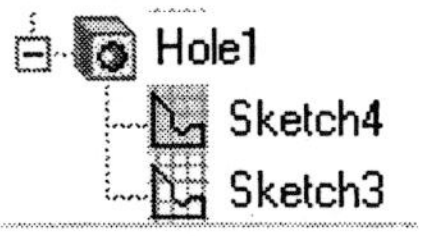

30) Display the Section view of BulbHole through the Right plane. Click the **Right** plane from the FeatureManager. Click **View** from the Main menu. Click **Display, SectionView**. Click the **Flip Side to View** check box. Click **OK**.

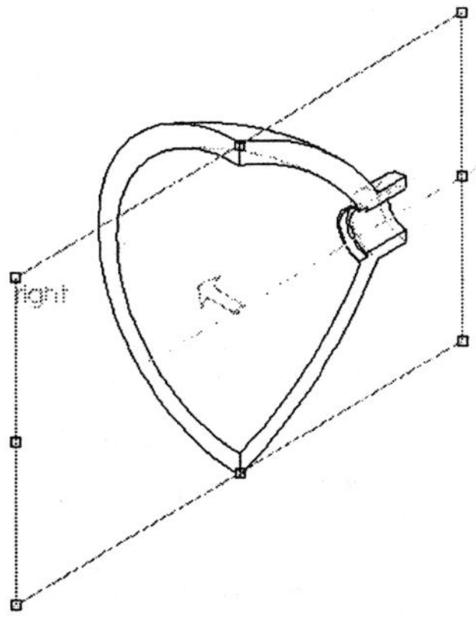

31) Display the Full view. Click **View, Display, SectionView.**

32) Display the Temporary Axis. Click **View, Temporary Axis.**

33) Rename **Cbore Hex Head Bolt** to **BulbHole.**

34) Save the LENS. Click **Save** .

Create the LENS - Boss Revolve Thin Feature

Create a Boss Revolve Thin feature. Rotate an open sketched profile around a centerline. The sketch profile must be open and cannot cross the centerline.

Use the Boss Revolve Thin feature to physically connect the LENS to the BATTERYPLATE in the FLASHLIGHT.

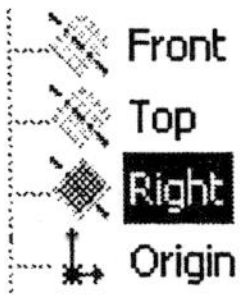

Create the Boss Revolve Thin feature.

35) Select the Sketch plane. Click the **Right** plane. Create the Sketch. Click **Sketch** .

36) Display the Right view. Click **Right** .

37) Sketch the centerline. Click **Centerline**. Sketch a **horizontal centerline** collinear to the Top plane through the Origin.

38) Select the right edge. **Right-click** in the Graphics window. Click **Select** from the Pop-up menu. Click the **right edge** of the Base feature.

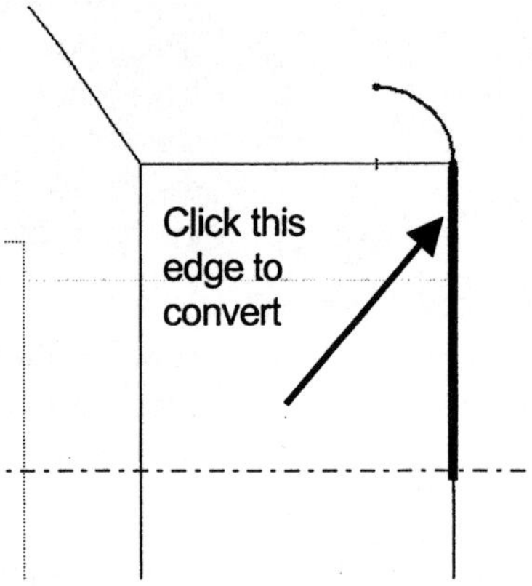

39) Click **Convert Entities**. Select the **edge**. Create an arc tangent to the extracted edge.

40) Click **TangentArc**. Click the **top point** of the vertical line. Drag the **mouse pointer** to the left. The mouse pointer displays a vertical line when the endpoint aligns with the arc center point. Create the 90° arc. Release the **left mouse button**.

Note: To create the 90° arc, the Snap to points in the Grid/Units must be unchecked.

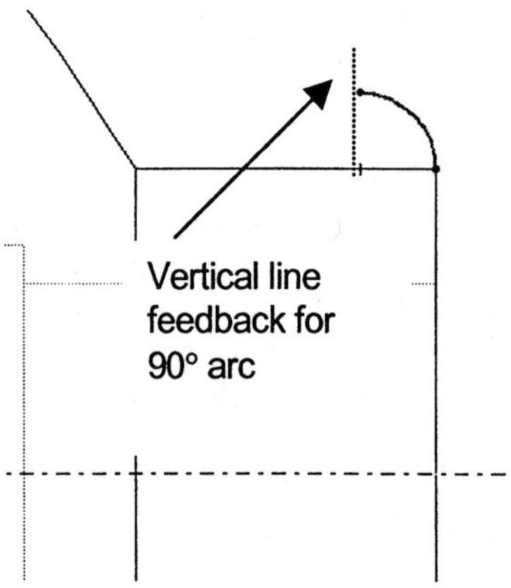

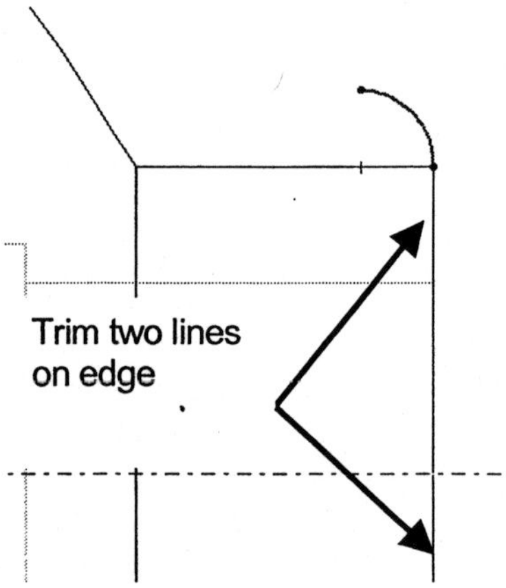

41) The vertical line segments are required to create the Tangent Arc. Remove the two line segments.

Click **Trim**. Click **both vertical edges**. The Sketch consists of an arc and a centerline.

42) Add a dimension. Click **Dimension**. Create a radial dimension. Enter **.100, [2.54]**.

The sketch arc requires a coincident relationship. This insures that the center point of the arc is coincident with the horizontal silhouette edge of the Base-Revolve feature.

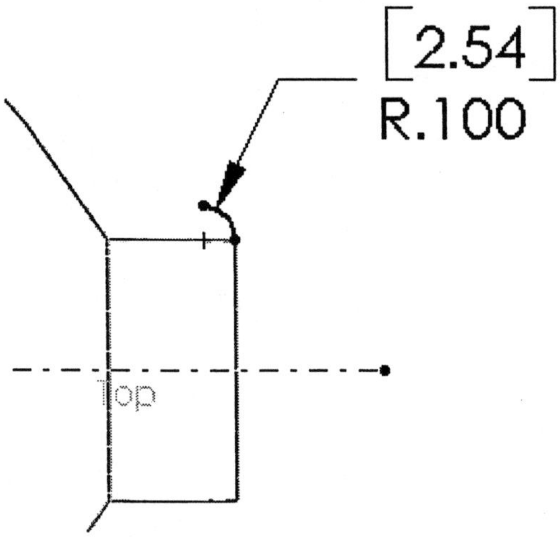

43) Add geometric relations.

Click **Add Relations**. Click the **arc center point**. Click the **horizontal line** (silhouette edge) of the Base-Revolve feature. Click the **Coincident** button. Click **Apply**. Click **Close**.

The black Sketch is fully defined.

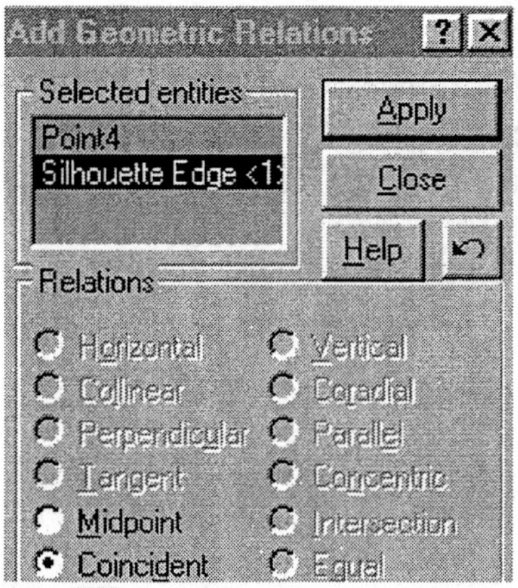

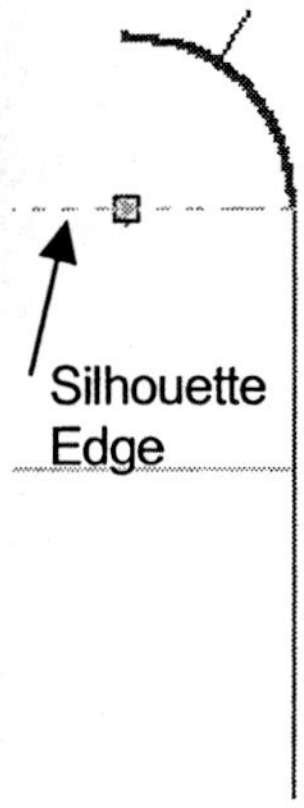

44) Revolve the Sketch. Click **Revolve**. A warning message appears:

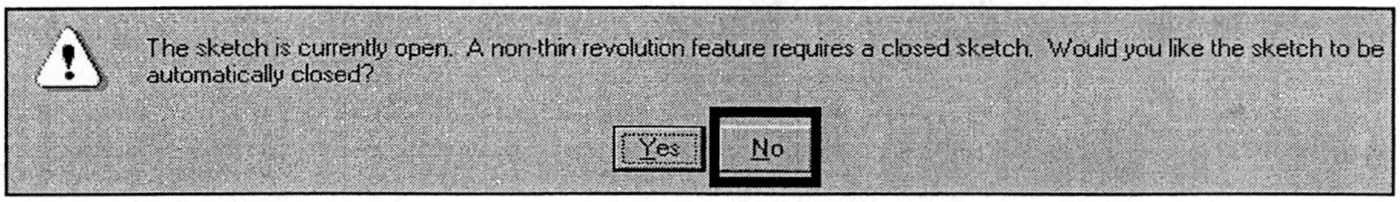

45) Keep the Sketch open. Click **No**. A second warning message appears:

46) Click **OK**. The Thin Feature check box is active.

47) Create the Thin-Revolved feature on both sides of the Sketch. Select **Mid-Plane** from the Type list box. Enter **.050**, [**1.27**] for Wall Thickness. Display the Boss-Revolve-Thin1 feature. Click **OK**.

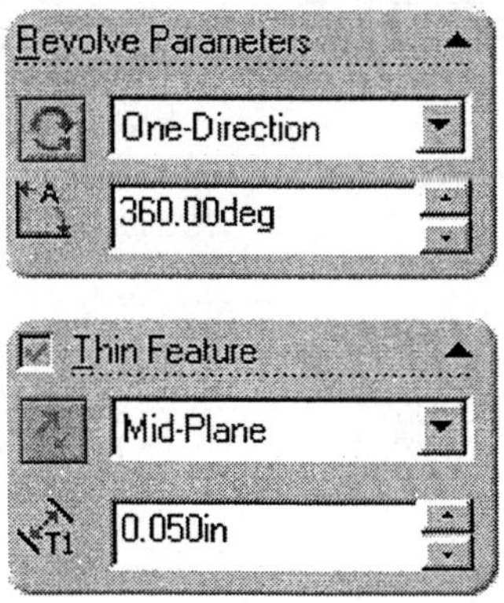

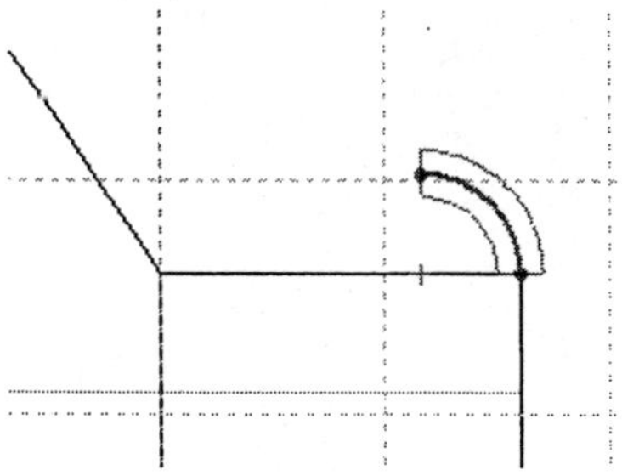

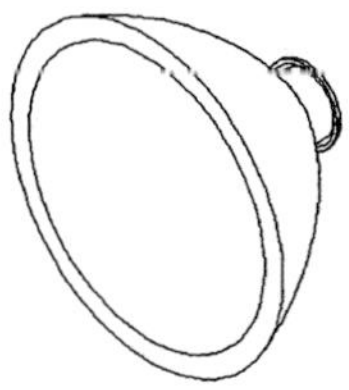

48) Rename **Boss-Revolve-Thin1** to **LensConnector**.

49) Save the LENS. Click **Save**.

50) Turn off the Temporary Axis. Click **View**, **Temporary Axis**.

Create the LENS - Extruded Boss Feature

Use the Extruded-Boss feature to create the front LensCover. The feature extracts the front outside circular edge from the Base-Revolve feature. The front LensCover is a key feature for designing the mating component. The mating component is the LENSCAP.

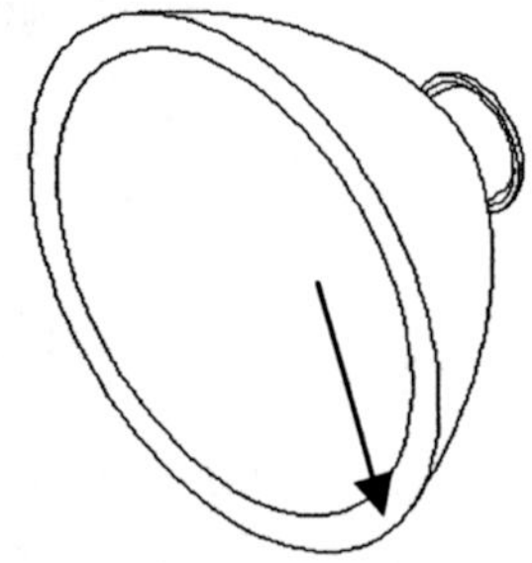

Create the Extruded Boss feature.

51) Select the Sketch plane. Click the **front circular face**.

52) Create the Sketch. Click **Sketch**.

53) Display the Front view. Click **Front**.

54) Click the **outside circular edge**. Click **Offset Entities**. Click the **Bi-directional** check box. Enter **.250, [6.35]**.

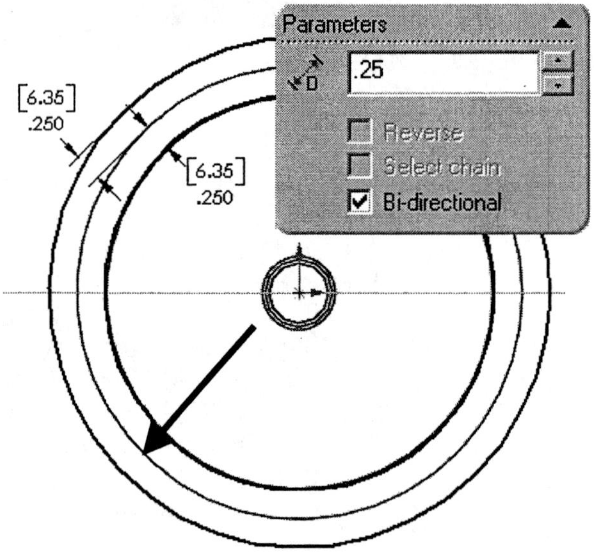

55) Extrude the Sketch. Click **Extrude Boss/Base**. Enter **.250**, **[6.35]** for Depth. Display the Boss-Extrude feature. Click **OK**.

56) Verify the position of the Boss Extrude. Click the **Top** view.

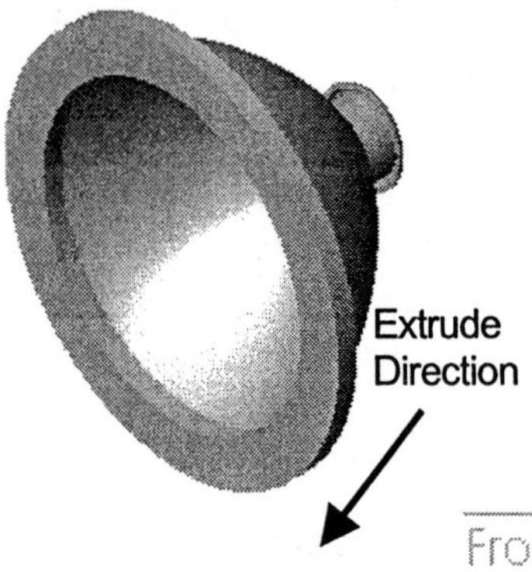

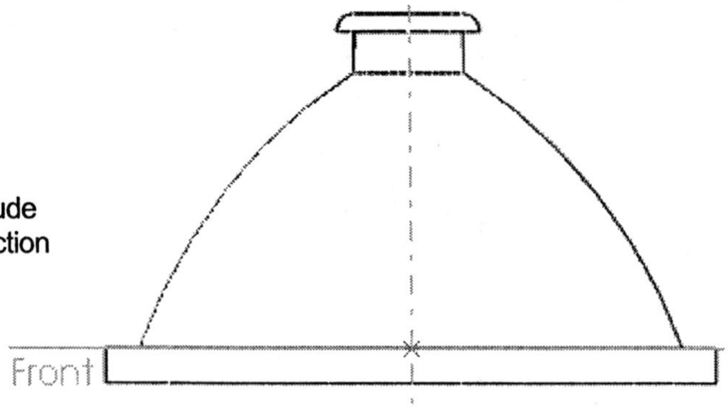

57) Rename **Boss-Extrude** to **LensCover**.

58) Save the LENS. Click **Save**.

Create the LENS - Extruded Boss Feature

An Extruded Boss feature is used to create the LensShield. The feature extracts the inside circular edge of the LensCover and places it on the Front plane. The LensShield feature is transparent in order to view the BULB and simulate clear plastic.

Create the Extruded Boss feature.

59) Select the Sketch plane. Click the **Front** plane.

60) Create the Sketch. Click **Sketch**.

61) Display the Front view. Click **Front**.

62) Sketch the profile. Click the **front inner circular edge** of the LensShield (Boss-Extrude2). Click **Convert Entities**. The circle is projected onto the Front Plane.

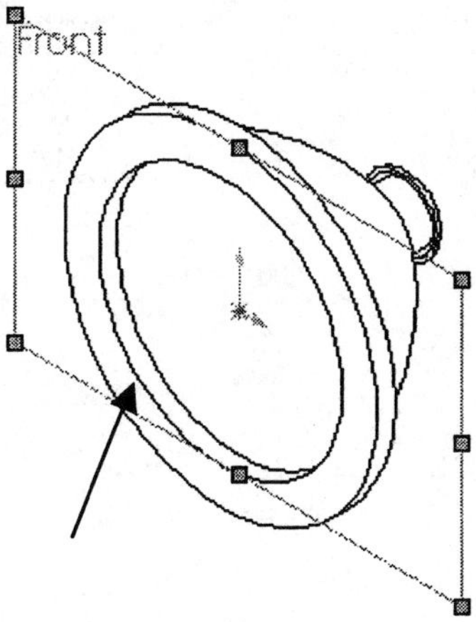

Front Plane

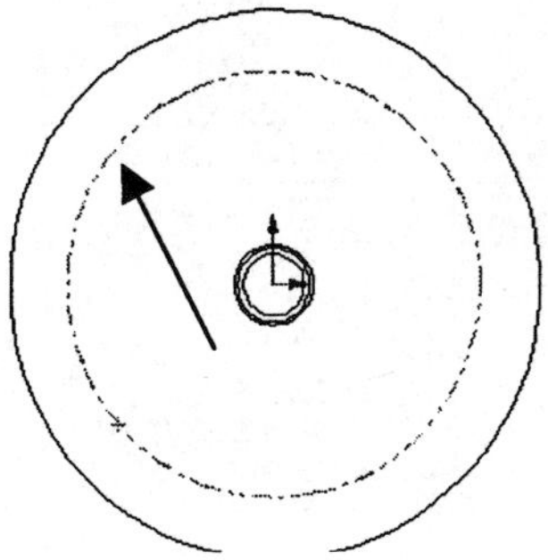

63) Extrude the Sketch. Click **Extruded Boss/Base**. Enter **.100, [2.54]** for Depth. Click **OK**.

Note: If you select the inside circular edge on Plane1, you will create a disjoint feature.

64) Rename **Boss-Extrude3** to **LensShield.**

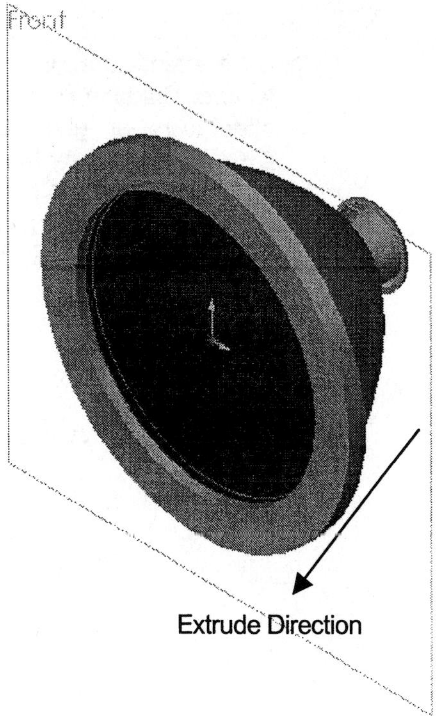

65) Add transparency to the LensShield. Right-click the **LensShield** in the Graphics window. Click **Feature Properties**. The Feature Properties dialog box is displayed.

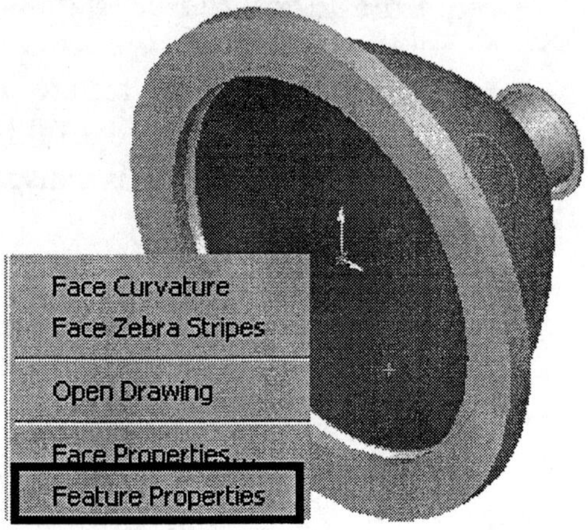

66) Click the **Color** button. The Entity Property dialog box is displayed. Click the **Advanced** button.

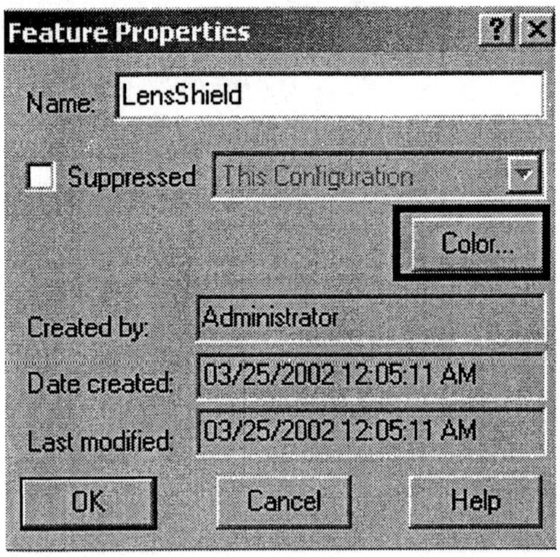

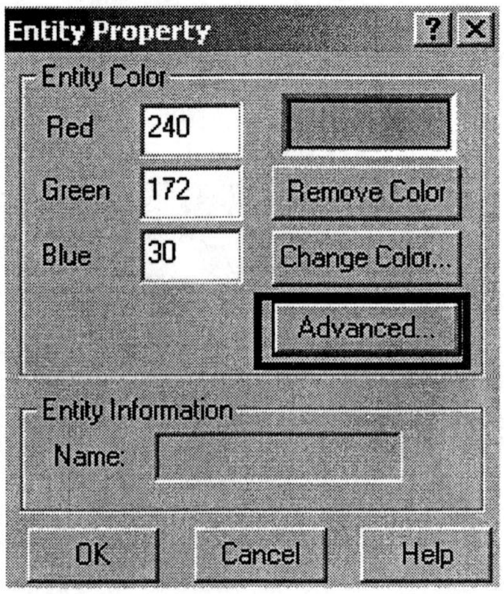

67) Set the transparency for the feature. Drag the **Transparency slider** to the far right side. Click **OK** from the Material Properties dialog box. Click **OK** from the Entity Property dialog box. Click **OK** from the Feature Properties box.

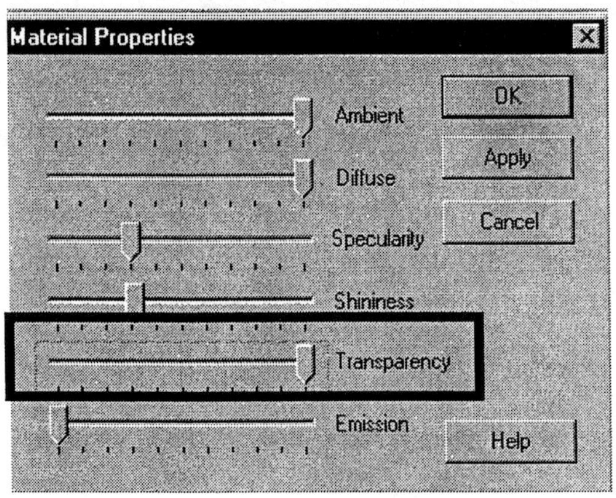

68) Display the transparent faces.

Click **Shaded**. When the LensShield is selected, the faces are not transparent. Click anywhere in the **Graphics window** to display the face transparency.

69) Save the LENS.

Click **Save**.

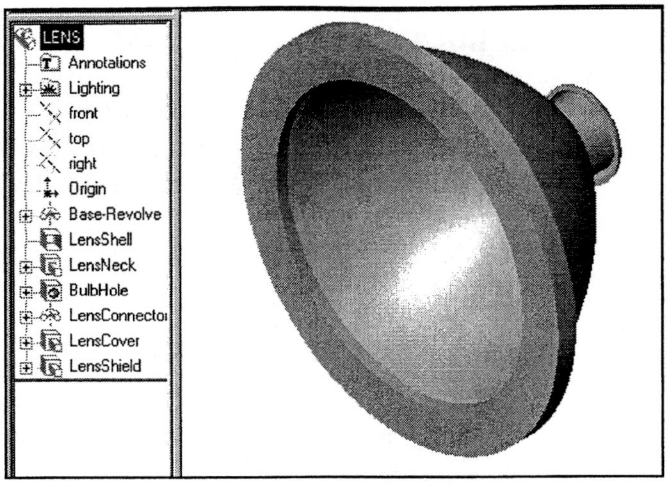

BULB

The BULB is contained within the LENS assembly. The BULB is a purchased part. The BULB utilizes the Revolved feature as the Base feature.

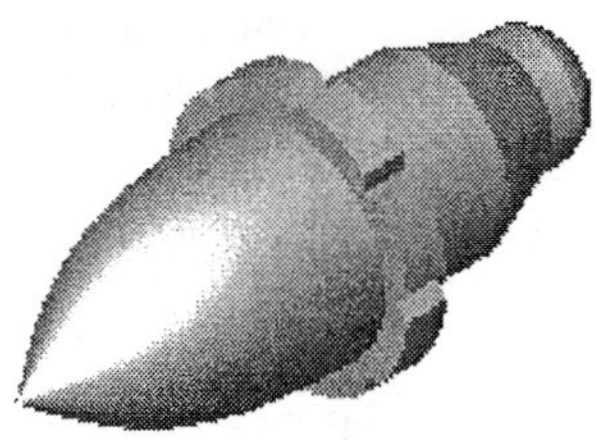

BULB Feature Overview

Create the Revolved Base feature from a sketched profile on the Right plane, Figure 2.11a.

Create a Revolved Boss feature using a B-Spline sketched profile. A B-Spline is a complex curve, Figure 2.11b.

Create a Revolved Cut Thin feature at the base of the BULB, Figure 2.11c.

Create a Dome feature at the base of the BULB, Figure 2.11d.

Create a Circular Pattern feature from an Extruded Cut, Figure 2.11e.

Figure 2.11a 2.11b 2.11c 2.11d 2.11e

Modify the BULB to practice Edit Definition and Edit Sketch after a design change.

Create the BULB - Revolved Base Feature

The solid Revolved Base feature requires a centerline and a sketched profile. The flange of the BULB is located inside the Counterbore Hole of the LENS. Align the bottom of the flange with the Front plane. The Front plane mates against the Counterbore face.

Create a Revolved Base feature.

70) Create the BULB. Click **New** . Click **PartEnglishTemplate**, **[PARTMETRICTEMPLATE]**. Click **OK**. Click **Save** . Select **My Documents** folder in the Save In text box. Enter the name of the part. Enter **BULB**. Click **Save**.

71) Select the Sketch plane. Click the **Right** plane. Create the Sketch. Click **Sketch** .

72) Show the three planes. Hold down the **Ctrl** key. Click **Front**, **Top** and **Right** from the FeatureManager. Right-click **Show**. Release the **Ctrl** key.

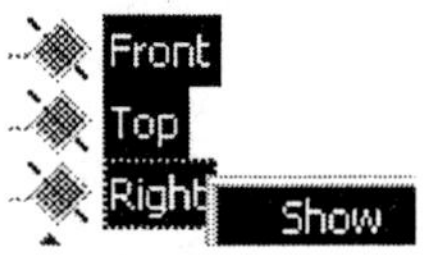

73) Display the Right view. Click **Right** .

74) Sketch the centerline. Click **Centerline** . Sketch a horizontal **centerline** collinear to the Top plane through the Origin .

75) Sketch the profile. Create six lines. Click **Line** .

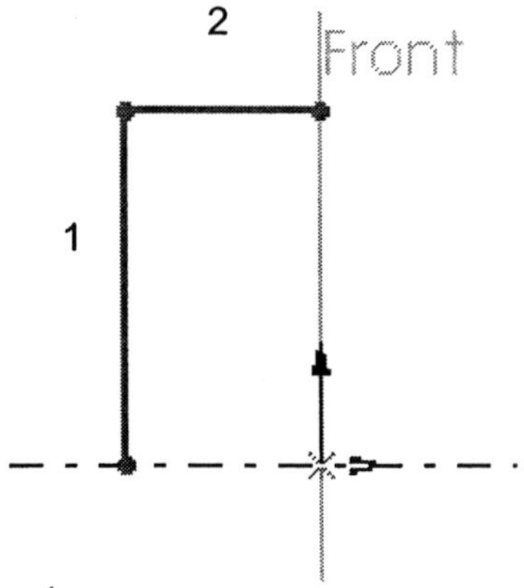

Create the first line. Sketch a **vertical line** to the left of the Front plane.

Create the second line. Sketch a **horizontal line** with the endpoint coincident to the Front plane.

Create the third line. Sketch a short **vertical line** towards the centerline, collinear with the Front plane.

Create the forth line. Sketch a **horizontal line** to the right.

Create the fifth line. Sketch a **vertical line** with the endpoint collinear with the centerline.

Create the sixth line. Close the Sketch. Sketch a **horizontal line.**

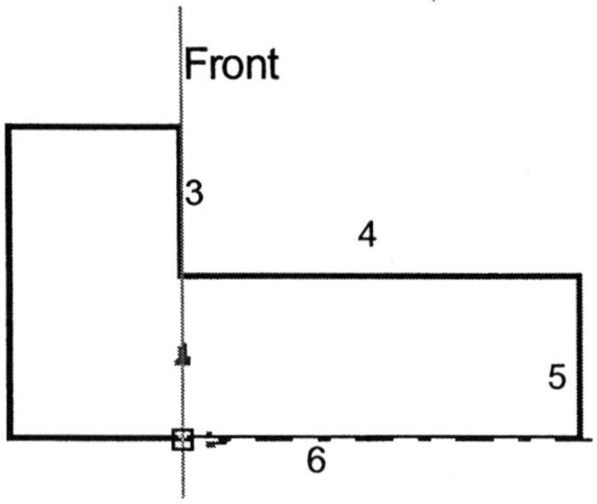

76) Add dimensions. Click **Dimension**.

Create a vertical linear dimension. Click the **right line**. Enter **.200**, **[5.08]**.

Create a vertical linear dimension. Click the **left line**. Enter **.295**, **[7.49]**.

Create a horizontal linear dimension. Click the **top left line**. Enter **.100**, **[2.54]**.

Create a horizontal linear dimension. Click the **top right line**. Enter **.500**, **[12.7]**.

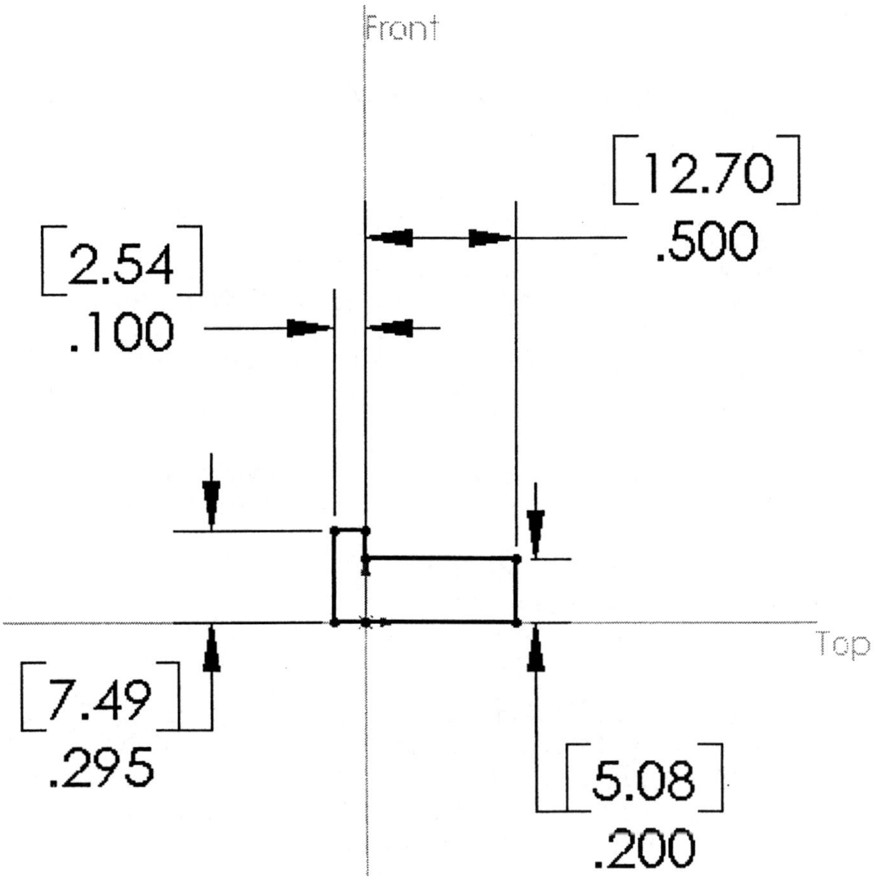

77) Revolve the Sketch. Click **Revolve** from the Feature toolbar. The Revolve Feature dialog box is displayed. Accept the default option values. Click **OK**.

78) Save the BULB. Click **Save**.

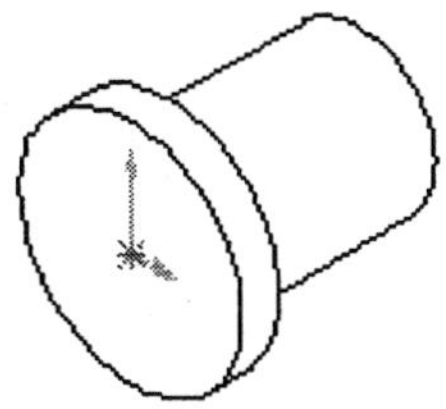

Create the BULB - Revolved Boss Feature

The bulb requires a second solid Revolve feature. The profile utilizes a complex curve called a B-Spline (Non-Uniform Rational B-Spline or NURB). B-Splines are drawn with control points. Adjust the shape of the curve by dragging the control points.

Create the Revolved Boss feature.

79) Turn the Grid Snap off. Click **Grid**. Uncheck the **Snap to points** check box.

80) Select the Sketch plane. Click the **Right** plane. Create the Sketch. Click **Sketch**. Display the Right view. Click **Right**.

81) Sketch the centerline. Click **Centerline**. Sketch a **horizontal centerline** collinear to the Top plane, coincident to the Origin.

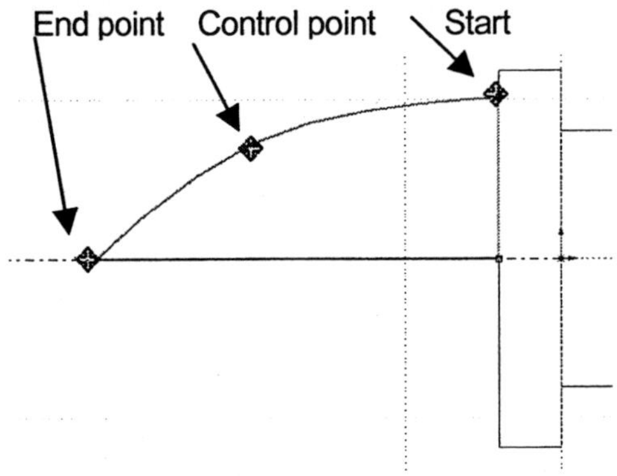

Sketch the profile. Click **B-Spline**. Sketch the start point. Click the **left vertical edge** of the Base feature.

Sketch the control point. Drag the **mouse pointer** to the left of the Base feature and below the first point. Release the **left mouse** button.

Sketch the end point. Click the **control point**. **Drag** the mouse pointer to the centerline. Release the **left mouse button.**

82) Adjust the B-Spline. Click **Select**. Position the **mouse pointer** over the B-Spline control point. Drag the **mouse pointer** upward. Release the **left mouse button.**

Note: SolidWorks does not require dimensions to create a feature.

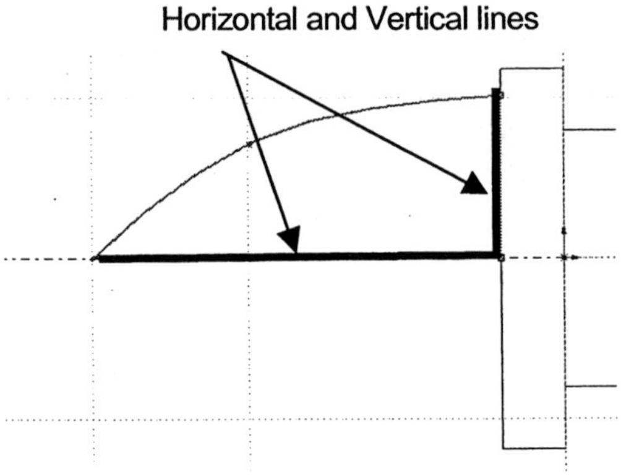

83) Complete the profile. Sketch two lines. Click **Line**. Create a horizontal line. Sketch a **horizontal line** from the B-Spline endpoint to the left edge of the Base-Revolved feature. Create a vertical line. Sketch a **vertical line** to the B-Spline start point, collinear with the left edge of the Base-Revolved feature.

84) Revolve the Sketch. Click **Revolve** from the Feature toolbar. The Revolve Feature dialog box is displayed. Accept the default options. Display the Revolve feature. Click **OK**.

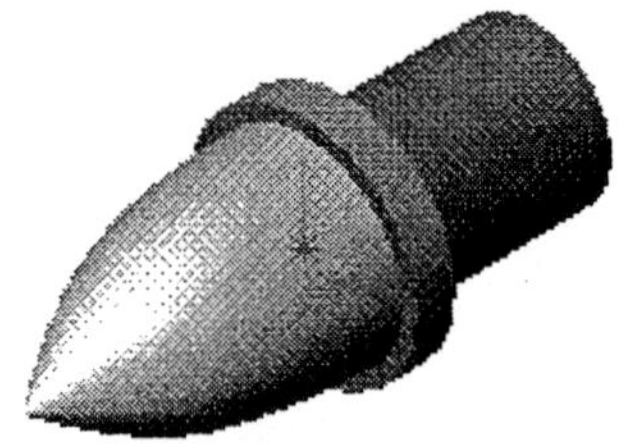

85) Save the BULB. Click **Save**.

Create the BULB - Revolved Cut Thin Feature

A Revolved Cut Thin feature removes material by rotating an open sketch profile around a centerline.

Create the Revolved Cut Thin feature.

86) Select the Sketch plane. Click the **Right** plane. Create the profile. Click **Sketch**. Display the Right view. Click **Right**.

87) Sketch the centerline. Click **Centerline**. Sketch a **horizontal centerline** collinear to the Top plane, coincident to the Origin.

88) Sketch the profile. Click **Line**. Sketch a line from the **midpoint** of the top silhouette edge downward and to the right. Sketch a horizontal line with the .260, [6.6] end point coincident with the vertical **right edge**.

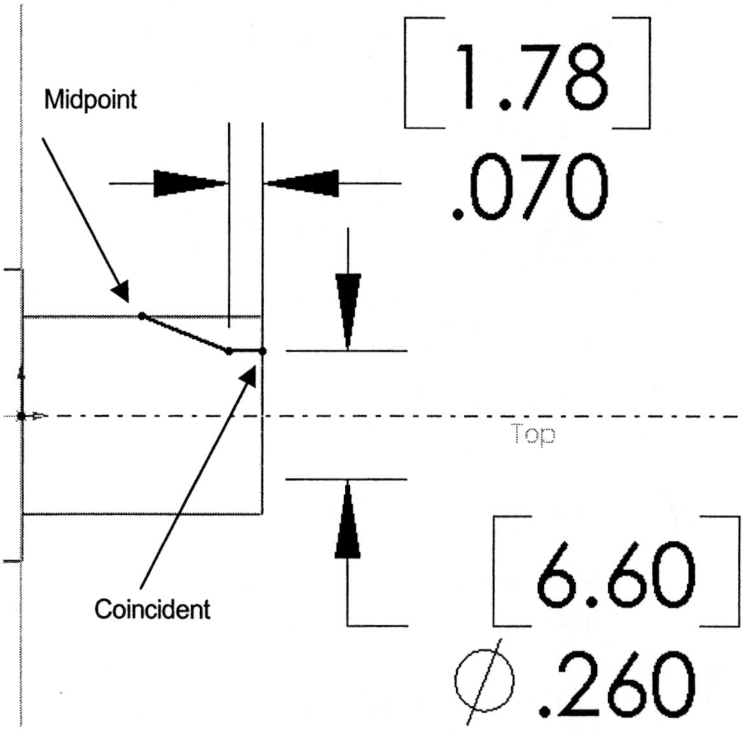

89) Add relations. Hold down the **Ctrl** key. Click the **start point** of the line. Click the top **Silhouette edge**. Release the **Ctrl** key. Click the **Midpoint** button. Click **OK**. Hold down the **Ctrl** key. Click the **end point** of the line. Click the right **vertical edge**. Release the **Ctrl** key. Click the **Coincident** button. Click **OK**.

90) Add dimensions. Click **Dimension**. Create the diameter dimension. Click the **centerline**. Click the **short horizontal line**. Enter **.260**, [**6.6**]. Add a horizontal dimension. Click the **short horizontal line**. Enter **.070**, [**1.78**]. The black Sketch is fully defined.

Note: The ∅.260 is displayed as a diameter dimension. Right-click Properties, uncheck the Display diameter check box to display a radius value.

91) Revolve the Sketch. Click **Revolved Cut** from the Feature toolbar. Click **No** to the Warning Message, "Would you like the sketch to be automatically closed?" Click **OK** to the Warning Message, "The profile is only suitable for a thin feature".

92) The Cut Revolve Thin Feature dialog box is displayed. The direction arrow points away from the centerline. Click the **Direction** button. Enter **.150**, **[3.81]** for Thickness. Display the Revolved Cut Thin feature. Click **OK**.

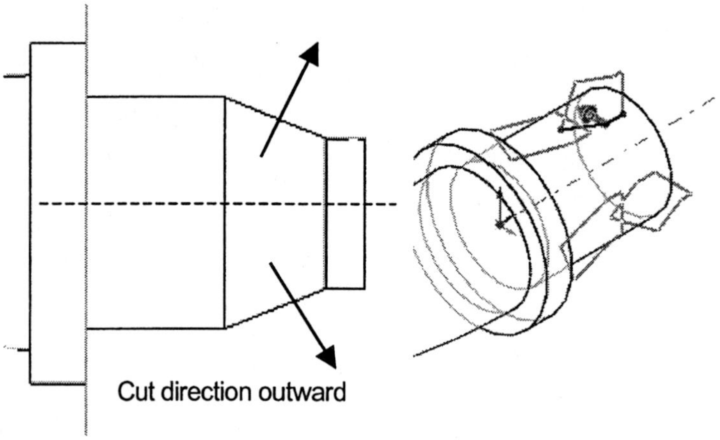

93) Save the BULB. Click **Save**.

Create the BULB - Dome Feature

A Dome feature creates spherical or elliptical shaped geometry. Use the Dome feature to create the Connector feature of the BULB.

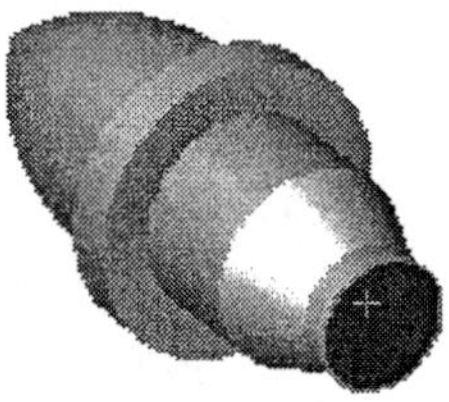

Create the Dome feature.

94) Select the Sketch plane. Click the **back circular face** of the Revolve Cut Thin.

95) Click **Insert** from the Main menu. Click **Features**, **Dome**. The Dome dialog box is displayed. Enter **.100**, **[2.54]** for Height. Display the Dome. Click **OK**.

96) Save the BULB. Click **Save**.

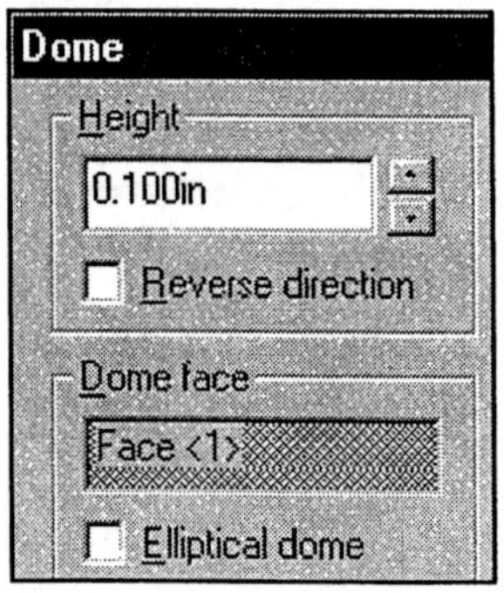

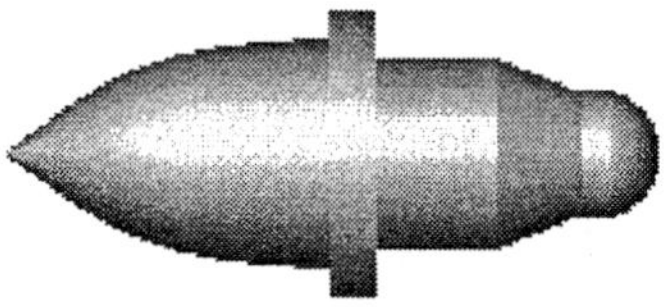

Create the BULB - Circular Pattern

The Pattern feature creates one or more instances of a feature or a group of features. The Circular Pattern feature places the instances around an axis of revolution.

The Pattern feature requires a seed feature. The seed feature is the first feature in the Pattern. The seed feature in this section is an Extruded-Cut.

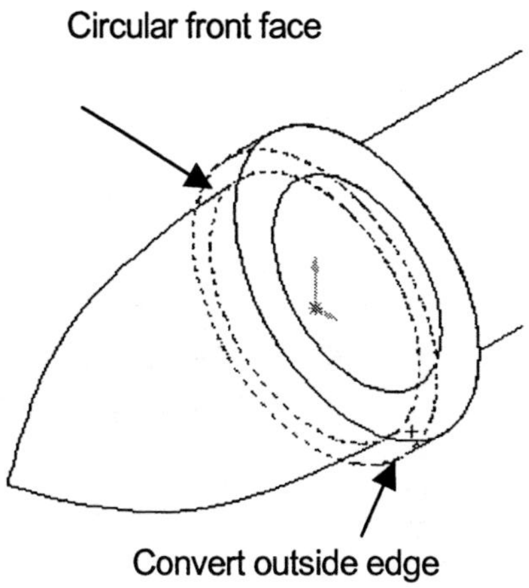

Create the Circular Pattern.

97) Select the Sketch plane. Click the **front circular face** of the Base feature.

98) Create the Sketch. Click **Sketch**.

99) Extract the outside circular edge. Click **Select**. Click the **outside circular edge**. Click **Convert Entities**.

100) Display the Front view. Click **Front**.

101) Show the Right plane. Click the **Right** plane in the FeatureManager. Right-click **Show.**

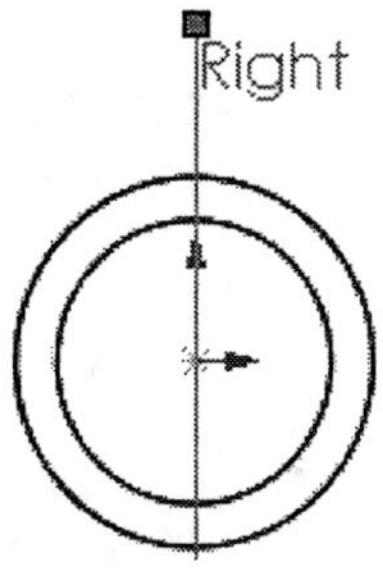

102) Sketch the centerline. Click **Centerline**. Sketch a **vertical centerline** coincident with the top and bottom circular circles and coincident with the Right plane.

103)Zoom to display the centerline and the outside circular edge.

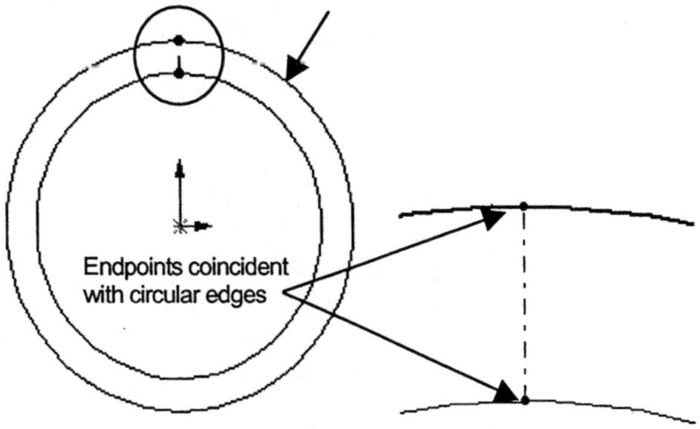

104) Sketch a V-shaped line. Click **Mirror**. Select the **centerline**. Click **Line**.

Create the first point. Click the **midpoint** of the centerline. Create the second point. Click the coincident **outside circle edge**. Turn the Mirror off.

Click **Mirror**.

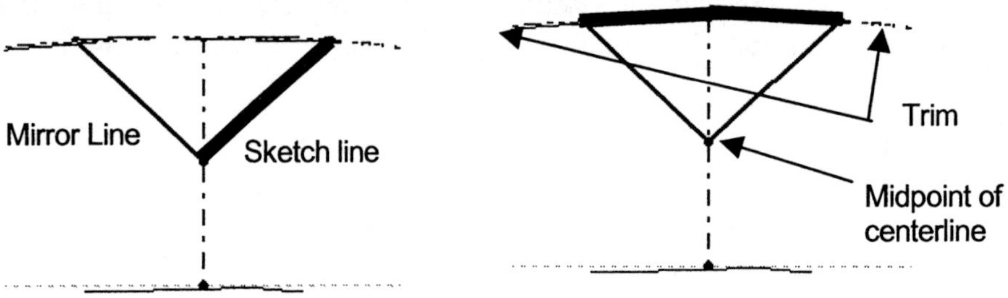

105) Trim the lines. Click **Trim**. Click the **circle** outside the V shape.

106) Add the geometry relations. Hold down the **Ctrl** key. Click the **two lines**. Click the **Perpendicular** button. Release the **Ctrl** key. The black Sketch is fully defined.

107) Extrude the Sketch. Click **Extruded Cut**. Click **Up to Next** from the Type list box. Display the Extruded Cut. Click **OK**.

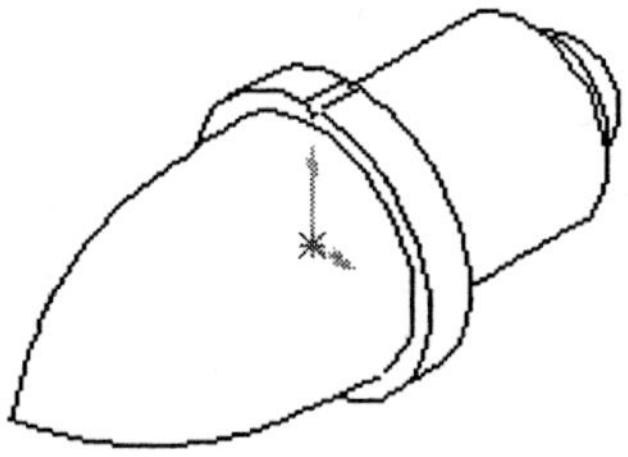

108) Display the Temporary axis. Click **View, Temporary Axis** from the Main menu.

The Cut-Extrude is the seed feature for the Pattern.

109) Create the Pattern. Click the **Cut-Extrude** feature. Click **Circular Pattern**. The Circular Pattern dialog box is displayed. Click the **Direction selected** text box. Click **Temporary Axis**. Create 4 copies of the Cut. Enter **4** in the Total Instances spin box. Click the **Equal spacing** check box. Click the **Geometry pattern** check box. Display the Pattern feature. Click **OK**.

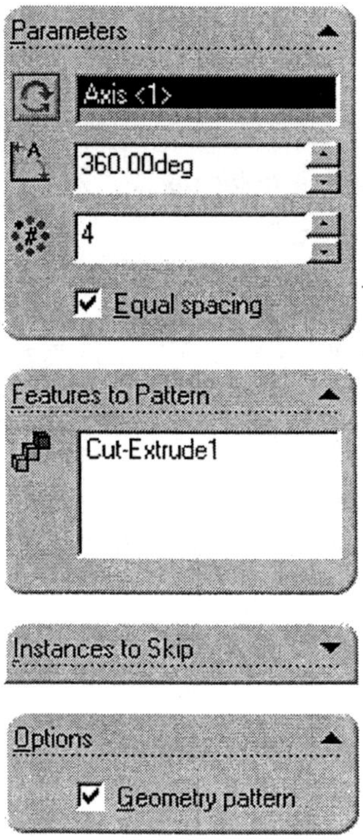

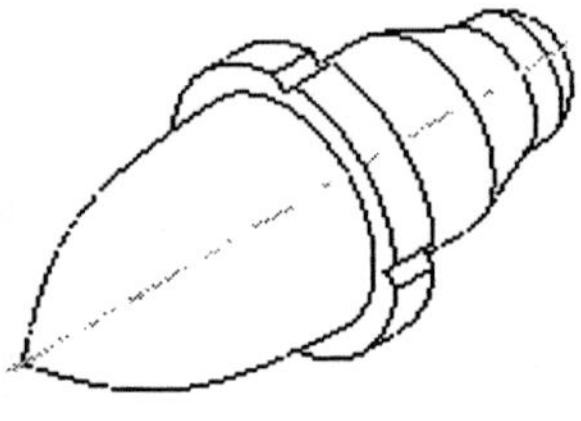

110) Edit the Pattern feature. Right-click on the **Circular Pattern** from the Feature Manager. Click **Edit Definition.** Enter **8** in the Total instances spin box. Display the updated Pattern. Click **OK.**

111) Hide the Temporary axis. Click **View** from the Main menu. Click **Temporary Axis.** Hide the Planes. Click **Planes** from the View menu.

112) Save the BULB. Click **Save.**

Design Change with Rollback

You are required to implement a design change for the BULB. The BULB requires a small fillet on the front outside circular face. The Fillet feature is created before the v-shaped Extruded Cut and Circular Pattern.

The Rollback and Edit Definition functions are used to implement the design change.

The Rollback function allows a feature to be redefined in any state or order. Implement the design change. Add the new Fillet feature before the Extruded Cut feature. Reorder the Fillet feature in the FeatureManager and view the results.

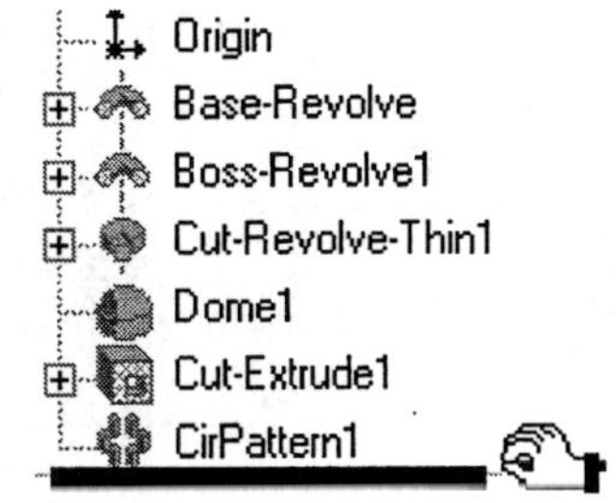

Create the Fillet.

113) Position the Rollback bar. Place the **mouse pointer** over the yellow Rollback bar at the bottom of the FeatureManager design tree. The mouse pointer displays a symbol of a hand. Drag the **Rollback** bar upward to below the Dome feature.

114) Click the outside **front circular edge** of the BULB. Click **Fillet.** Enter **0.01** for Fillet Radius. Click **OK.**

115) Position the Rollback bar. Drag the **Rollback** bar to the bottom of the FeatureManager.

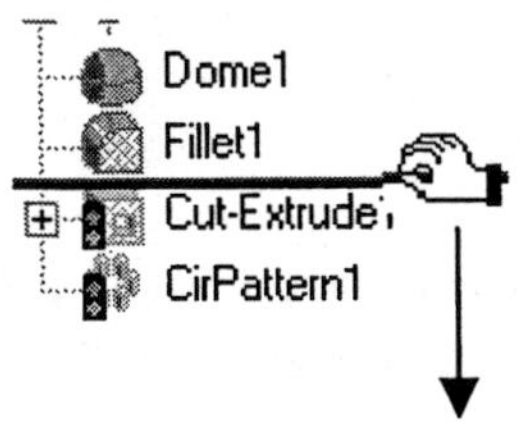

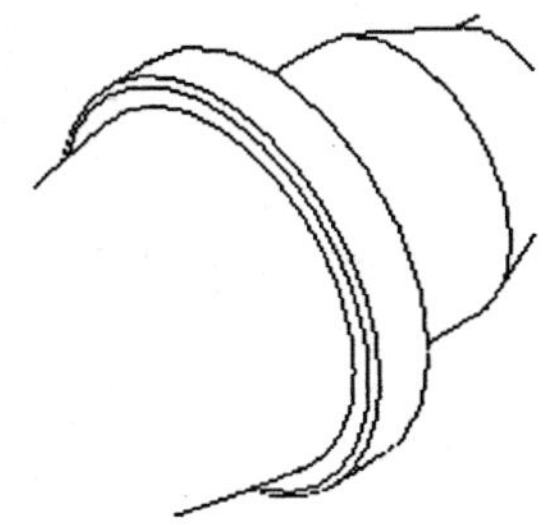

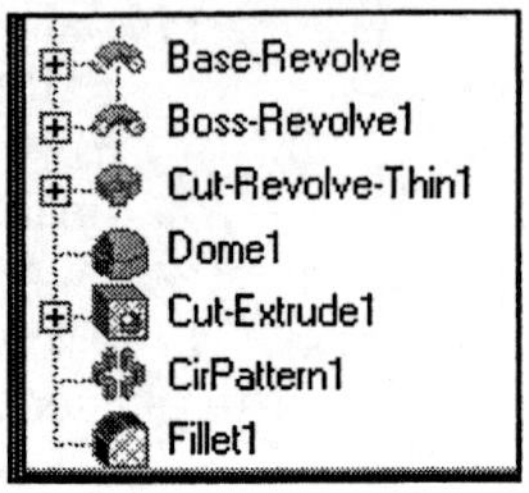

116) Reorder Fillet1. Drag the **Fillet1 text** to the bottom of the FeatureManager.

117) A Fillet Error message occurs. The circular edge is no longer valid. This is not the design intent. Drag the **Fillet1 text** before the Cut-Extrude1 text.

118) Suppress the Circular Pattern. Right-click **CirPattern1** in the FeatureManager. Click **Suppress**.

119) Save the BULB. Click **Save**.

The v-shape Extruded Cut requires a 2D sketch plane. The Extruded Cut fails when the Fillet radius becomes too large and removes the sketch plane. Creating features on curved surfaces with reference planes is discussed in the next project.

A suppressed feature is a feature that is not displayed. It is useful to hide features for clarity. Suppressed features improve model Rebuild time.

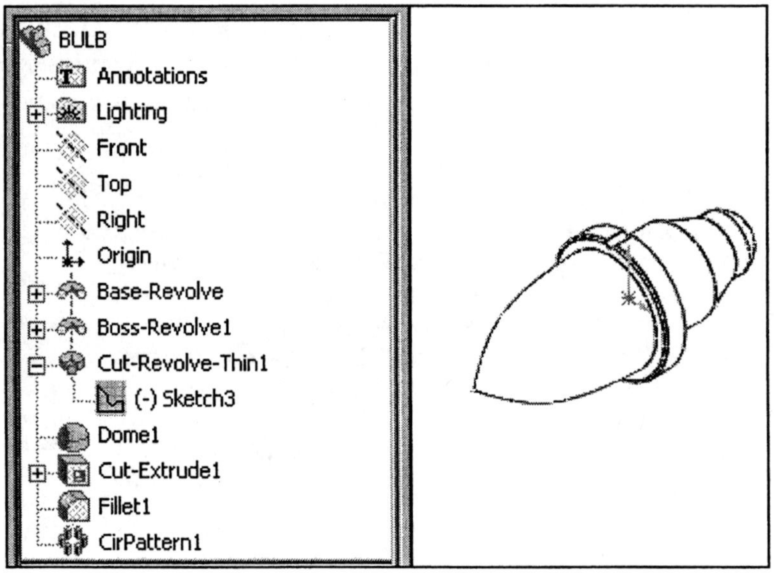

Customizing Toolbars

The default Toolbars contains numerous icons that represent basic functions. SolidWorks contains additional features and functions not displayed on the default Toolbars.

Customize the Toolbar.

120) Place the Dome icon and the Rib icon on the Features Toolbar. Click **Tools** from the Main menu. Click **Customize**. The Customize dialog box is displayed.

121) Click the **Commands** tab. Click **Features** from the category text box. Drag the **Dome** icon into the Features Toolbar.

You have just created two parts:

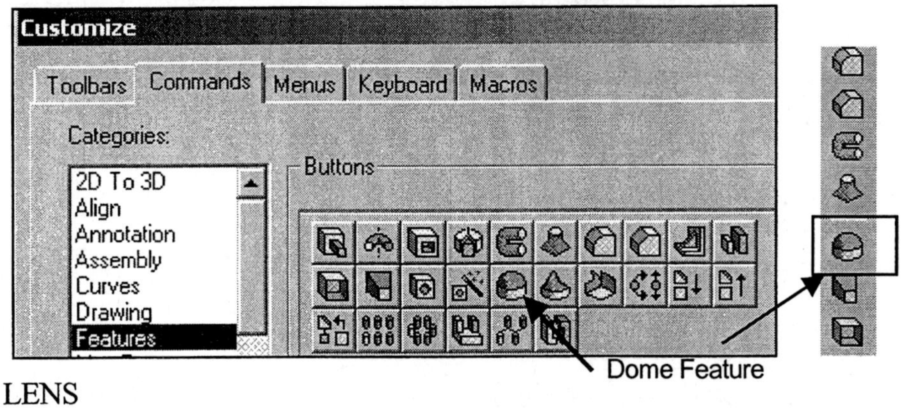

- LENS
- BULB

Practice the exercises before moving onto the next section.

Questions

1. Identify the function of the following features:

   Revolved Base

   Revolved Cut Thin

2. Name the two line types required in the sketch of a Revolved feature.
3. What is the function of the Shell feature?
4. An arc requires _______ points?
5. Name the required points of an arc?
6. When do you use the Hole Wizard feature?
7. Describe the Mid Plane option for a Revolved Thin feature.
8. What is a B-Spline?
9. Identify the required information for a Circular Pattern?
10. How do you control the display of the Temporary Axis?
11. Define Rollback in the FeatureManager.
12. How do you add the Dome feature icon to the Feature Toolbar?

Exercises

Create the following Parts:

Revolved

Exercise 2.1: Create the SIMPLE SCREW.

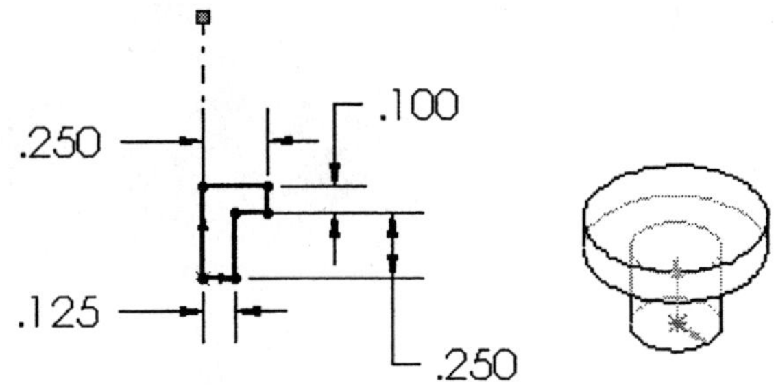

Exercise 2.1

Exercise 2.2: Create the SIMPLE CAP SCREW.

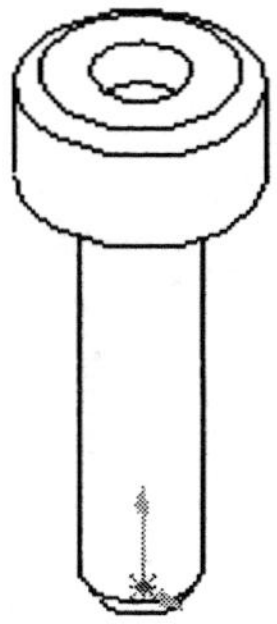

Exercise 2.2

Exercise 2.3: Create the SPOOL.

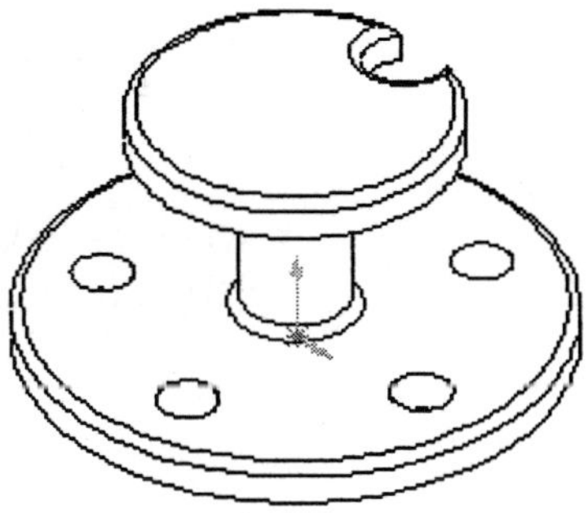

Exercise 2.3

Exercise 2.4: Create a TRAY and GLASS. Use real objects to determine the overall size and shape of the Base feature. Below are a few examples.

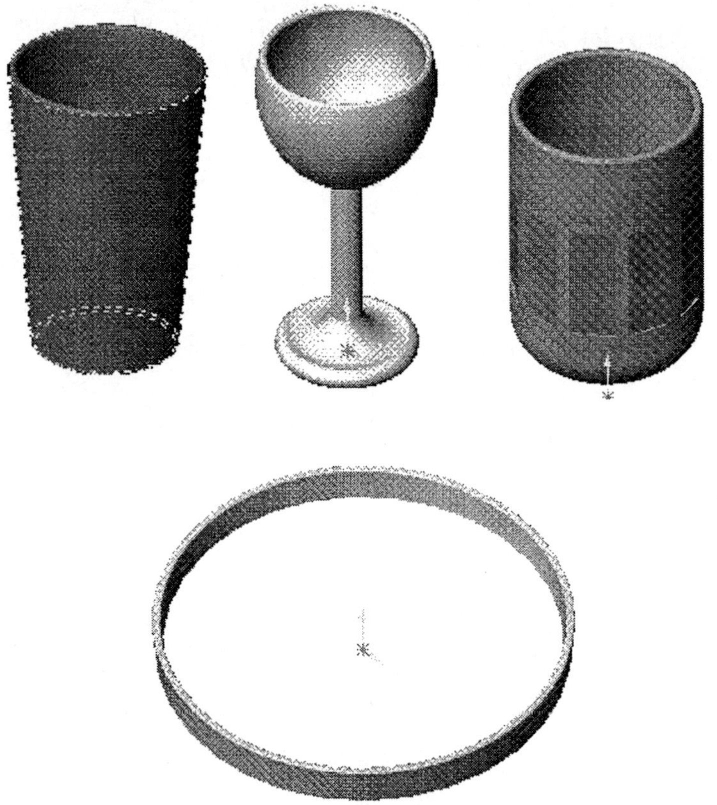

Exercise 2.4

Exercise 2.5: Create a JAR-BASE. Use the dimensions from the JAR-BASE to determine the size of the JAR-COVER.

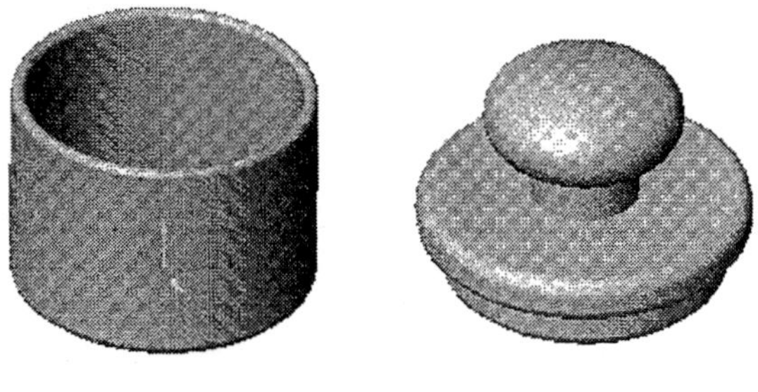

Exercise 2.5

Exercise 2.6: Industry Collaborative Exercise.

Note: PhotoWorks Add-In is required to complete this exercise. Select Tools, Add-Ins. Select PhotoWorks. PhotoWorks toolbar and On-Line Help are added to the Main menu.

Create picture images of the BATTERY and LENS parts. Incorporate the new images into a PowerPoint presentation.

PhotoWorks software application renders photo-realistic images of SolidWorks 2001Plus parts and assemblies. PhotoWorks generates an image directly from the view in the active SolidWorks Graphics window. Rendering effects include materials, lighting, image background, image quality and image output format. A scene contains a combination of these rendering effects.

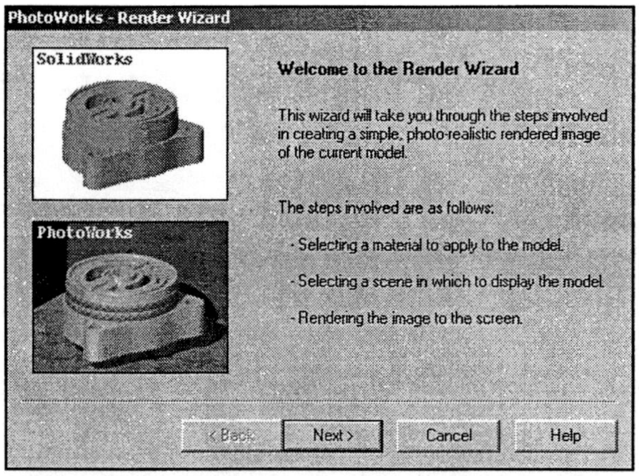

The Render Wizard  steps you through the process to apply a material to the model, select a scene to display the model and render the image to the screen.

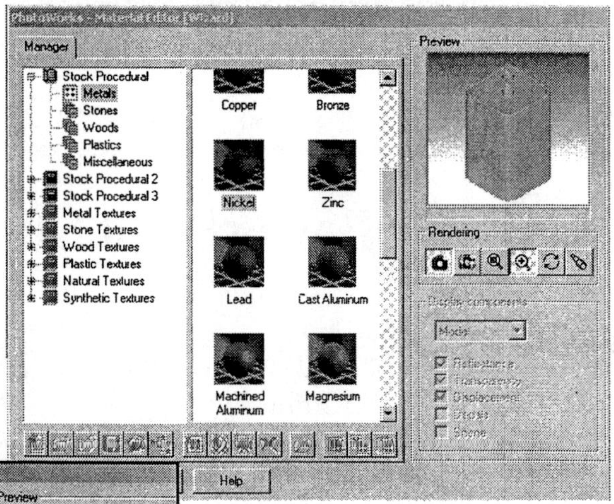

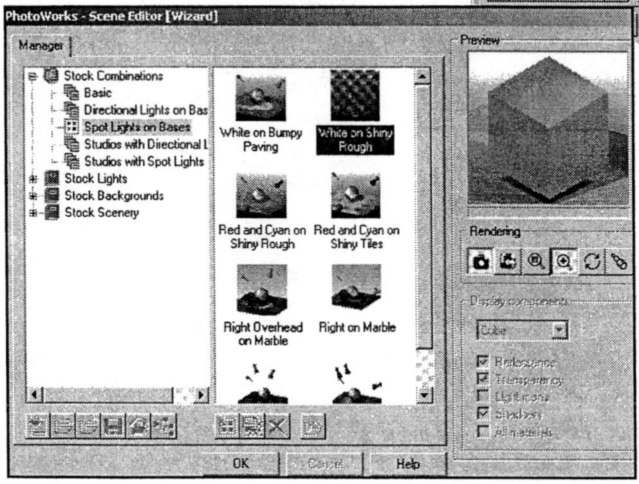

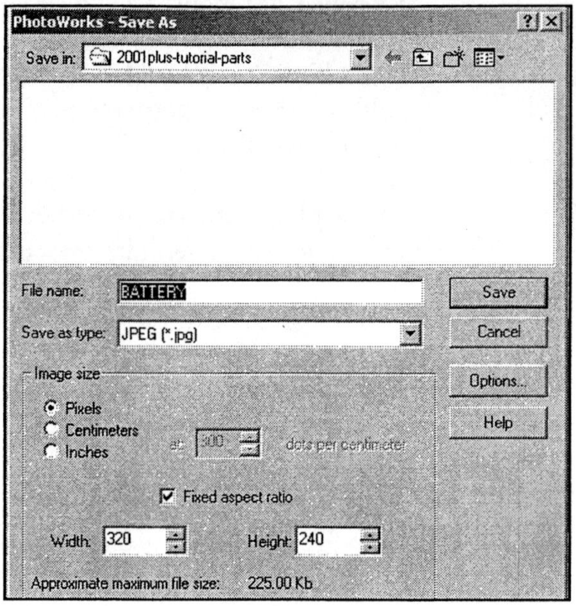

Save the image in JPEG format.

Click Render Image to File. Select JPEG for Save as Type. The default picture size is 320 x 240 pixels.

Insert the BATTERY.JPG file as a Picture into a new PowerPoint document.

Repeat the above procedure for the LENS. Select a different material and scene.

Additional PhotoWorks examples are found in Help, Getting Started and Help PhotoWorks Help Topics.

Note: Image size and file type affect file size and render time. Images below were created with 320 x 240 pixels and cropped. JPEG format creates a smaller file size than Bitmap for PowerPoint and web-based presentations. The Lens Shield feature is suppressed.

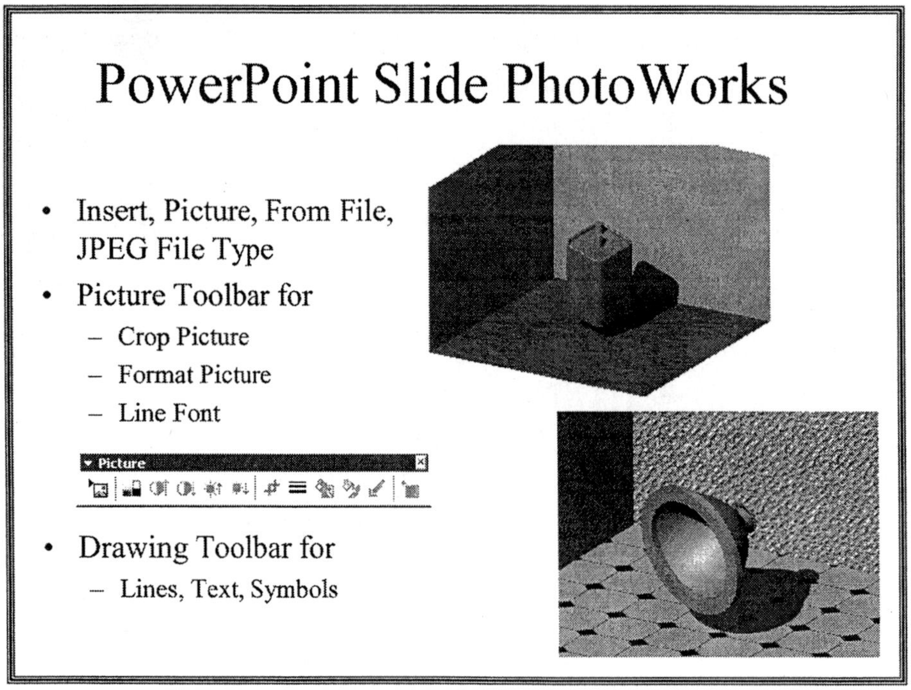

Exercise 2.7: Industry Collaborative Exercise.

Note: SolidWorks Animator Add In is required for this exercise. Select Tools, Add-Ins, SolidWorks/Animator.

Create a rotate model animation of the LENS. Incorporate the new animation into a PowerPoint presentation.

SolidWorks Animator add-in software application captures motion and animates SolidWorks assemblies. SolidWorks Animator generates Windows-based animations (**.avi** files) that you can support on any Windows-based computer.

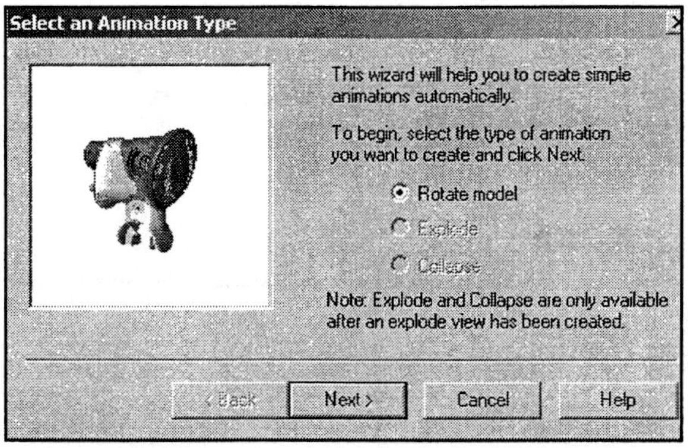

Start with the LENS part. Click Tools, Add Ins. Check the SolidWorks Animator check box. Create a rotate model animation of the LENS. Click the Animator Wizard . Click Rotate Model. Click Next. Select Y-axis for axis of roation. Click Next. Enter 20 seconds for Duration. Enter 0 for Start Time. Click Finish to view the rotation.

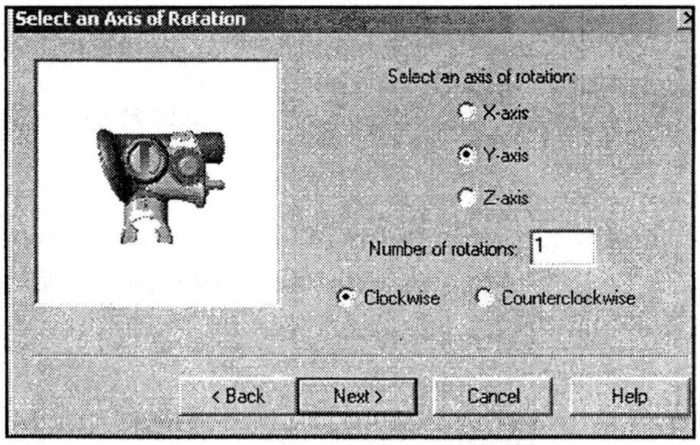

Record the animation and create the .AVI file. Click Record Animation to File . The Renderer is the SolidWorks screen. Microsoft Video 1 is the default Video Compression. The time required to create the .AVI file is greater than the time required to create the SolidWorks/Animator screen file.

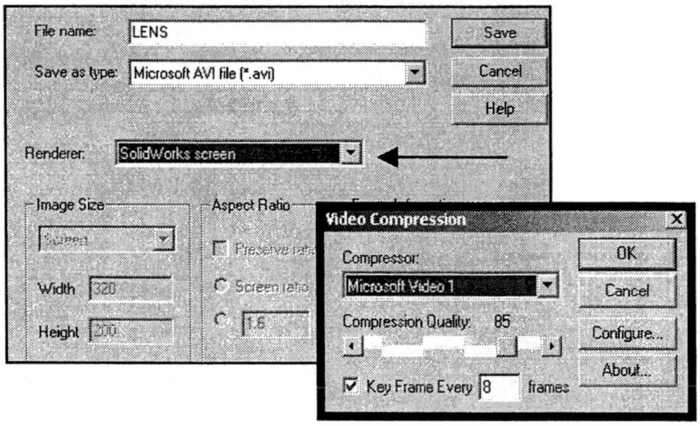

Play the .AVI file with the Windows Media Player.

Insert the LENS.AVI file as a Movie into PowerPoint and create a presentation.

Note: LENS.AVI create with PhotoWorks Renderer. Image size 320 x 240 pixels.

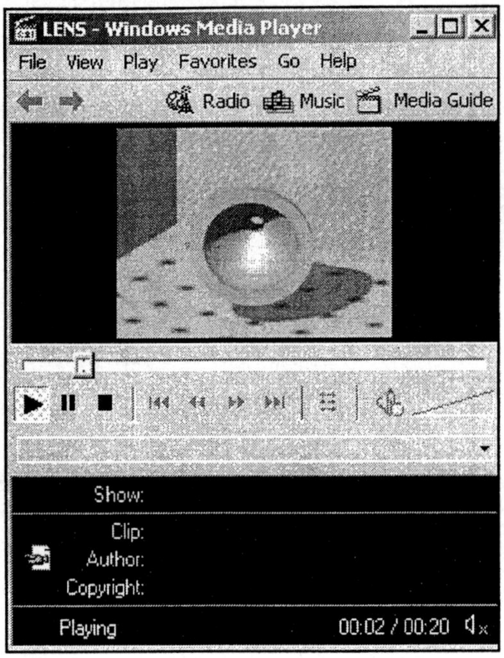

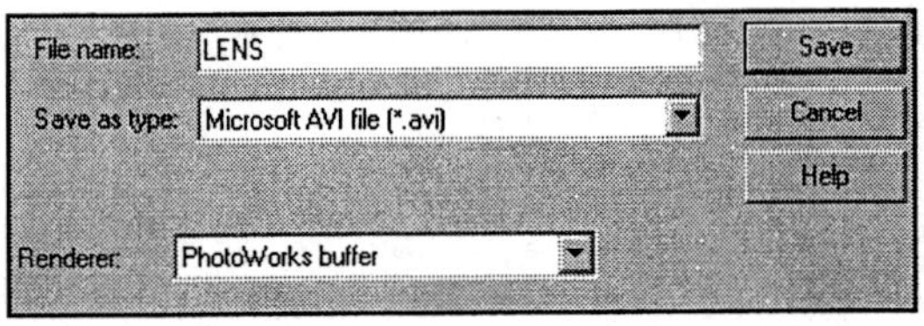

PowerPoint with SolidWorks/Animator

Insert, Movie, Movie From File

Crop the .avi file to fit onto the PowerPoint slide

Add text and border

Change animation timing with Slide Show, Custom Animation

Double-click to run .avi

Time = 2 seconds

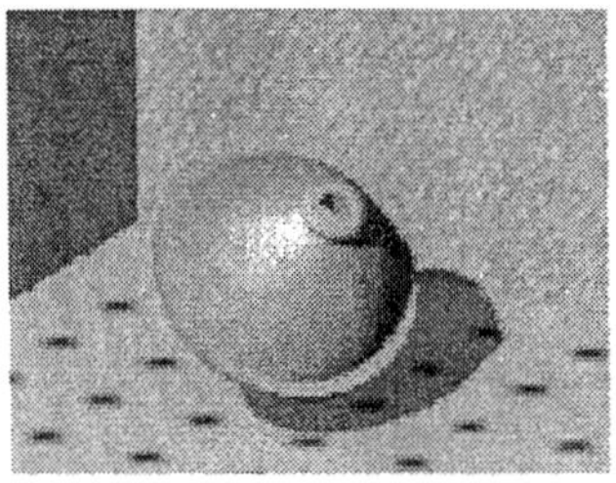

Time = 10 seconds

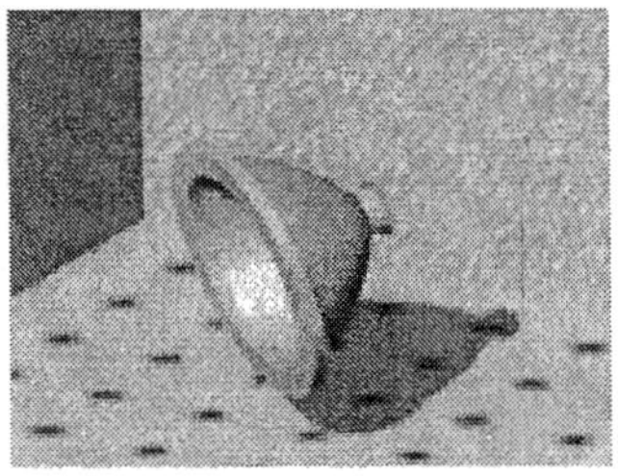

Time = 18 seconds

Project 3

Sweep and Loft Features

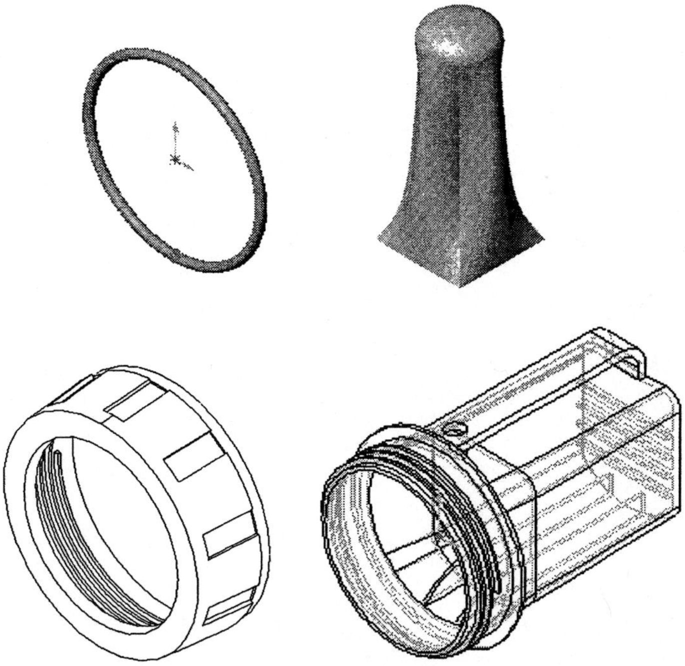

Below are the desired outcomes and usage competencies based upon the completion of Project 3.

Project Desired Outcomes	Usage Competencies
Create four key FLASHLIGHT parts: • O-RING • SWITCH • LENSCAP • HOUSING	Comprehension of the fundamental definitions and process of Feature-Based 3D Solid Modeling using Sweeps and Lofts.
	Ability to combine multiple features to create parts.

Notes:

Project 3 – Sweep and Loft Features

Project Objective

Create four parts of the FLASHLIGHT. Create the O-RING, SWITCH, LENSCAP and HOUSING parts.

Project Situation

Communications is a major component to a successful development program. Provide frequent status reports to the customer and to your team members.

Communicate with suppliers. Ask questions. Check on details. What is the delivery time for the BATTERY, LENS and SWITCH?

Talk to colleagues to obtain useful manufacturing suggestions and ideas.

Your team decided that plastic injection molding is the most cost effective method to produce large quantities of the desired part.

Investigate surface finishes that support the customers advertising requirement. You have two fundamental choices:

- Adhesive label
- Silk-screen

There are time, quantity and cost design constraints. For the prototype, use an adhesive label. In the final product, create a silk screen.

Investigate your options on O-Ring material. Common O-Ring materials for this application are Buna-N (Nitrile®) and Viton®.

Note: Be cognizant of compatibility issues between O-Ring materials and lubricants.

Project Overview

Create four parts in this project:

- O-RING
- SWITCH
- LENSCAP
- HOUSING

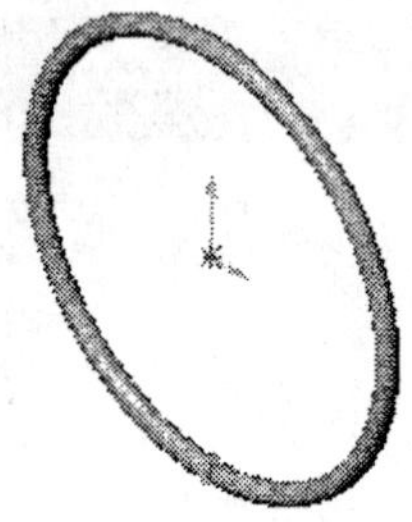

Figure 3.1a

Two Base feature types are discussed in this project:

- Sweep – O-RING
- Loft – SWITCH

Use the Sweep feature as the O-RING Base feature, Figure 3.1a. Create the O-RING Sweep feature by moving a small circular profile along a large circular path. The diameter of the large circular path of the O-RING is determined by the diameter of the LENS. Position the O-RING between the LENS and the LENSCAP.

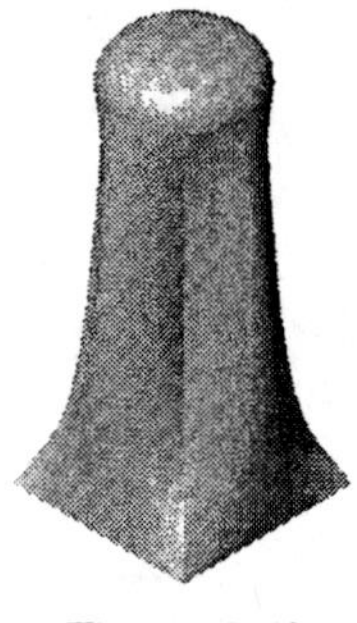

Figure 3.1b

The SWITCH is a purchased part. The SWITCH is a complex assembly. Use the Loft feature to create the outer SWITCH, Figure 3.1b.

The LENSCAP and HOUSING are designed plastic parts. The LENS and BATTERY is enclosed by the LENSCAP and HOUSING respectively. How do you design the LENSCAP and HOUSING to ease the transition from development to manufacturing? Answer: Review the fundamental design rules behind the plastic injection manufacturing process:

- Maintain a constant wall thickness. Inconsistent wall thickness creates stress.
- Create a radius on all corners. No sharp edges. Sharp edges create vacuum issues when removing the mold.
- Allow a minimum draft angle of 1 degree. Draft sides and internal ribs. Draft angles assist in removing the part from the mold.

Obtain additional information on material and manufacturing from material suppliers or manufacturers. Example: GE Plastics of Pittsfield, MA (www.geplastics.com & www.gepolymerland.com) provides design guidelines for selecting raw materials and creating plastic parts and components.

Review the three developed parts in Project 1 and Project 2:

- BATTERY
- LENS
- BATTERYPLATE

The BATTERY and LENS are purchase parts. The BATTERYPLATE is a created part.

Position the BATTERY next to the LENS, Figure 3.2a. The diameter of the LENSCAP must be slightly larger than the diameter of the LENS.

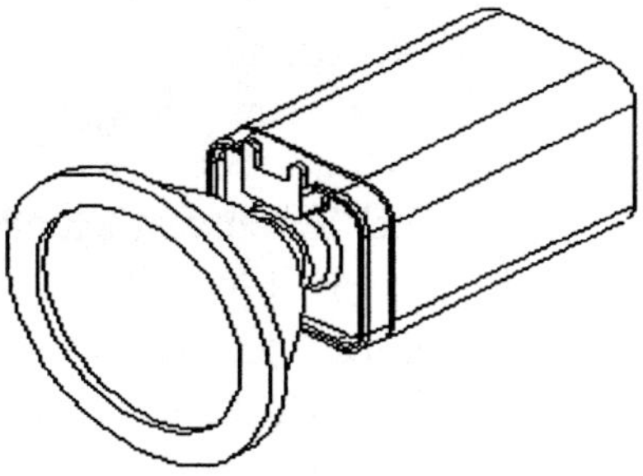

Figure 3.2a

The overall height of the LENSCAP and HOUSING is approximately 7.0 inch, [177.8mm]. The dimension of the LENS and BATTERY determines the size of the HOUSING.

Create the LENSCAP. Use the Base-Extrude feature with a draft angle, Figure 3.2b.

Create an internal thread in the LENSCAP. Use the Sweep feature and a sketched Helix/Spiral curve. Revolve a Boss feature around the outside of the LENSCAP. Use a Circular Pattern feature.

Figure 3.2b

Create the HOUSING. The Base Extrude feature dimensions are determined from the LENS and LENSCAP. The Loft feature is used to transition a circular profile of the LENSCAP to a square profile of the BATTERY.

Use the Extrude feature to enclose the BATTERY. Create a constant wall thickness with the Shell feature. Create the handle. Use the Sweep feature.

The Rib feature provides extra internal support and rigidity, Figure 3.2c. Ribs maintain the proper alignment of the BATTERY and LENS.

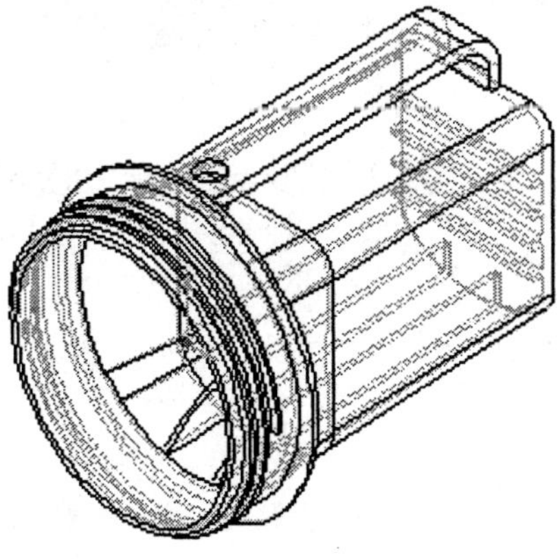

Figure 3.2c

Note: In this project, wall thickness, clearance fit and thread size are increased for improve illustration. Parts are simplified.

O-RING

The O-RING captures the LENS to the LENSCAP. Create the O-RING. Use the Sweep feature as the Base feature. A Sweep feature adds material by moving a profile along a path.

Figure 3.3

Note: Sweep features require a cross section and a path. Example: Think of a doughnut with a bite, Figure 3.3. The cross section is the small circle, "bite" of the doughnut.

The path is the large circle that constitutes the circumference of the donut.

This is your O-RING.

O-RING Feature Overview

Create the O-RING. The default orientation of the O-RING is based on the orientation in the assembly.

Sketch a large circular path on the Front plane. Sketch the small cross section on the Right plane, Figure 3.4a. The path and the cross section are combined to create the Sweep feature, Figure 3.4b.

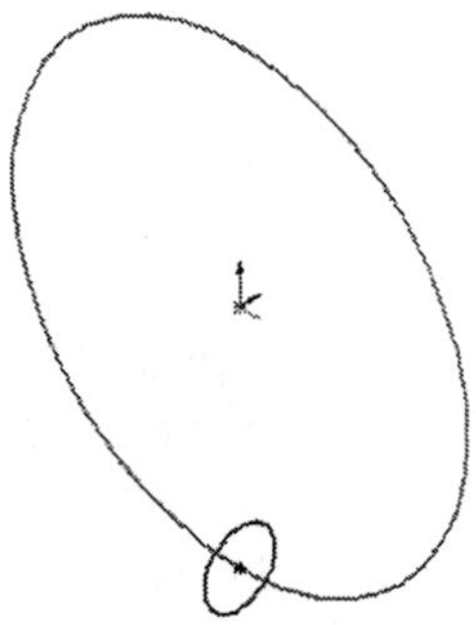

Figure 3.4a

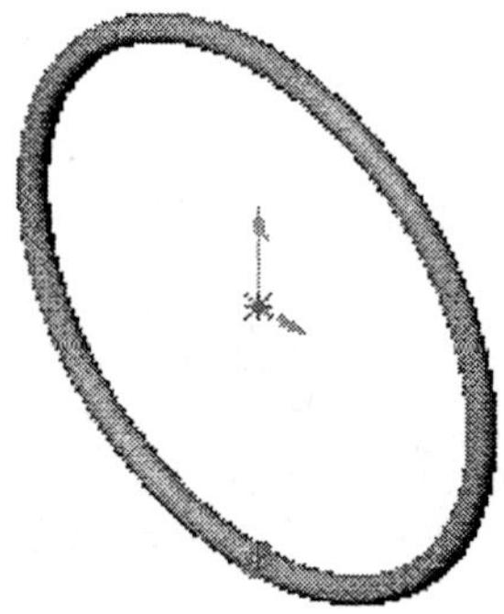

Figure 3.4b

The diameter of the large circular path is 4.350 inch, [110.49mm]. The outside diameter of the LensCover feature is 4.466 inch, [113.44mm], Figure 3.4c.

Let's start.

Note: Use the PartMetricTemplate for metric parts. Millimeter dimensions are provided in brackets[x].

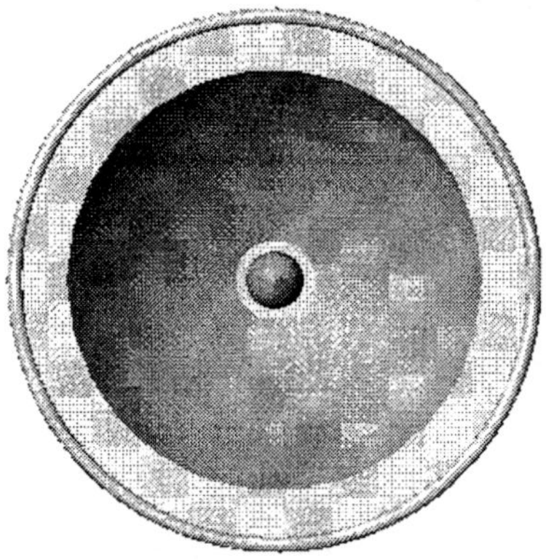

Figure 3.4c

Create the O-RING

Create the O-RING with the Sweep Base feature.

The Sweep-Base feature uses:

- A circular path sketched on the Front plane.
- A small cross section sketched on the Right plane.

Create the O-RING.

1) Click **New**. Click **PartEnglishTemplate**, [**PartMetricTemplate**]. Click **OK**.

Click **Save**. Select **My Documents** folder in the Save In text box. Enter the name of the part. Enter **O-RING**. Click the **Save** button.

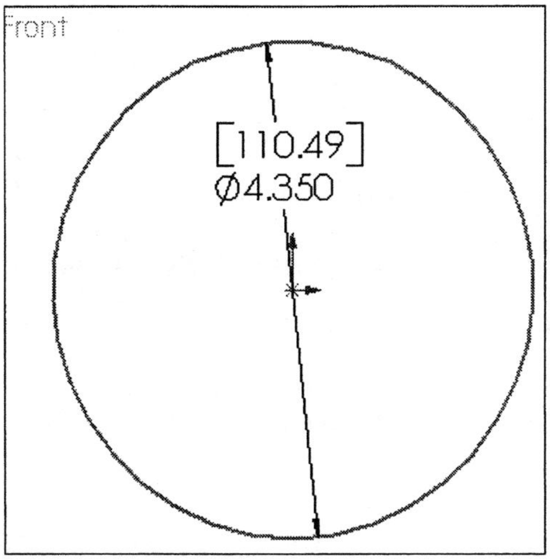

Create the Sweep path.

2) Select the Sketch plane. The Front plane is the default Sketch plane.

3) Sketch the path. Click **Sketch**. Click **Circle**. Sketch a circle centered at the **Origin**.

4) Add a dimension. Click **Dimension**. Dimension the circumference. Enter **4.350**, [**110.49**].

5) Close the Sketch. Click **Sketch**.

6) Rename **Sketch1** to **sketch-path**.

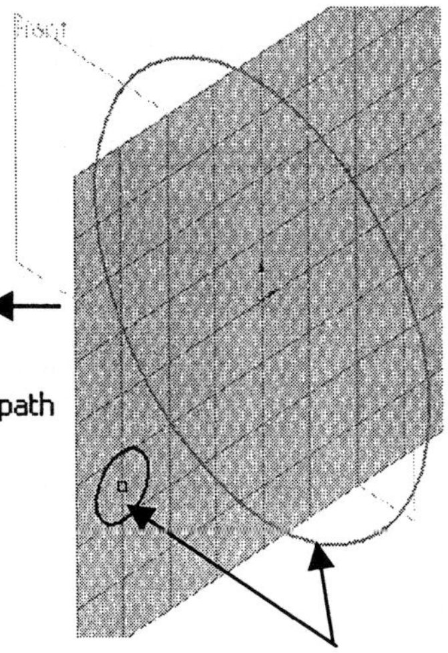

Create the Sweep profile.

7) Display the Isometric view. Click **Isometric**.

8) Select the Sketch plane. Click the **Right** plane from the FeatureManager.

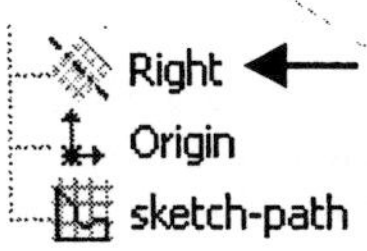

9) Sketch the profile (cross section). Click **Sketch**. Click **Circle**. Create a small **circle** left of the sketch-path.

The Pierce geometric relation positions the center of the cross section on the sketched path.

10) Add geometric relations. Right-click **Select** in the Graphics window. Hold down the **Ctrl** key. Click the **small circle center point**. Click the **large circle circumference**. Click **Pierce**. Release the **Ctrl** key. The center point of the small circle pierces the sketch-path (large circle).

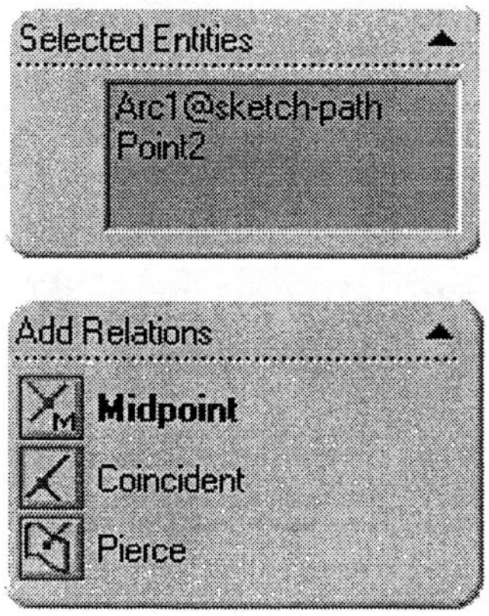

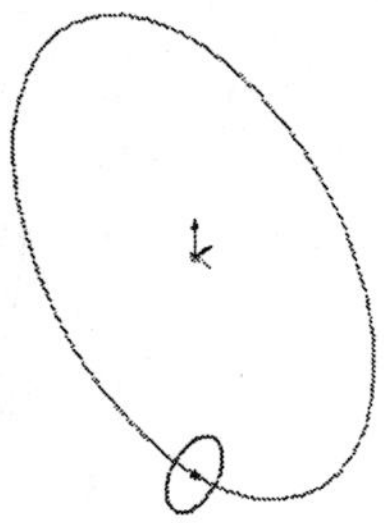

11) Add a dimension. Click **Dimension** . Dimension the small circle. Enter diameter. Enter **.125**, [**3.18**].

12) Close the Sketch. Click **Sketch** .

13) Rename **Sketch2** to **sketch-profile**.

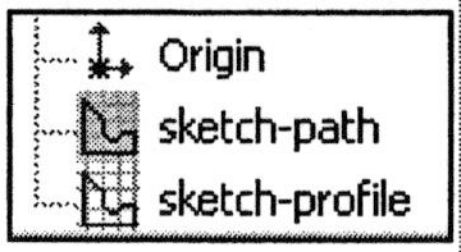

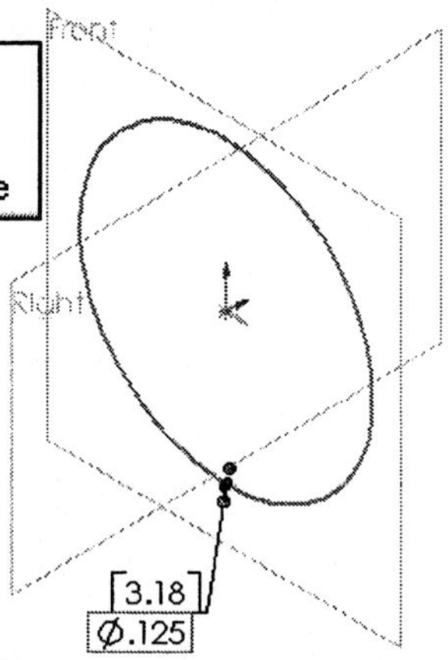

For detailed cross sections, perform the following steps:

- Create a large cross section.
- Pierce the section to the path.
- Add dimensions to create the true size.

Sweep the Profile.

14) Click **Sweep** from the Feature toolbar. The Sweep dialog box is displayed. Split the FeatureManager. Position the **mouse pointer** at the top of the FeatureManager. The mouse pointer displays the Split bar . Drag the **Split bar** half way down to display the FeatureManager and the Base-Sweep PropertyManager.

15) Select the Sweep profile. Click the **Sweep profile** text box. Click the **sketch-profile** from the FeatureManager. Click the **Sweep path** text box. Click the **sketch-path** from the FeatureManager. Display the Sweep. Click **OK**.

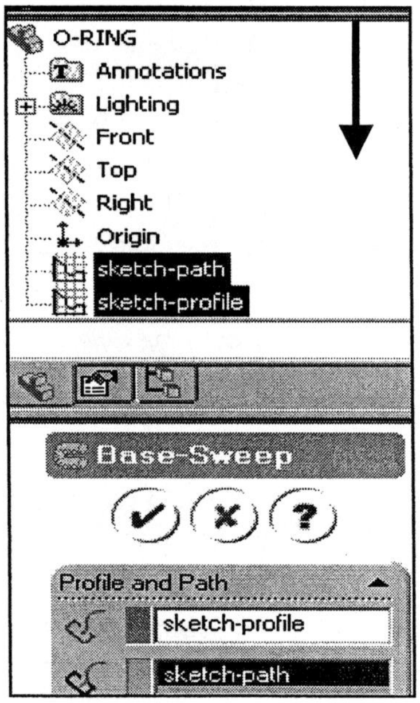

16) Expand the Base-Sweep feature. Click the **Plus Sign** in the FeatureManager. The Base-Sweep is composed of the Sketch-section and the Sketch-path. Collapse the Base-Sweep feature. Click the **Minus Sign**.

17) Save the O-RING. Click **Save**.

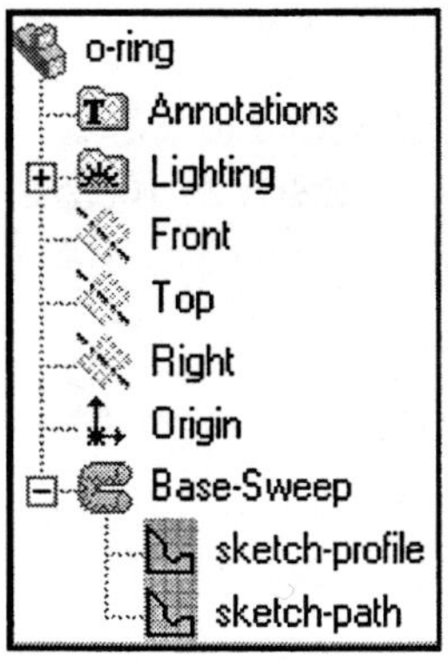

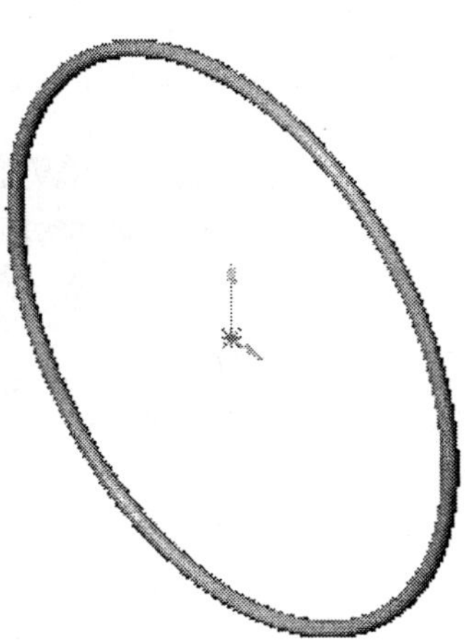

SWITCH

The SWITCH is a purchased part. The SWITCH is a complex assembly. Create the outside casing. Create the SWITCH with the Base-Loft feature, Figure 3.5.

A Loft feature uses two or more cross sections to define geometry contours. Sketch each cross section on a plane. The path connects the cross sections to create the solid Loft feature.

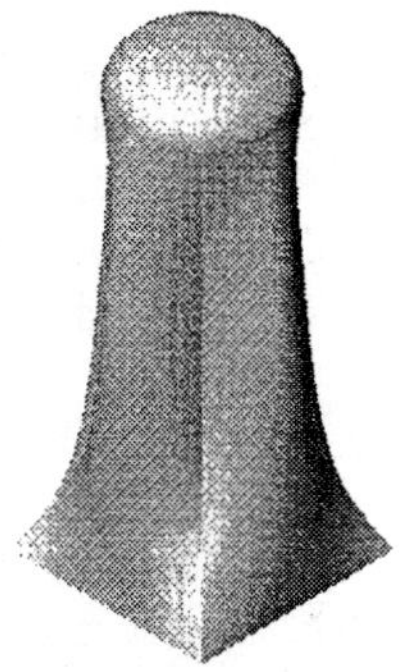

Figure 3.5

SWITCH Feature Overview

The orientation of the SWITCH is based on its position in the assembly. The SWITCH is comprised of three cross sections. Each cross section is sketched on a different plane, Figure 3.6a. Create the Base-Loft feature, Figure 3.6b. Create a Dome feature to the top face of the Loft, Figure 3.6c.

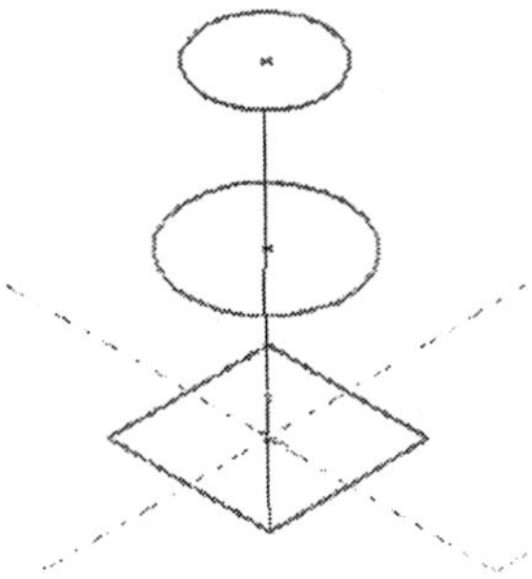

Figure 3.6a

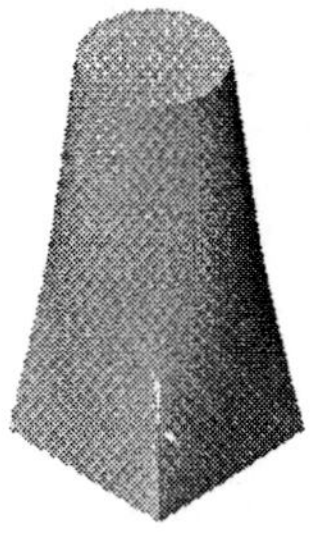

Figure 3.6b

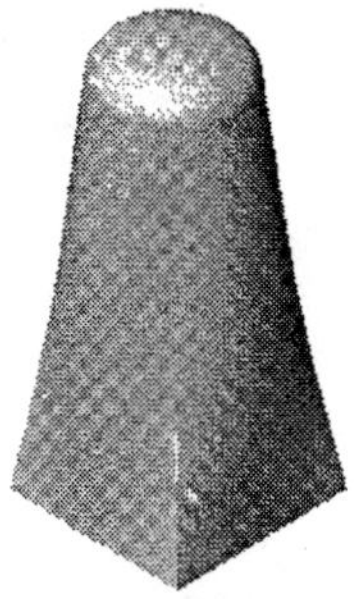

Figure 3.6c

Create the SWITCH – Loft Feature

Create the SWITCH with a Loft Base feature. The Loft Base feature requires three cross sections. Sketch each cross section on a unique plane.

Create the Sketch planes.

18) Click **New**. Click **PartEnglishTemplate**, **[PartMetricTemplate]**. Click **OK**. Click **Save**. Select **My Documents** folder in the Save In text box. Enter the name of the part. Enter **SWITCH**. Click the **Save** button.

19) Create the first Sketch plane. The first Sketch plane is the Top plane. Right-click the **Top** plane from the FeatureManager. Click **Show**. Display the Isometric view. Click **Isometric**. Turn Grid and Snap to Points off. Click **Grid**. Uncheck all **Grid** and **Snap check boxes.**

20) Create the second Sketch plane. Copy the Top plane. Click the **edge** of the Top plane. Hold the **Ctrl** key down. Drag the **Top** plane upward. Create the offset Sketch plane. Release the **mouse button**. Release the **Ctrl** key. Enter **.5**, **[12.7]** for Distance. Click **OK**.

Hold Ctrl key and drag Top plane upward

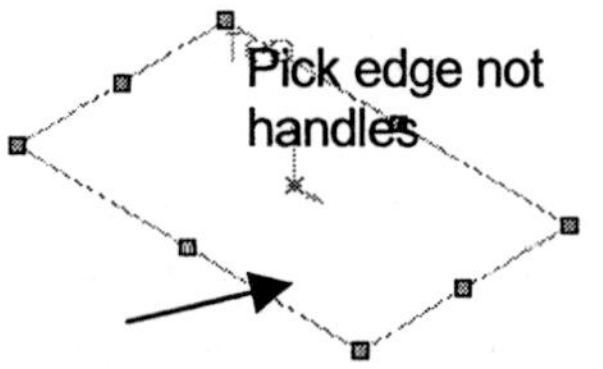

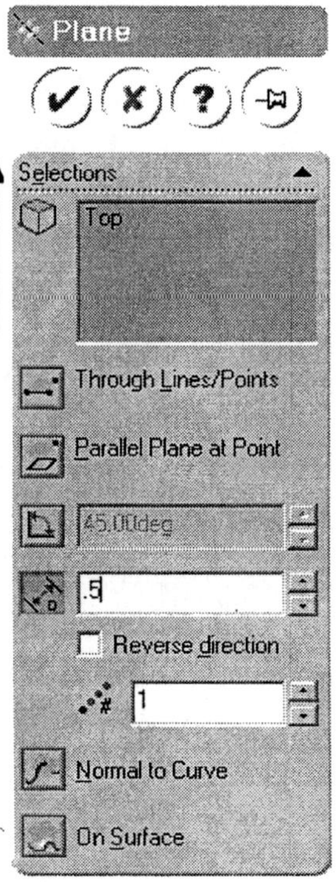

21) Create the third Sketch plane. Copy the Top plane. Click the **edge** of the Top plane. Hold the **Ctrl** key down. Drag the **Top** plane upward above Plane1. Release the **mouse button**. Release the **Ctrl** key. The default name for the third Sketch plane is Plane2. Enter **1.0**, **[25.4]** for Distance. Click **OK**.

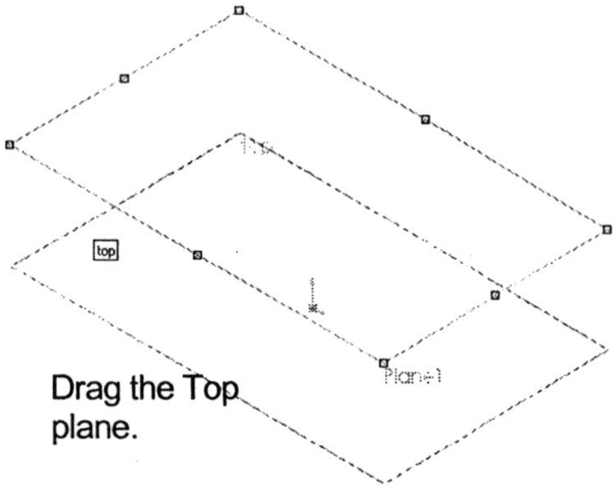

Drag the Top plane.

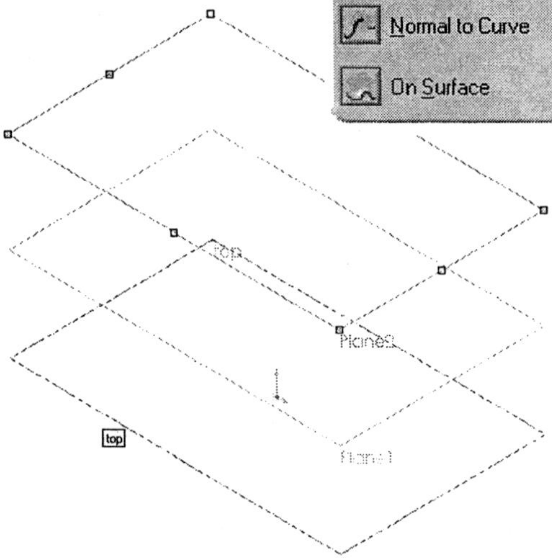

22) Display Sketch planes from the Front view. Click **Front** .

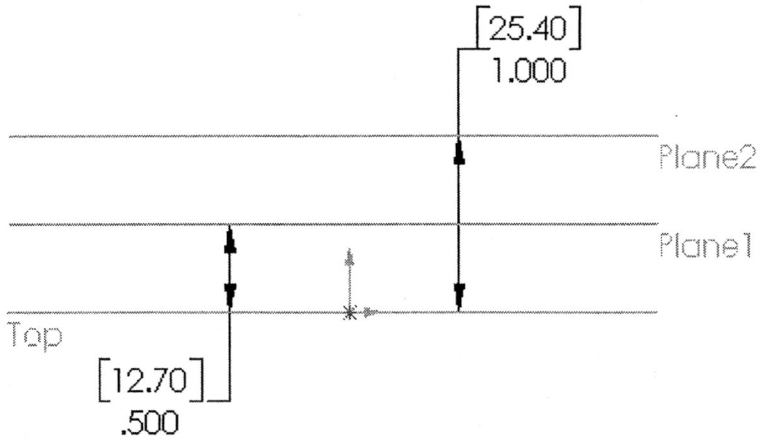

23) Display the Sketch planes. Click **Isometric** . Use the view icons to verify the location of the planes and cross sections. Click the **Front** plane from the FeatureManager. Click the **Right** plane.

Create Sketch1. Sketch1 is a square on the Top plane.

24) Select the Sketch plane. Click the **Top** plane.

25) Create the Sketch. Click **Sketch** .

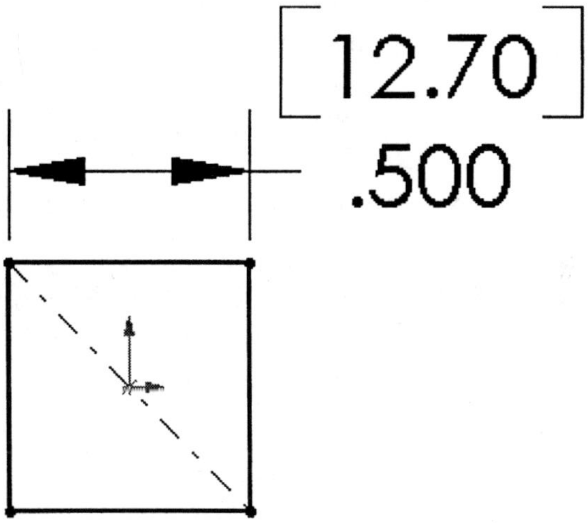

26) Display the Top view. Click **Top** .

27) Sketch the Profile. Click **Rectangle** . Sketch a **Rectangle** centered about the Origin.

28) Click **Centerline** . Create a diagonal centerline between two opposite corner points of the rectangle. Click the two **corner points** of the rectangle.

29) Add geometric relations. Right click **Select**. Add a **Midpoint** relation between the diagonal **centerline** and the **Origin**.

30) Add an **Equal** relation between the **left vertical line** and the **top horizontal line**. All four sides are equal. Add a dimension. Click **Dimension** . Dimension one line of the rectangle. Enter **.500**, **[12.7]**.

31) Close Sketch1. Click **Sketch**. Display the Isometric View. Click **Isometric**.

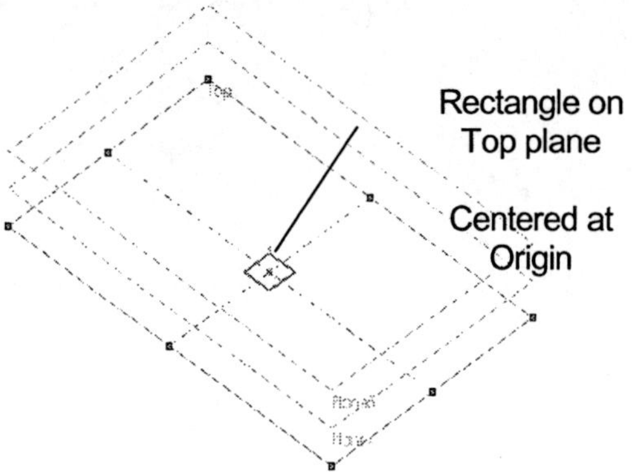

Create Sketch2. Sketch2 is a circle on Plane1.

32) Select the Sketch plane. Click **Plane1**.

33) Create the Sketch. Click **Sketch**. Click **Circle**. Create a **Circle** centered at the Origin.

34) Display the Top view. Click **Top**.

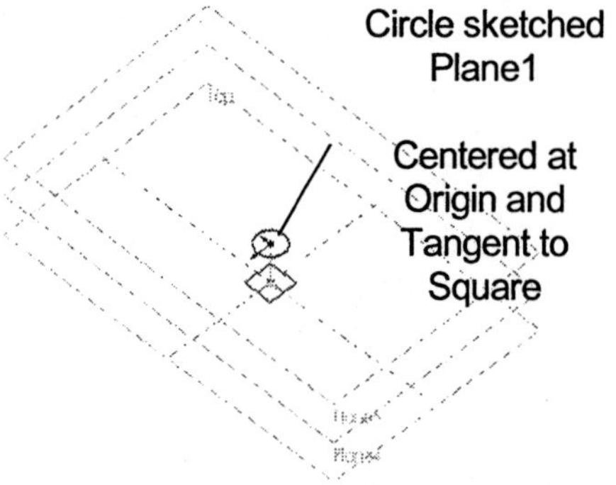

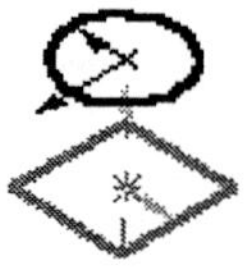

35) Add geometric relations. Right click **Select**. Hold down the **Ctrl** key. Click the **circumference** near the tangent. Click the Sketch1 **line**. Click **Tangent**. Release the **Ctrl** key. Click **Apply**.

Create a Tangent relation between the circle and line.

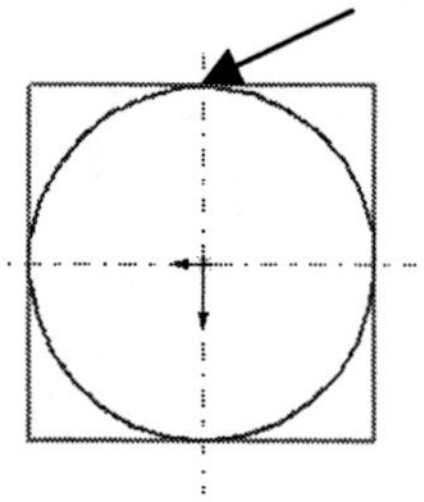

36) Close Sketch2. Click **Sketch**.

Create Sketch3. Sketch3 is a circle on Plane2.

37) Select the Sketch plane. Click **Plane2**. Create the Sketch. Click **Sketch**. Display the Top view. Click **Top**.

38) Click **Centerline**. Sketch a **centerline** coincident with the Origin and the upper right corner point.

39) Click **Point** from the Sketch Tools toolbar. Click the **midpoint** of the diagonal centerline.

40) Click **Circle**. Create a **Circle** centered at the **Origin**. Click the **midpoint** of the centerline.

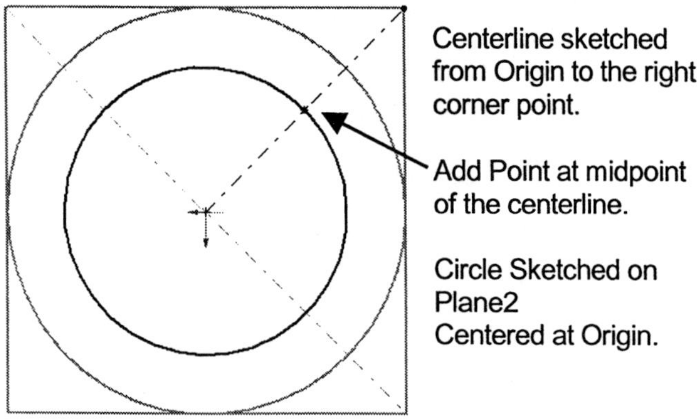

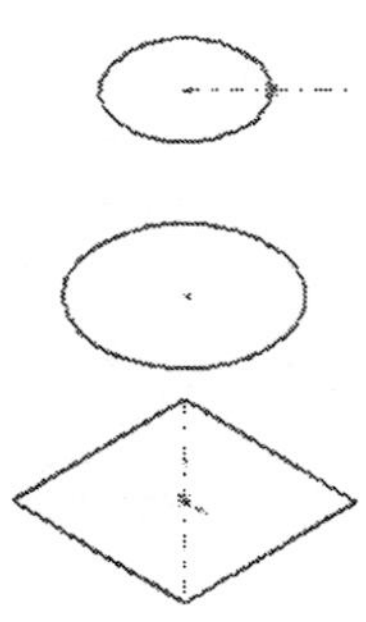

41) Close Sketch3. Click **Sketch**.

42) Hide the Planes. Click **View** from the Main menu. Click **Planes**. Fit the model to the Graphics window. Press the **f** key.

Create the Loft.

43) Click **Loft** from the Feature toolbar. The Loft dialog box is displayed. Display the Isometric view. Click **Isometric**.

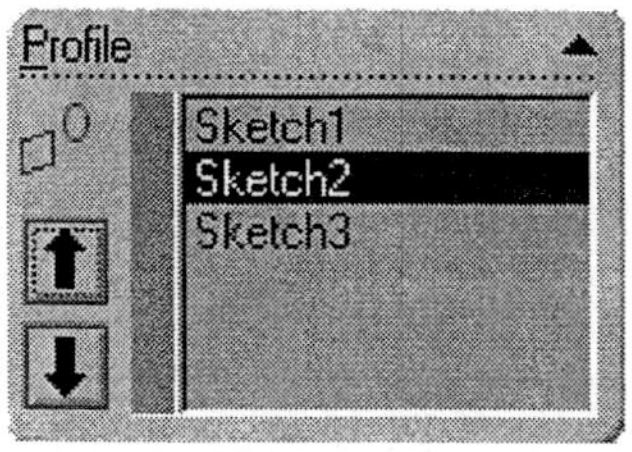

44) Select the Profiles. Display the FeatureManager.

Click **FeatureManager**. Click the **front corner** of Sketch1. Click **Sketch2**. Click **Sketch3**.

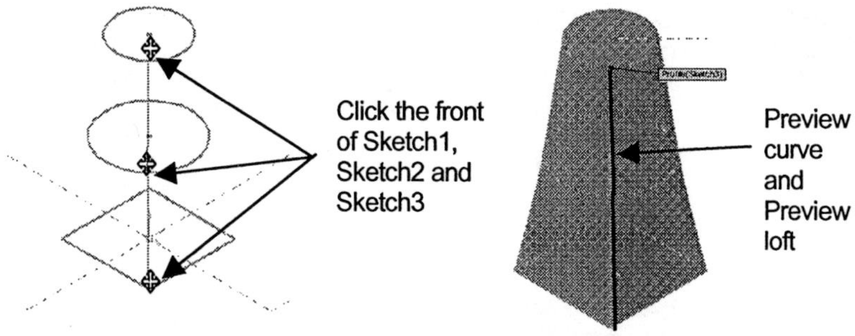

Caution should be used when selecting the profiles and location. The Loft feature is created by connecting the clicked locations of each profile. The system displays a preview curve and preview loft as the profiles are selected.

The Up button and Down button in the Loft dialog box is used to rearrange the order of the profiles. The correct order for the profiles is:

- Sketch1
- Sketch2
- Sketch3

Note: If you select an incorrect location on the Sketch, Right-click in the Graphics window. Click Clear Selections. Select the profiles again.

45) Display the Loft Base feature. Click **OK**.

46) Save the SWITCH. Click **Save**.

Create the SWITCH – Dome feature

Create the Dome feature on the top face of the Loft Base feature.

Create the Dome.

47) Select the face. Click the **top face** of the Base-Loft feature.

48) Click **Dome** from the Feature toolbar. The Dome dialog box is displayed. Enter **.100**, **[2.54]**. Display the Dome. Click **OK**.

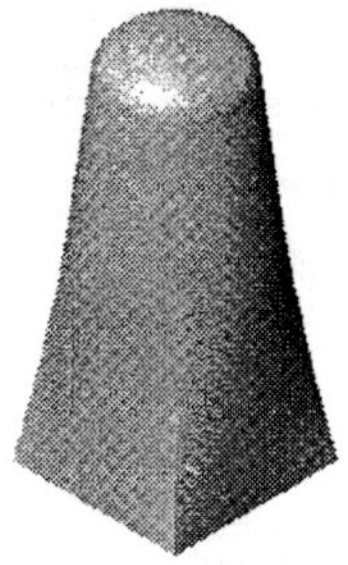

49) Expand the Base-Loft feature. Click the **Plus Sign** in the FeatureManager. The Base-Loft is composed of Sketch1, Sketch2 and Sketch3.

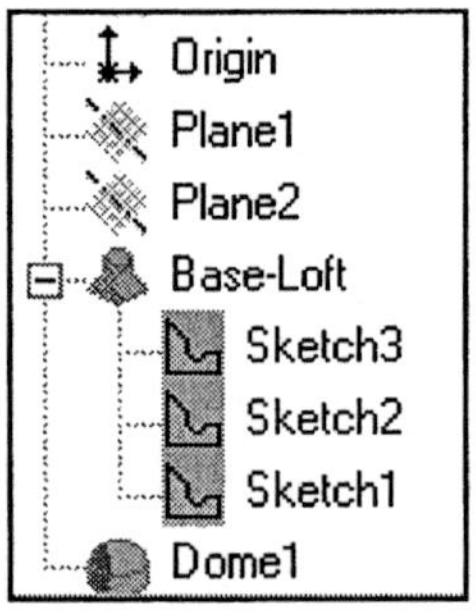

50) Modify the Base-Loft feature. Double-click on the **Base-Loft**. Double-click on the Plane1 offset dimension, **.500**, **[12.70]**. Enter **.125**, **[3.18]**.

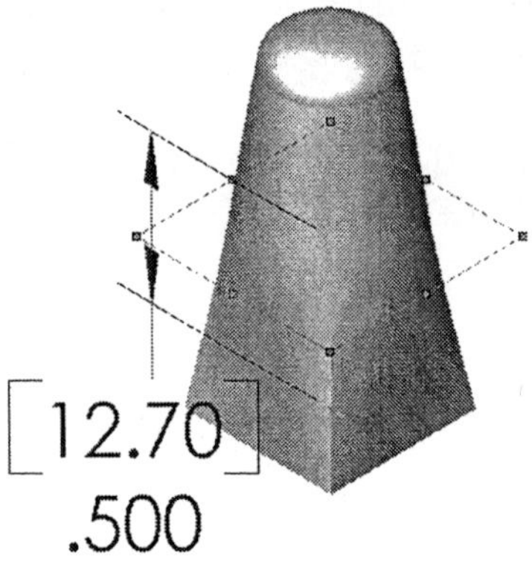

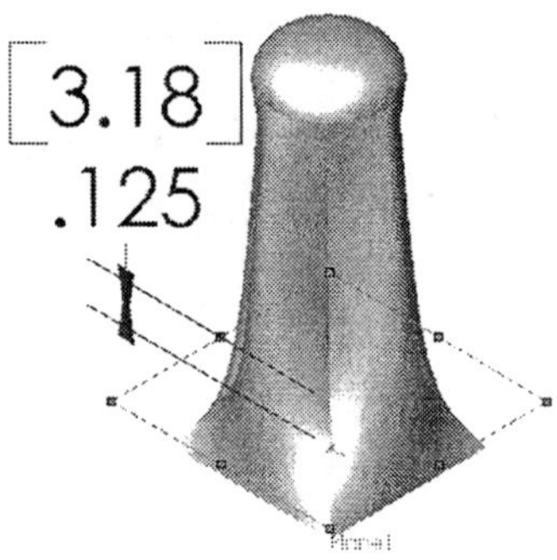

51) Click **Rebuild**. Click **OK**.

52) Save the SWITCH. Click **Save**.

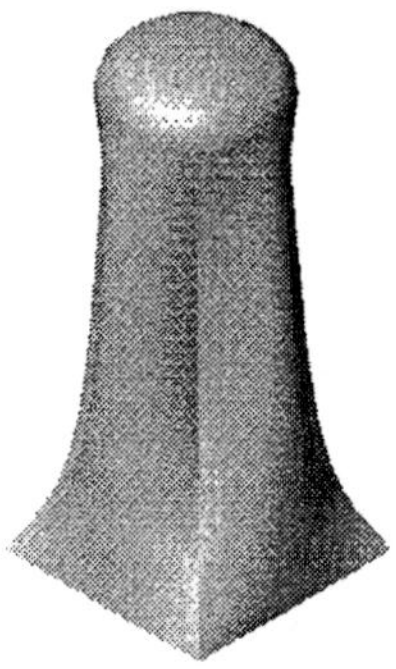

LENSCAP

The LENSCAP is a plastic part used to position the LENS to the HOUSING. An O-RING is positioned between the LENS and LENSCAP, Figure 3.7.

Figure 3.7

How should you sketch the LENSCAP? Sketch the LENSCAP on the Front plane.

What key dimensions from the LENS are required to create the LENSCAP?

The diameter and depth are the key dimensions to create the LENSCAP.

How do you determine the LENSCAP depth? Measure the depth of the LENS. The LENS is approximately 3inchs, [76.2mm] deep.

Position half of the LENS inside the LENSCAP. Position the other half within the HOUSING, Figure 3.8.

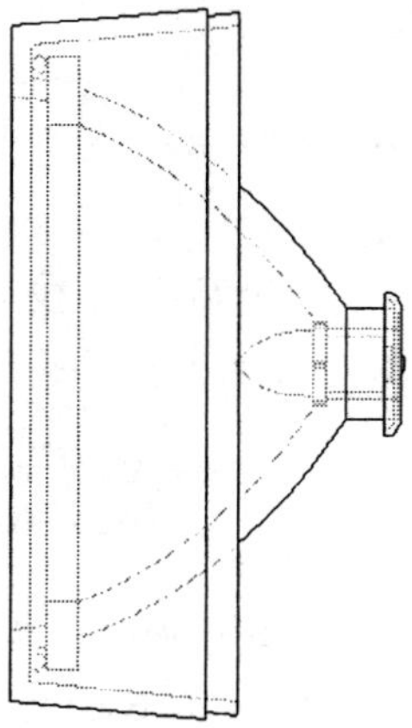

Figure 3.8

LENSCAP Feature Overview

Create an Extruded Base feature with a circular profile on the Front plane, Figure 3.9a. The LENSCAP is a plastic part. For plastic parts use a Draft angle.

Add an Extruded Cut feature. The Extruded Cut feature should be equal to the diameter of the LENS Revolved-Base feature, Figure 3.9b. Create a Shell feature. Use the Shell feature for a constant wall thickness, Figure 3.9c.

Add a Revolved-Cut feature on the back face, Figure 3.9d.

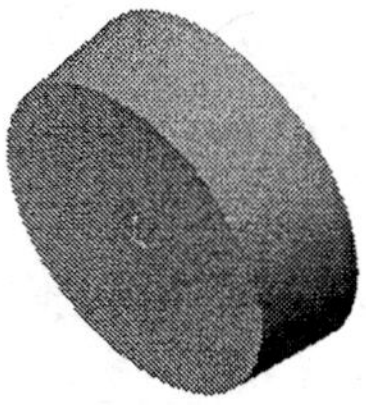

Figure 3.9a

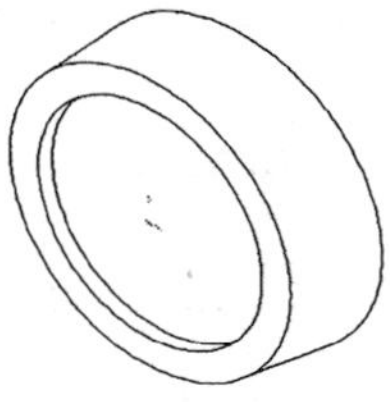

Figure 3.9b

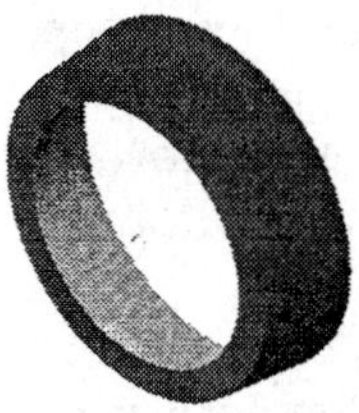

Figure 3.9c

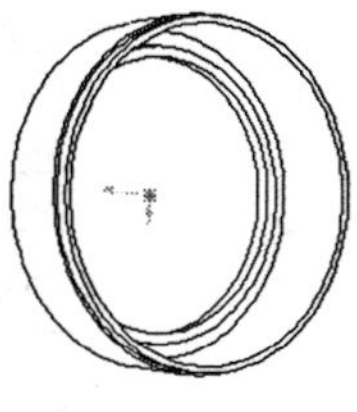

Figure 3.9d

Create an Extruded Thin Cut feature. The Extruded Thin Cut feature is used to grip the outside of the LENSCAP, Figure 3.9e. Create a Pattern feature. The Pattern feature is used to create multiple instances of the Extruded Cut feature, Figure 3.9f. Create a thread using a sketched Helical Curve feature and a Sweep feature, Figure 3.9g.

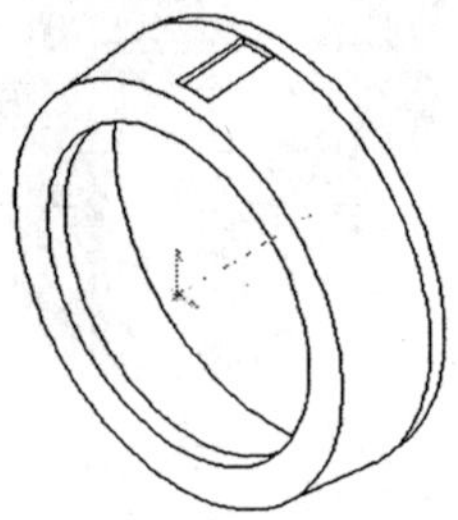

Figure 3.9e

Figure 3.9f

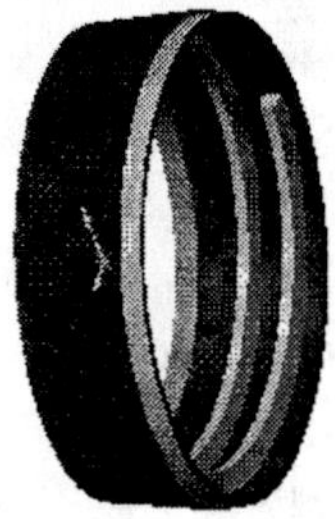

Figure 3.9g

Create the LENSCAP

Create the Extruded Base feature.

53) Click **New**. Click **PartEnglishTemplate**, **[PartMetricTemplate]**. Click **OK**. Click **Save**. Select **My Documents** folder from the Save In text box. Enter the name of the part. Enter **LENSCAP**. Click the **Save** button.

54) Select the Sketch plane. The Front plane is the default Sketch plane. Turn the Grid and Snap to Points off. Click **Grid**. Uncheck all **Grid** and **Snap check boxes**.

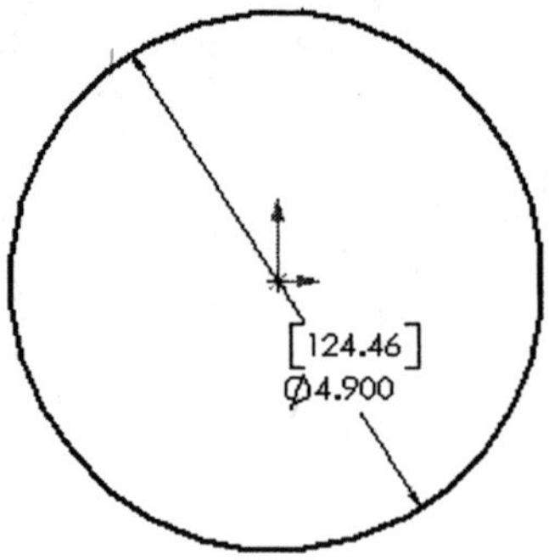

55) Create the Sketch. Click **Sketch**. Click **Circle**. Create a circle centered at the **Origin**. Add the dimension. Click **Dimension**. Click the **circumference** of the circle. Enter **4.900**, **[124.46]**.

56) Extrude the Sketch. Click **Base-Extrude**. Blind is the default Type option. Enter **1.725**, **[43.82]** for Depth. Click the **Reverse Direction** button. Click the **Draft While Extruding** checkbox. Enter **5** in the Angle text box. Click the **Draft Outward** check box. Display the Extruded-Base feature. Click **OK**.

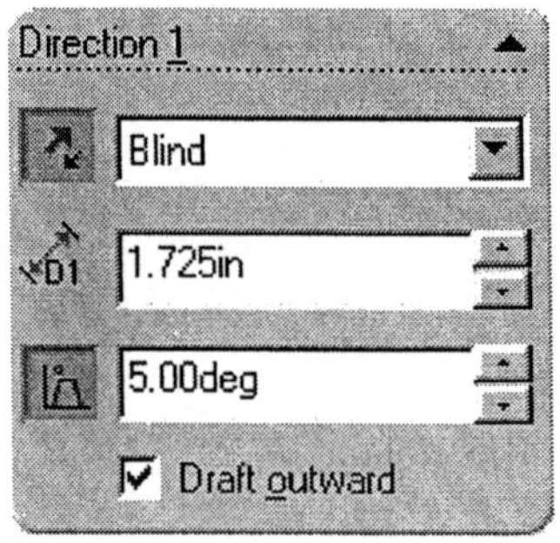

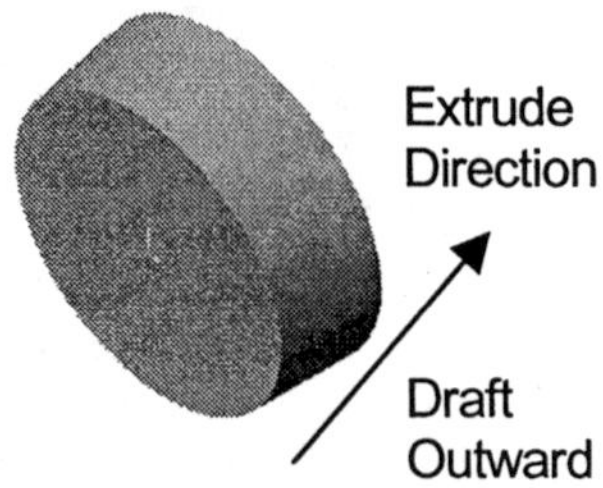

57) Save the LENSCAP. Click **Save**.

Create the LENSCAP – Extruded Cut Feature

Create an Extruded Cut feature on the front face of the Base feature. The diameter of the Extruded Cut equals the diameter of the Base-Revolved feature of the LENS.

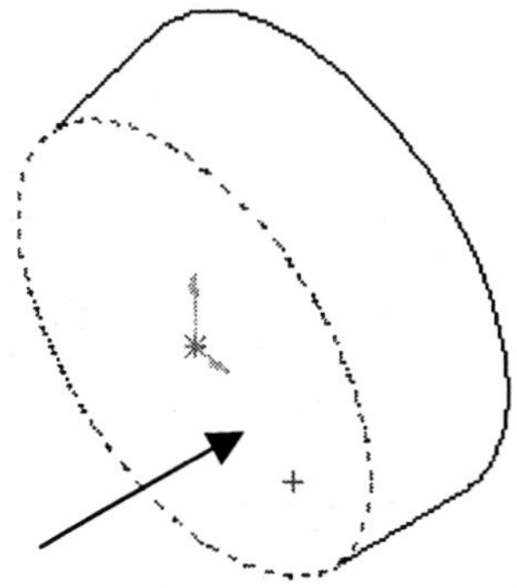

Create the Extruded Cut Feature.

58) Select the Sketch plane. Click the **front face**. Create a Sketch. Click **Sketch**. Click **Circle**. Create a circle centered at the Origin.

59) Add a dimension. Click **Dimension**. Click the **circumference** of the circle. Enter **3.875**, **[98.43]**.

60) Extrude the Sketch. Click **Extruded Cut**. Blind is the default Type option. Enter **.275**, **[6.99]** for Depth. Click the **Draft While Extruding** check box. Enter **5** for Angle. Display the Cut-Extrude feature. Click **OK**.

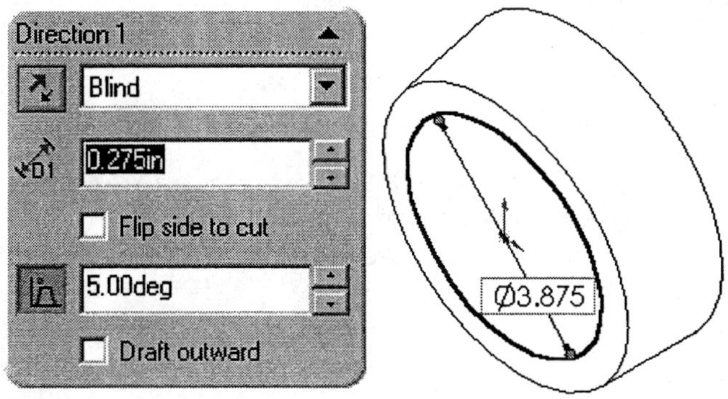

61) Rename **Cut-Extrude1** to **Front-Cut**.

62) Save the LENSCAP. Click **Save**.

Create the LENSCAP – Shell Feature

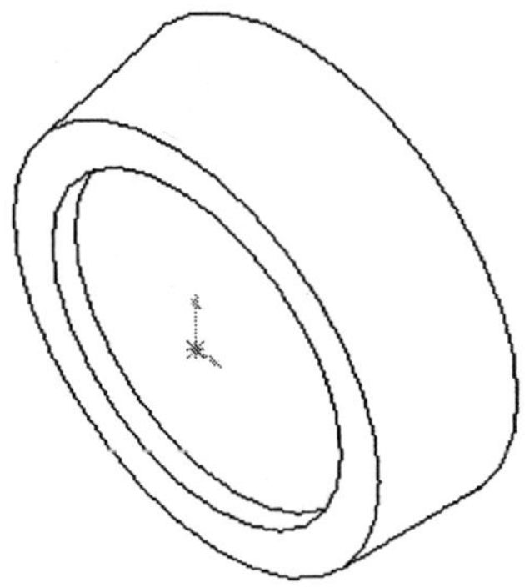

The Shell feature removes face material from the solid LENSCAP.

63) Rotate the LENSCAP to view the part. Click **Rotate**.

Create the Shell.

64) Click **Shell** from the Feature toolbar. The Shell Feature dialog box is displayed. Click the **front face** of the Front-Cut feature. Click the **back face** of the Extruded-Base feature. Enter **.150**, [**3.81**] for Thickness.

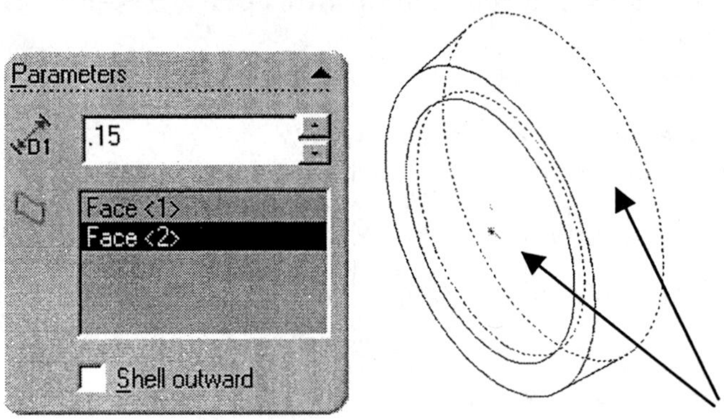

65) Display the Shell. Click **OK**.

66) Display the inside of the Shell. Click **Right**.

Click **Hidden In Gray**.

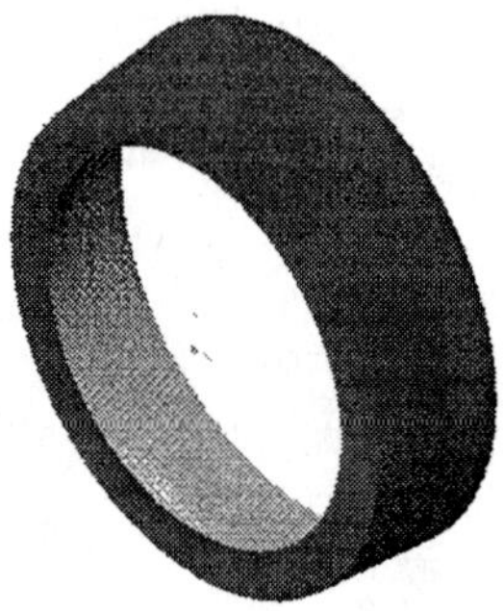

67) The inside gap created by the Shell feature is used to seat the O-RING in the assembly.

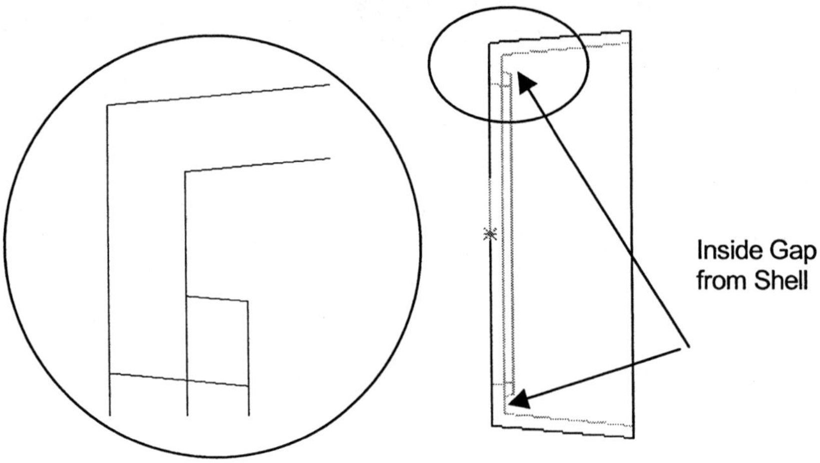

68) Save the LENSCAP. Click **Save**.

Create the LENSCAP - Revolved Cut Thin Feature

The Revolved Cut Thin feature removes material by rotating a sketched profile around a centerline. A new edge on the conical face is required to create the profile. Create the edge with a Split curve. A Split curve divides a face and creates an edge.

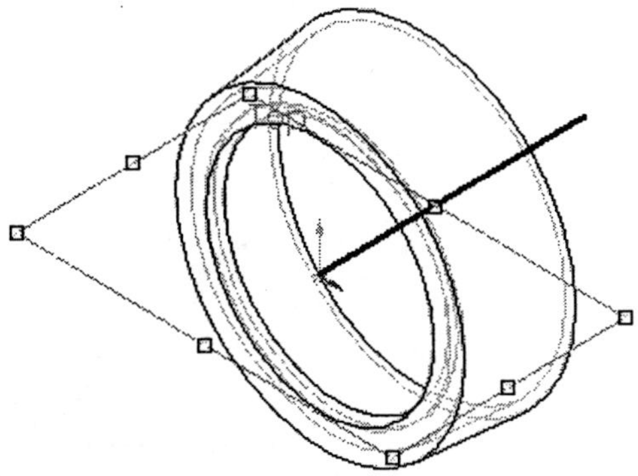

Create the Split Line.

69) Select the Sketch plane. Click the **Top** plane.

70) Create the Sketch. Click **Sketch**. Sketch a **vertical line** from the Origin to a point in space. The line must be longer than the Extruded Base feature.

71) Click **Insert**, **Curve**, **Split Line**. Click **Projection** from the Types of Split. Click **Next**. Select the **top outside conical face** for the Faces to Split text box. Click **OK**.

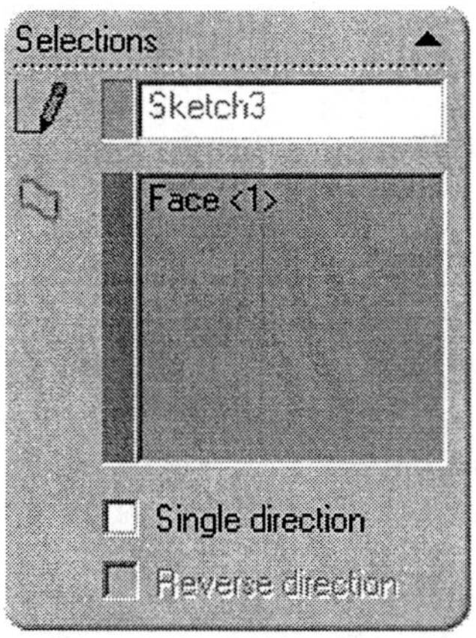

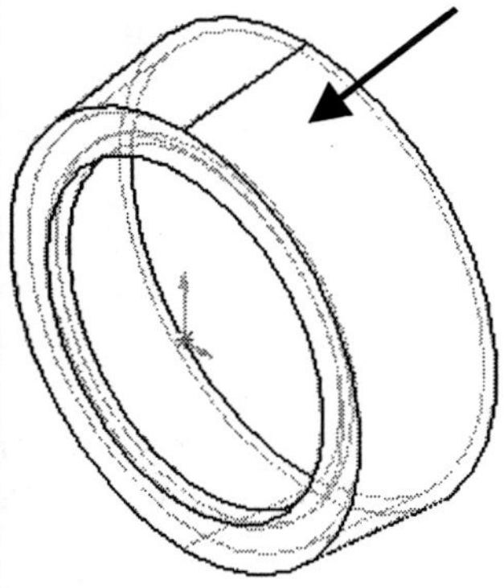

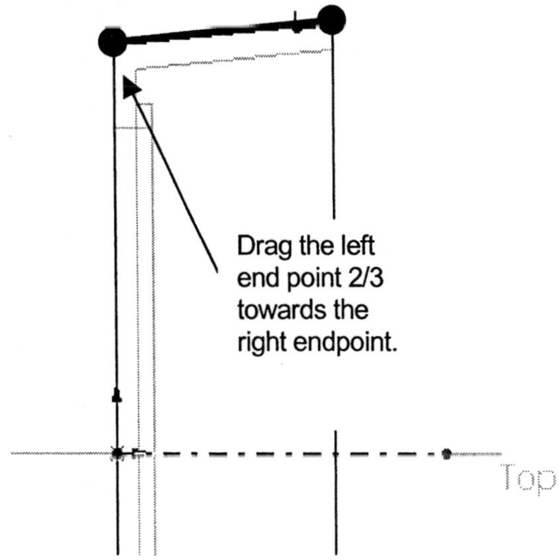

Create the Revolved Cut Thin feature.

72) Select the Sketch plane. Click the **Right** plane.

73) Create the Sketch. Click **Sketch**.

74) Display the Normal to view. Click **NormalTo** from the Standard View toolbar.

75) Sketch the centerline. Click **Centerline**. Sketch a **horizontal centerline** collinear to the Top plane coincident to the Origin.

76) Sketch the profile. Right-click in the **Graphics window**. Click **Select**. Select the **top outside edge** of the Split curve. Click **Convert Entities**.

77) Create a short line. Drag the **left endpoint** 2/3 towards the right endpoint. Release the **mouse button**.

78) Add a dimension. Click **Dimension**. Click the **line**. Create an aligned dimension. Enter **.250**, **[6.35]**.

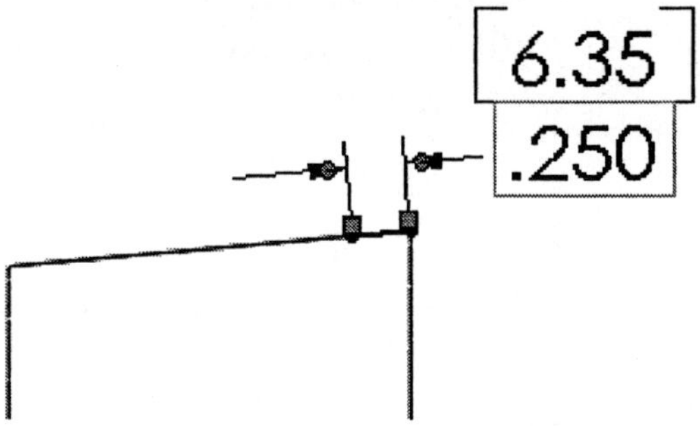

79) Revolve the Sketch. Click **Revolve Cut** from the Feature toolbar. Do not close the Sketch. The warning message states; "The sketch is currently open." Click **No**. The warning message states, "The open profile is only suitable for a thin feature." Click **OK**. Create the feature on one side of the Sketch. Click the **Thin Feature checkbox**. OneDirection is the default Type. Click the **Reverse** checkbox. Enter **.050**, **[1.27]** for Wall Thickness.

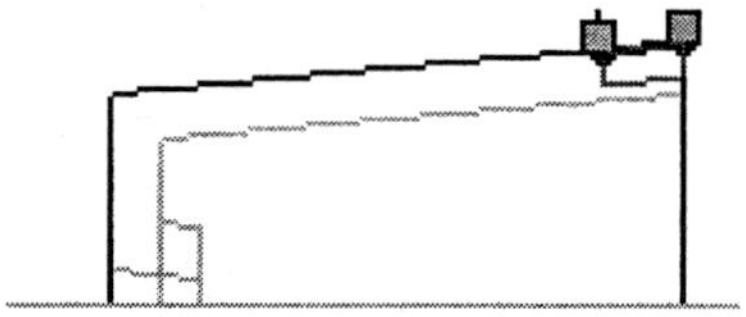

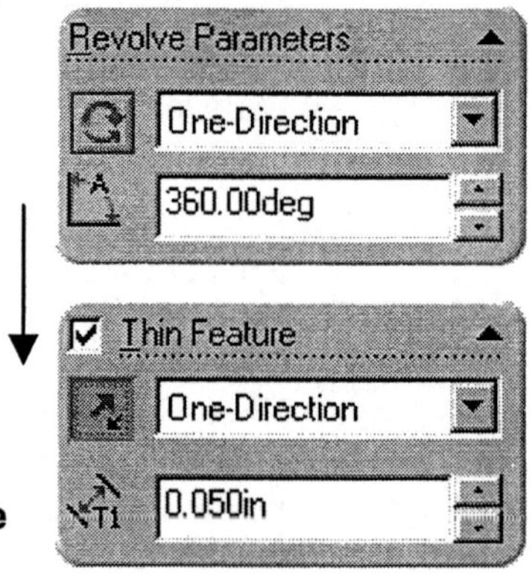

80) Display the Cut-Revolve-Thin feature. Click **OK**. Display the backside of the Revolve-Thin feature. **Rotate** the part.

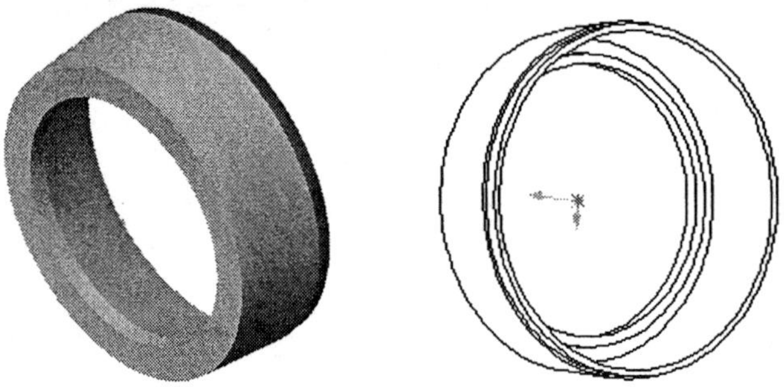

81) Rename **Cut-Revolve-Thin1** to **BackCut**.

82) Save the LENSCAP. Click **Save**.

Create the LENSCAP – Circular Pattern

A Pattern creates one or more instances of a feature or group of features. A Circular Pattern requires a seed feature and an axis of revolution. The seed feature in this example is an Extruded Cut Thin feature created on a new Surface Reference plane. Create the following:

- A Surface Reference plane for the Sketch plane.
- The seed feature as an Extrude-Thin feature.
- The Circular Pattern.

Create the Surface Reference Plane.

83) Display the planes in the Isometric view. Click **Isometric**.

84) Create a Surface Reference plane for the Sketch plane. Select the Plane Option. Click **Insert** from the Main menu. Click **Reference Geometry**. Click **Plane**. Click the **On Surface** button. Drag the **Split bar** half way down to display the FeatureManager and the Plane PropertyManager.

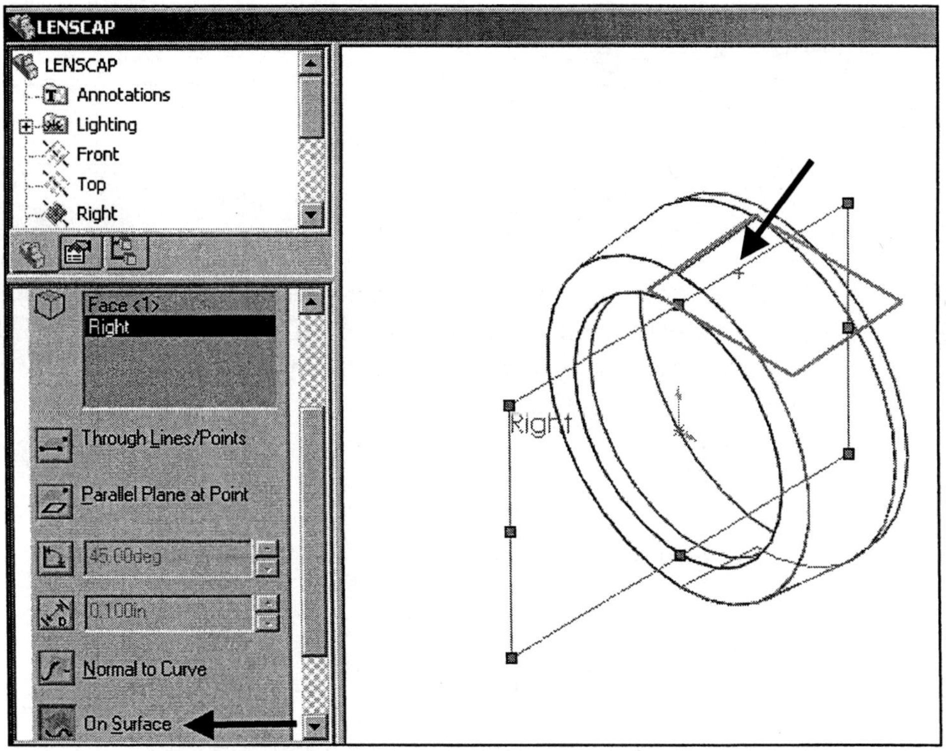

85) Select the Surface. Click the **top outside conical face** of the Extruded-Base feature. Select a plane perpendicular to the surface. Click the **Right** plane from the FeatureManager. Click **OK**.

86) Resize Plane1. Drag the center **left handle** to the left. Rename **Plane1** to **SurfacePlane.**

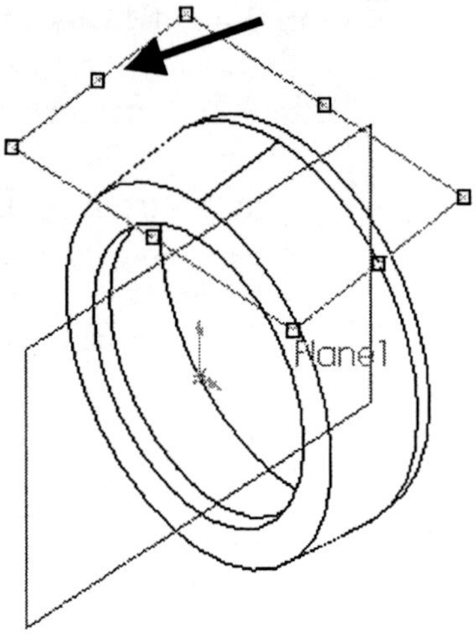

87) Save the LENSCAP. Click **Save.**

Create the seed feature.

88) Select the Sketch plane. Click the **SurfacePlane.**

Click **NormalTo** .

89) Create the Sketch. Click **Sketch** . Select the **Split curve**. Click **Convert Entities** . Drag the end point downward to the **intersection of the Right plane** and the **FrontCut.**

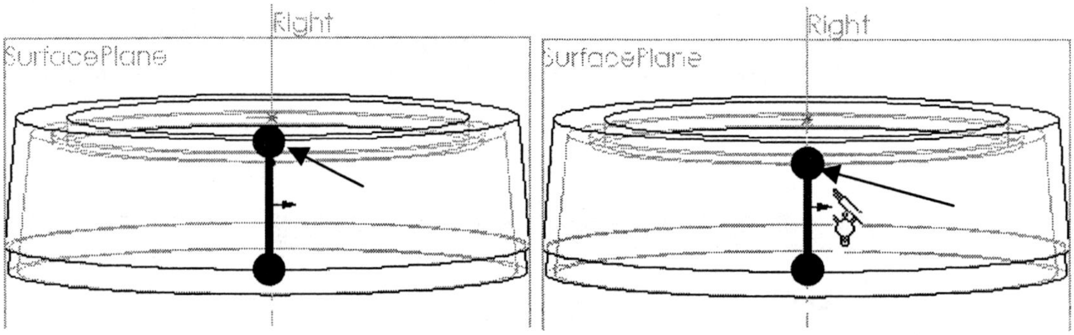

90) Add a **Coincident** relation between the **endpoint** of the line and the circular **edge** of the FaceCut.

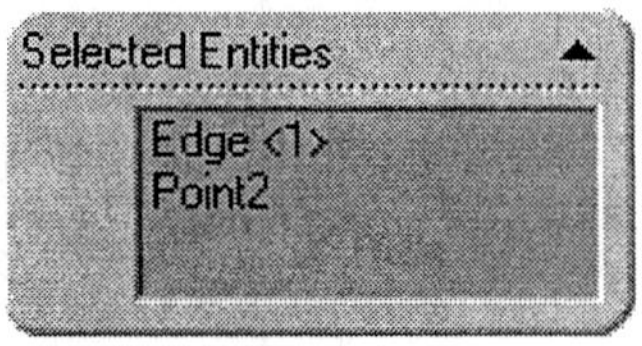

Note: Do not add a geometry relation if the sketched line and endpoints are displayed in black. If the vertical line is fully defined and displayed in black, no dimensions are required. The FrontCut and BackCut features control the length of the seed feature.

91) Extrude the Sketch. Click **Extruded Cut** . Click the **Thin Feature** check box. Select **Mid-Plane** for Thin Feature Type. Enter **.500, [12.7]** for Wall Thickness. Select **Blind** from the Type drop down list. Enter **.100, [2.54]** for Depth. Display the Extruded Cut Thin feature. Click **OK**.

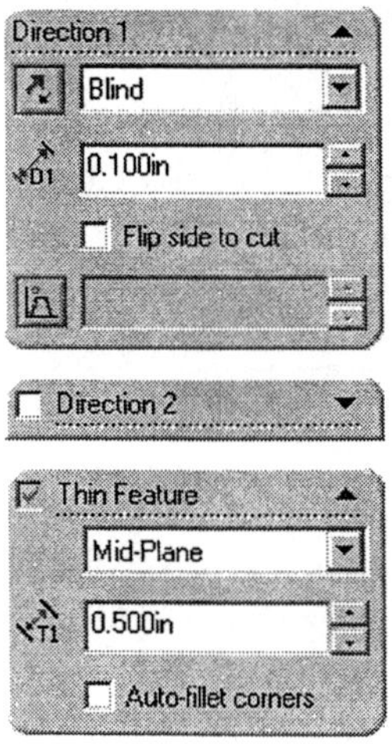

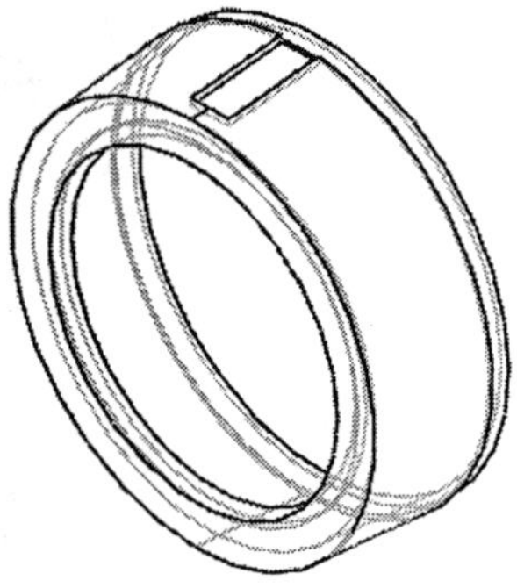

92) Rename **Cut-Extrude-Thin** to **Seed-Cut**.

93) Save the LENSCAP. Click **Save**.

Create the Circular Pattern.

94) Display the Temporary Axis. Click **View** from the Main menu. Click **Temporary Axis** from the View menu.

95) Create the Pattern. Click **Circular Pattern** . The Circular Pattern dialog box is displayed. Click the **Temporary Axis** for Direction 1. Create 10 cuts, equally spaced around the entire LENSCAP. Enter **360** for Total angle. Enter **10** for Total instances. Click the **Equal spacing** check box. Click **Seed Cut** (Cut Extrude-Thin1) for Feature to Pattern.

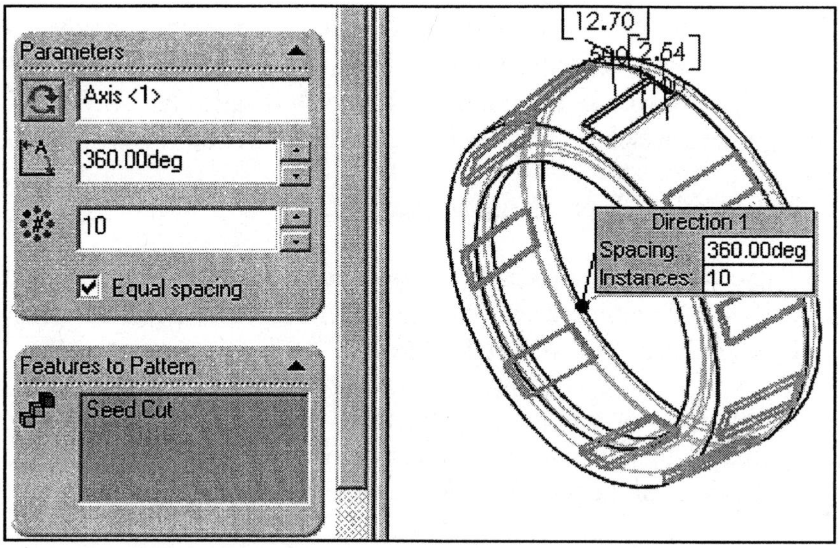

96) Create the Circular Pattern. Click **OK**.

97) Hide the Temporary axis. Click **View** from the Main menu. Click **Temporary axis**. Hide the Planes. Click **Planes**.

98) Save the LENSCAP. Click **Save**.

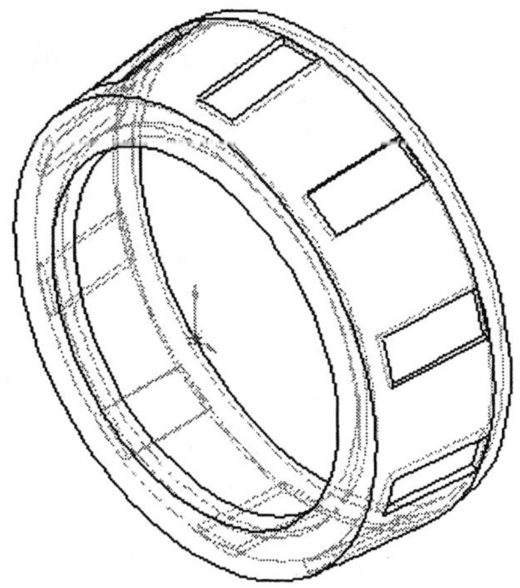

Suppress Feature

A suppressed feature is a feature that is not displayed. It is useful to hide features for clarity. Suppressed features improve model Rebuild time.

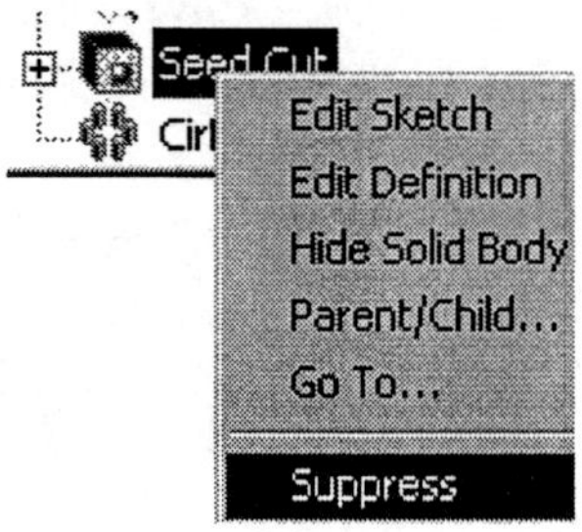

99) Suppress the SeedCut feature. Right-click on the **SeedCut** feature in the FeatureManager. Click **Suppress**. The SeedCut and the CirPattern features are displayed in gray in the FeatureManager.

The Circular Pattern feature is suppressed. The CirPattern feature is a child of the SeedCut feature.

Create the LENSCAP – Sweep Feature

The LENSCAP requires threads. Use the Sweep feature to create threads. The O-RING Base-Sweep feature required a circular path and a sketched cross section. The thread requires a spiral path. This path is called the Threadpath. The thread requires a sketched cross section. This cross section is called the Threadsection.

Note: Coils and springs use helical curves.

There are numerous steps required to create a thread. The plastic thread on the LENSCAP requires a smooth lead in. The thread is not flush with the back face. Use an offset plane to start the thread.

Create a new offset Sketch plane, ThreadPlane. Thread cross sections are normally small and detailed. Create a thread. Construction lines control the dimensions of the thread. Below are the following steps:

- Create a new plane for the start of the thread.
- Create the thread path.
- Create a large thread cross section for improve visibility.
- Create the Sweep feature.
- Reduce the size of the thread cross section.

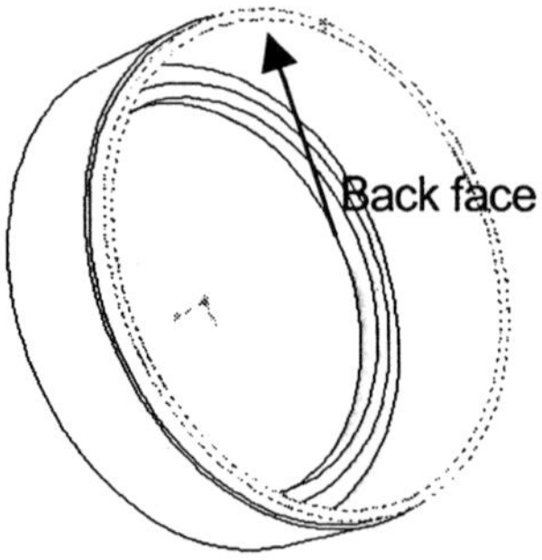

Create a new plane for the start of the thread

100) Display the back face of the LENSCAP. **Rotate** and **Zoom** the LENSCAP. Click the **narrow back face** of the BackCut feature.

101) Display the Right view. Click **Right** 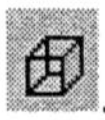.

102) Select the Plane Option. Click **Insert** from the Main menu. Click **Reference Geometry**. Click **Plane**. Click the **Offset** option. Click **Next**. The Offset Plane dialog box is displayed. Enter **.450**, **[11.43]** for Distance. Click the **Reverse** checkbox. Create the plane. Click **Finish**.

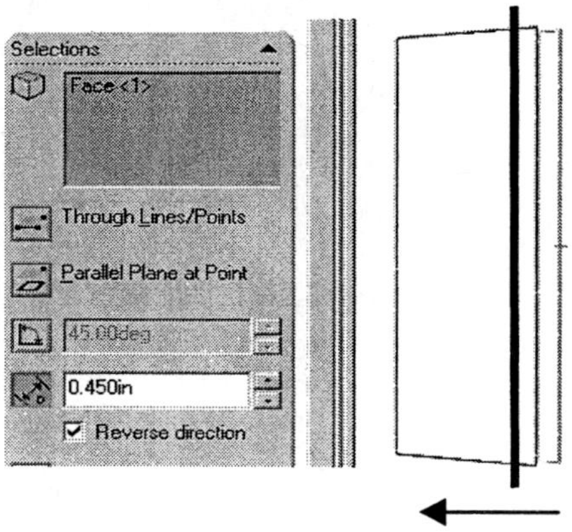

103) Rename **Plane2** to **Threadplane**. Display the Isometric view with the hidden lines removed. Click **Isometric**. Click **HiddenLinesRemoved**. The current Threadplane is displayed in green.

Create the Thread path.

104) Create the Sketch. Click **Sketch**. Extract the edge to the Threadplane. Click the **back inside circular edge** of the Shell. Click **Convert Entities**.

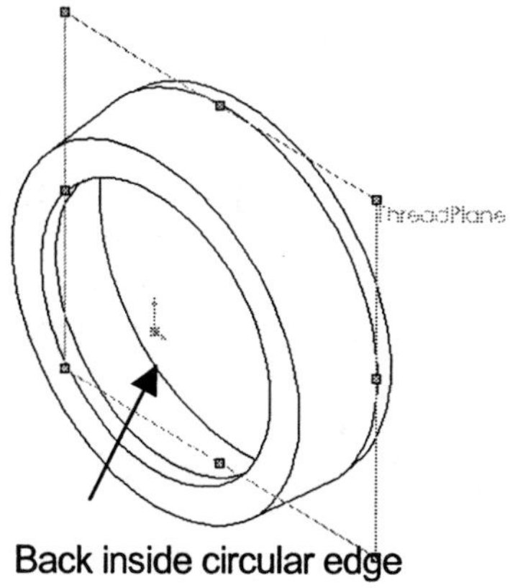

105) The circular edge is displayed on the Threadplane. Verify the Sketch position. Click **Top**. Click **Isometric**.

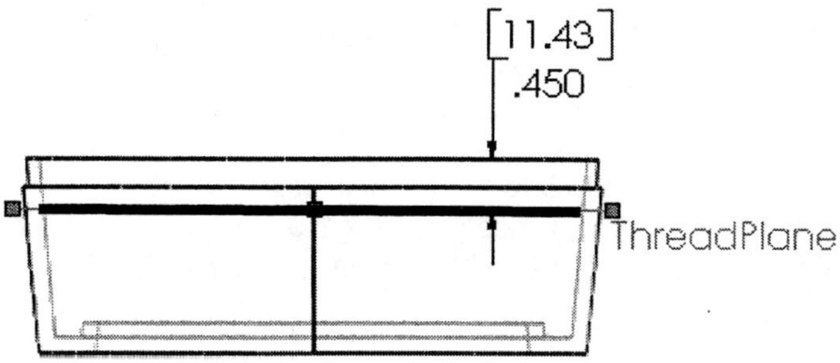

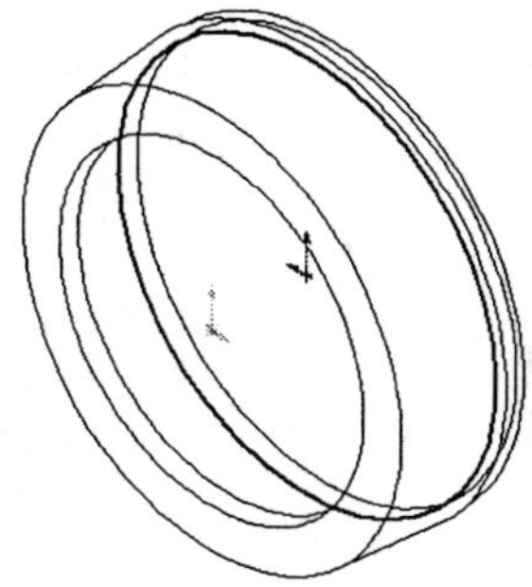

106) Create the Helix/Spiral curve path. Click **Insert** from the Main menu. Click **Curve**. Click **Helical/Spiral**. The Helix Curve dialog box is displayed. Enter **.250**, **[6.35]** for Pitch. Enter **2.5** for Revolutions. Click the **Taper Helix** check box. Enter **5** for Angle. Enter **0** for Starting angle. Click the **Reverse direction** checkbox. Do not check the Taper outward check box. Display the Helical/Spiral path. Click **OK**.

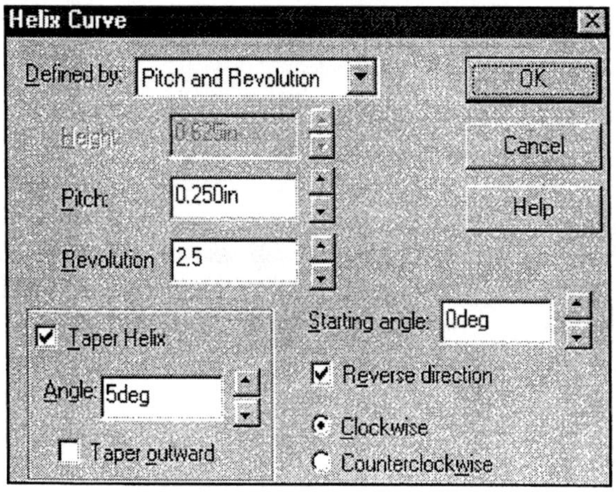

107) Rename **Helix1** to **ThreadPath**.

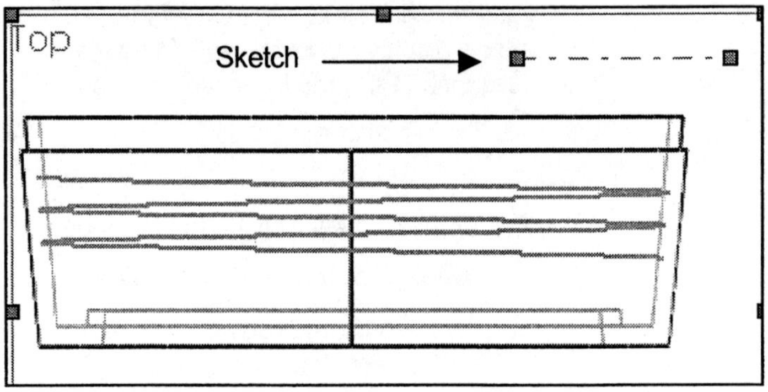

Create a large thread cross section for improve visibility.

108) Select the Sketch plane. Click the **Top** plane. Create the Sketch. Click **Sketch** . Display the Top view. Click **Top** 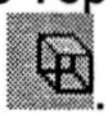. Sketch the profile to the top right of the LENSCAP.

109) Click **Centerline** 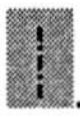. Create a short horizontal **centerline**. Create the first **vertical centerline** through the right endpoint.

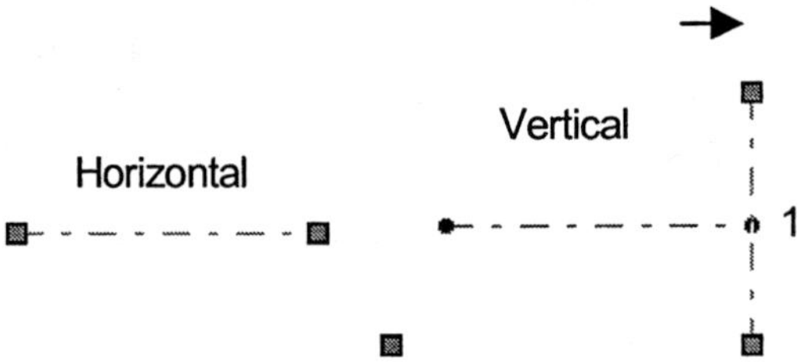

110) Add a Midpoint relation. Right click **Select**. Hold down the **Ctrl** key. Click the right **endpoint** of the horizontal line. Click the **Midpoint** button. Release the **Ctrl** key.

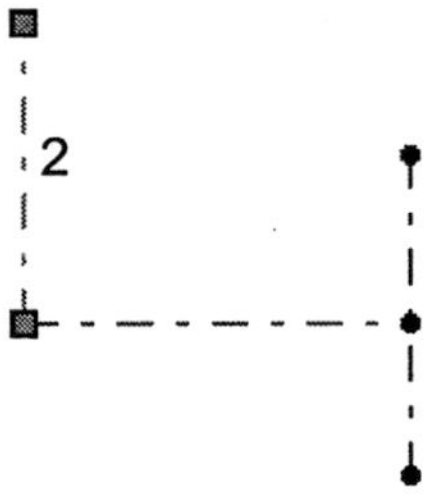

111) Create the second **vertical centerline** with the start point coincident with the left horizontal endpoint. Drag the **centerline** upward until it is approximately the same size as the left vertical line.

112) Create the third **vertical centerline** with the start point coincident with the left horizontal endpoint. Drag the **centerline** downward until it is approximately the same size as the left vertical line.

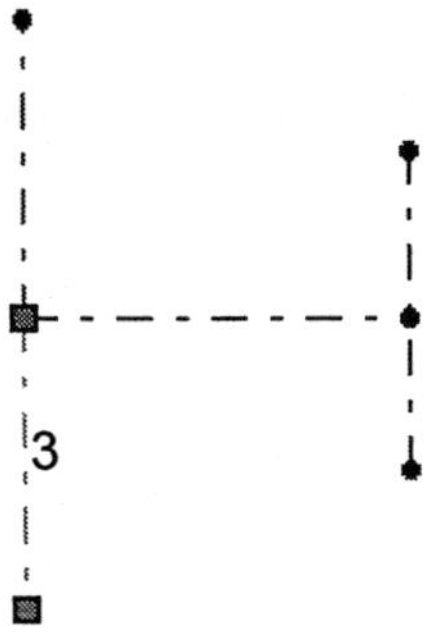

113) Add an Equal relation. Right click **Select**. Hold down the **Ctrl** key. Click the **three vertical lines**. Click the **Equal** button. Release the **Ctrl** key

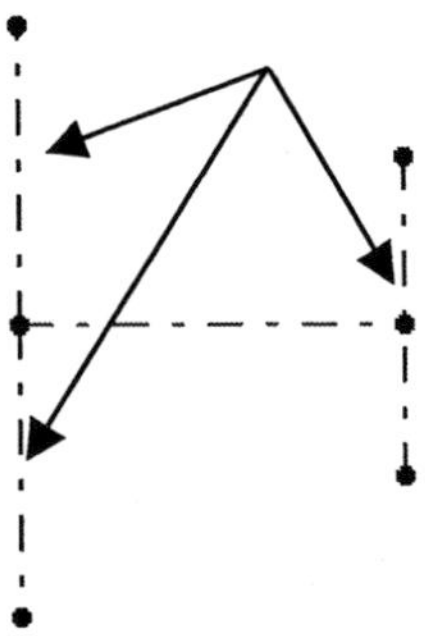

114) Add dimensions.

Click **Dimension**. Click the **left vertical endpoints**. Enter **.500**, [**12.7**].

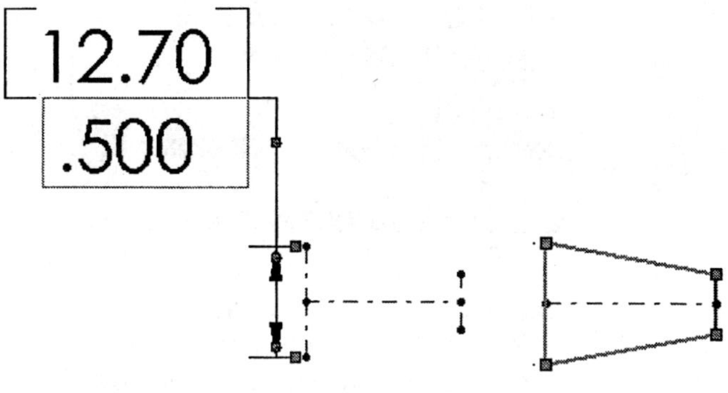

115) Sketch the profile.

Click **Line**. The profile is a trapezoid. Click the 4 **endpoints** of the vertical centerlines to create the trapezoid. Maintain the centerlines as part of the Sketch with geometric relationships.

116) Add an Equal relation. Right click **Select**. Hold down the **Ctrl** key. Click the **left**, **top** and **bottom lines** of the trapezoid. Click the **Equal** button. Release the **Ctrl** key.

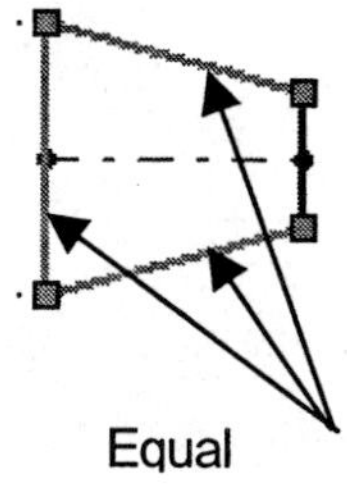

117) Add a **Pierce** relation between the **left midpoint** of the cross section and the starting **back edge** of the Threadpath. The Pierce relation positions the midpoint of the cross section to the path.

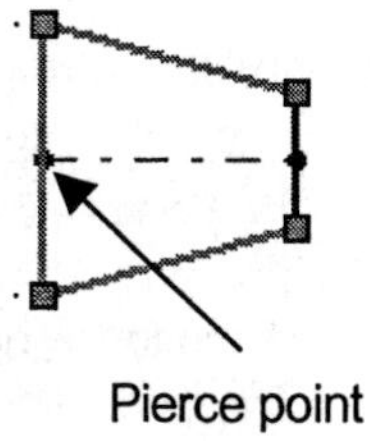

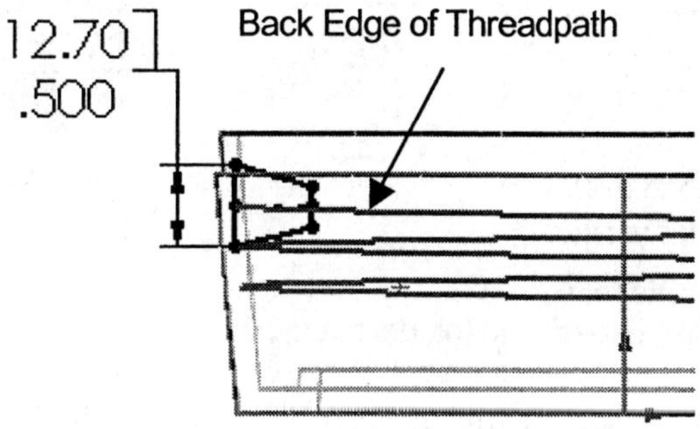

118) Click **Isometric** to verify your selection. Display the Sketch. Click **Apply**.

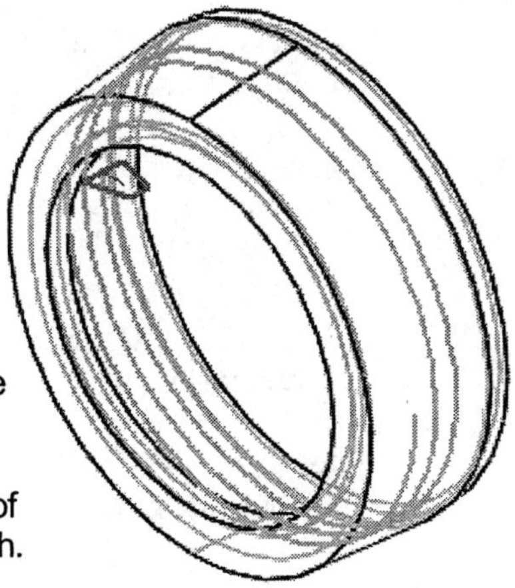

Pierce the left midpoint of the cross section and the left starting edge of the Threadpath.

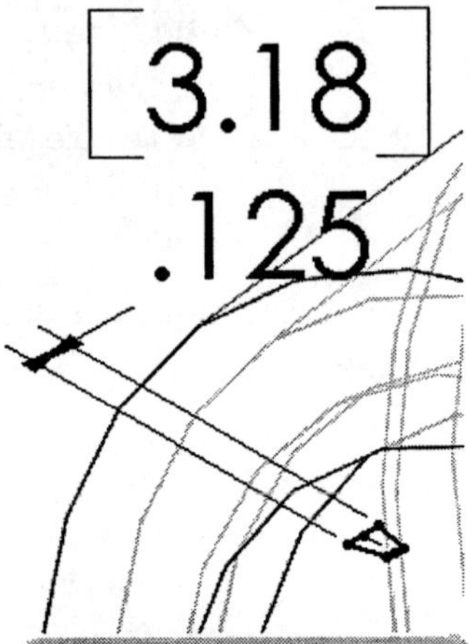

119) The thread cross section is too large. Modify the left line. Double click the **dimension text**. Enter **.125**, [3.18]. **Rebuild** the sketch.

120) Close the Sketch. Click **Sketch**.

121) Rename **Sketch6** to **Threadsection**.

122) Create the Sweep. Click **Sweep** from the Feature toolbar. The Sweep dialog box is displayed. This step should not have a number: Create the Sweep. Then the steps.

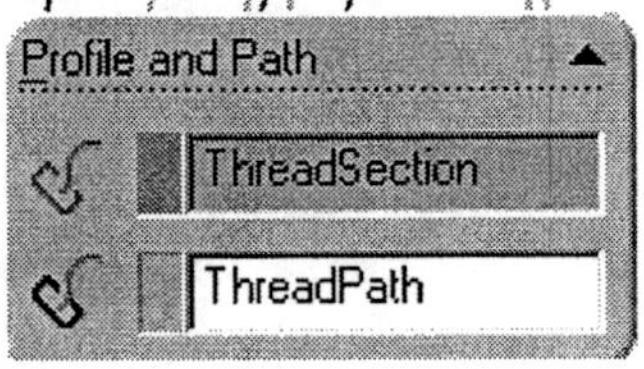

123) Select the cross section. Click the **Sweep section** text box. Click the **Threadsection** from the FeatureManager. Click the **Sweep path** text box. Click the helical **Threadpath** from the FeatureManager. Display the Sweep. Click **OK**.

124) Rename **Boss-Sweep1** to **Thread**.

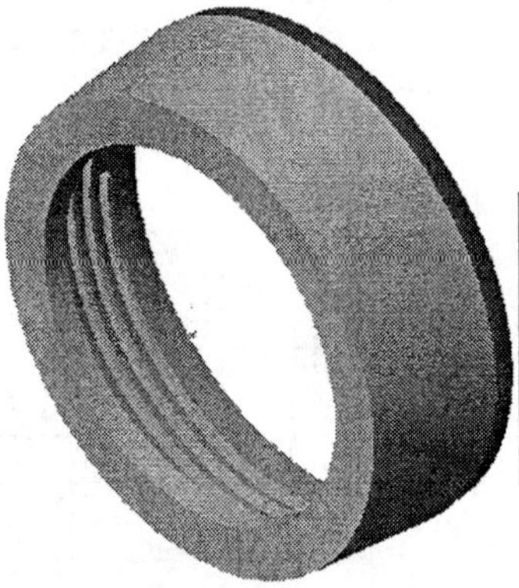

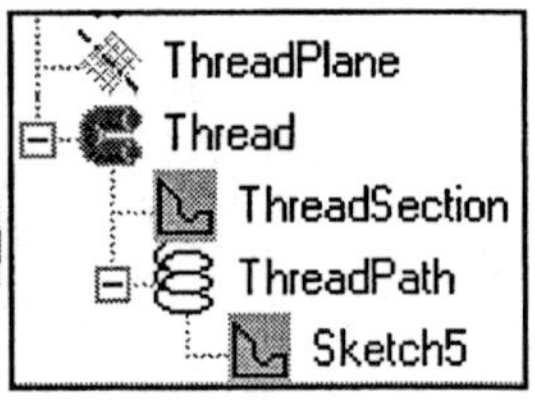

Expand the Thread feature.

125) Click the **Plus Sign** in the FeatureManager. The Thread feature is composed of the following: Threadsection and Threadpath. Expand the ThreadPath. Click the **Plus Sign**. The Threadpath contains the circular Sketch and the definition of the Helical curve.

Note: If the Threadsection geometry intersects itself, the cross section is too large. Reduce the cross section size and create the Sweep feature again.

126) Restore the CirPattern feature. Right-click on the **CirPattern** feature. Click **Suppress**.

127)Rebuild the LENSCAP. View the Thread. **Rotate** the LENSCAP. The LENSCAP is complete.

128) Save the LENSCAP. Click **Save**.

HOUSING

The HOUSING provides storage and support for the BATTERY. The SWITCH and LENSCAP connect to the HOUSING.

Do you remember the original customer requirements?

- An inexpensive reliable flashlight.
- Available advertising space of 10 square inches, [64.5 square centimeters].
- A light weight semi indestructible body.
- Self standing with handle.

The HOUSING must meet the above customer requirements. In a design situation, you may not have information on all of the required dimensions. Where do you start?

The LENS, BATTERYPLATE and BATTERY are internal to the HOUSING, Figure 3.11.

The LENSCAP thread fastens to the HOUSING thread.

Create a circular Extruded Base feature that fits inside the LENSCAP and contains the LENS.

In this exercise, create a loose fit between the BATTERY and the HOUSING. Review the mating parts and determine dimensions that require modification.

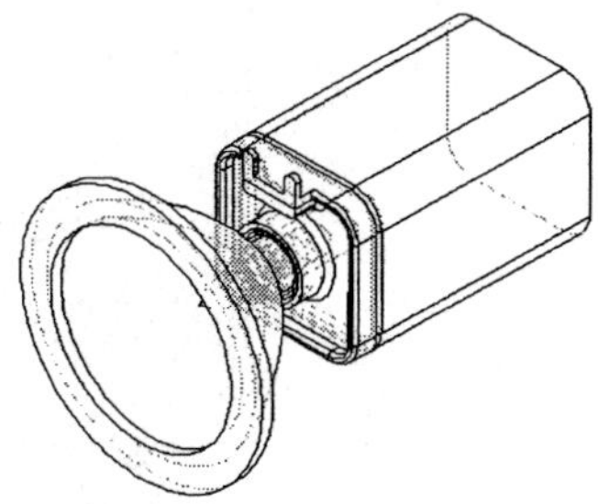

Figure 3.11

The LENS and the BATTERY are purchased parts. You cannot modify their size or shape! How do you transition the HOUSING from a rectangular shape to a circular shape? Answer: Create a Loft feature, Figure 3.12.

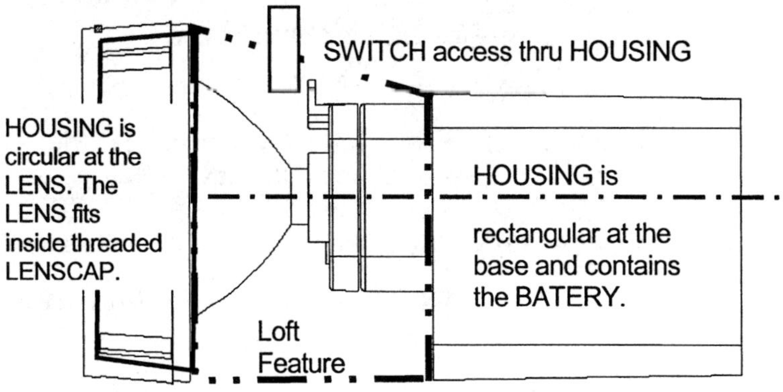

Figure 3.12

The HOUSING supports the threaded LENSCAP and LENS. The SWITCH fits through an opening in the HOUSING.

Internal Ribs are added to the HOUSING for structural integrity.

HOUSING Feature Overview

The HOUSING is composed of the following features:

- Extruded Base feature, Figure 3.13a.
- Loft feature, Figure 3.13b.
- Boss Extrude feature and Shell feature, Figure 3.13c.
- Handle Sweep and Thread-Sweep features, Figure 3.13d.
- Rib feature and Linear Pattern, Figure 3.13e.

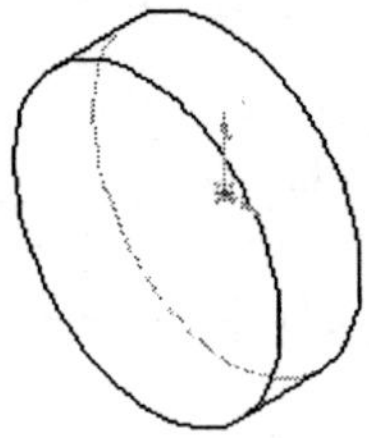

Figure 3.13a

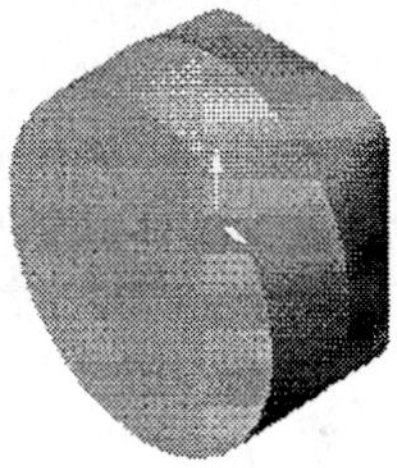

Figure 3.13b

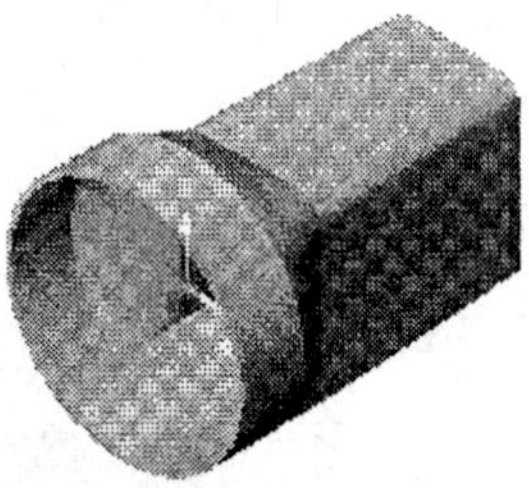

Figure 3.13c

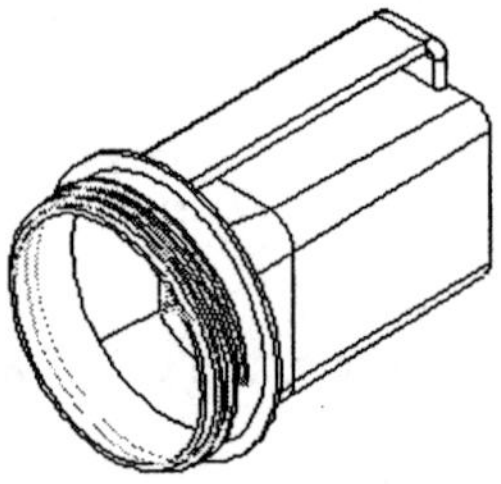

Figure 3.13d

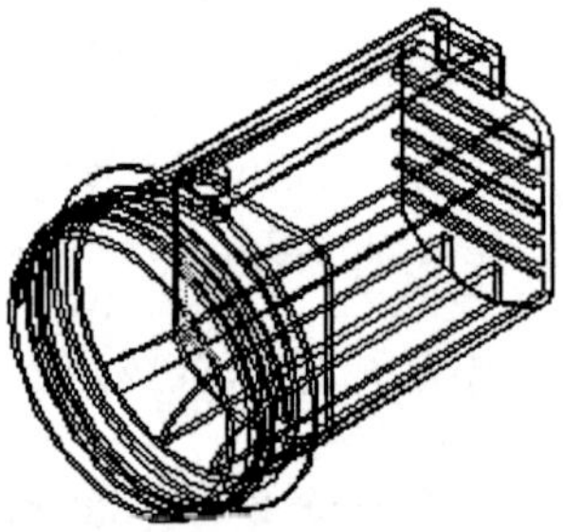

Figure 3.13e

Create the HOUSING – Extruded Base Feature

Create the HOUSING part. The Extruded Base feature is the first feature.

Create the Extruded Base feature.

129) Click **New** . Click **PartEnglishTemplate**, **[PartMetricTemplate]**. Click **OK**. Click **Save** . Select **My Documents** folder from the Save In text box. Enter the name of the part. Enter **HOUSING**. Click the **Save** button.

130) Select the Sketch plane. The Front plane is the default Sketch plane. Turn the Grid and Snap off. Click **Grid** . **Uncheck** all **Grid** and **Snap** checkboxes.

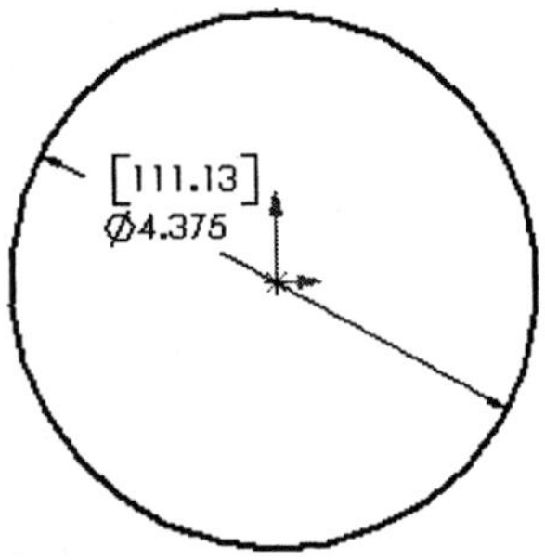

131) Create the Sketch. Click **Sketch** . Click **Circle** . Create a circle centered at the **Origin** .

132) Add a dimension. Click **Dimension** . Dimension the diameter of the circle. Enter **4.375**, **[111.13]**.

133) Extrude the Sketch. Click **Extruded-Base**. The Extrude Feature dialog box is displayed. Blind is the default Type option. Enter **1.300**, [**33.02**] for Depth. Display the Base-Extrude feature. Click **OK**.

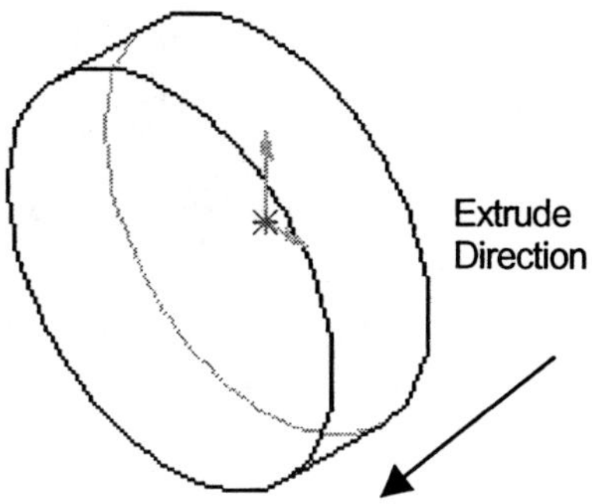

134) Save the HOUSING. Click **Save**.

Create the HOUSING - Loft Feature

The Loft feature is composed of two profiles. Create the first profile from the back face of the Base-Extrude feature. Create the second profile on an Offset plane. Close the Sketch.

Create the first profile.

135) Select the Sketch plane. Click the **back face** of the Base-Extrude feature.

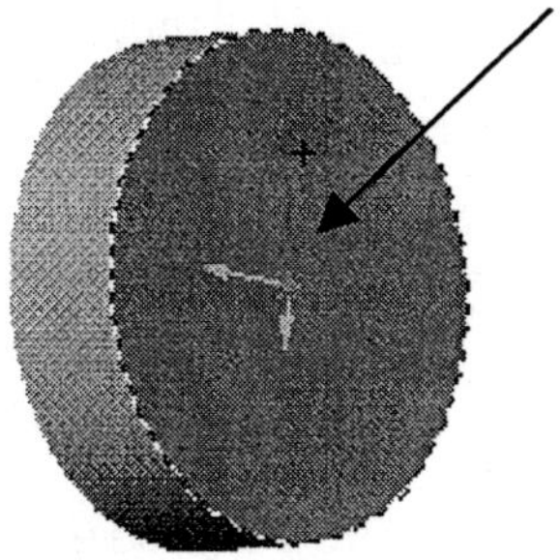

136) Click **Sketch**. Extract the entire face. Click **Convert Entities**.

137)Close the Sketch.

138) Rename **Sketch2** to **Sketch Circle**.

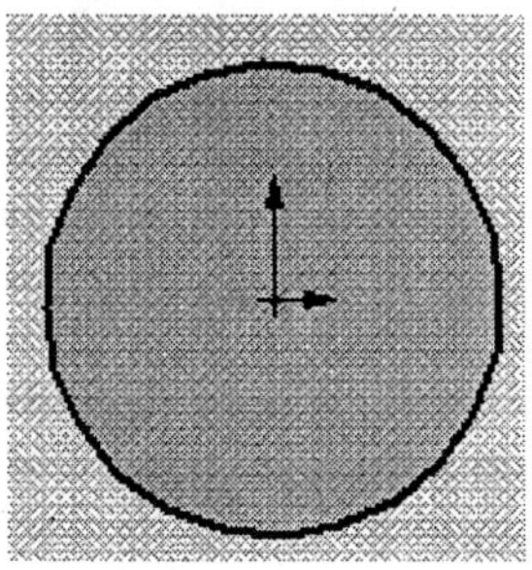

Create the second profile.

139) Create the Sketch plane. Click the **back face**. Click **Insert** from the Main menu. Click **Reference Geometry, Plane**. Create an **Offset** plane. Enter **1.300**, [**33.02**]. Verify the plane position. Click **Top**. Click **OK**.

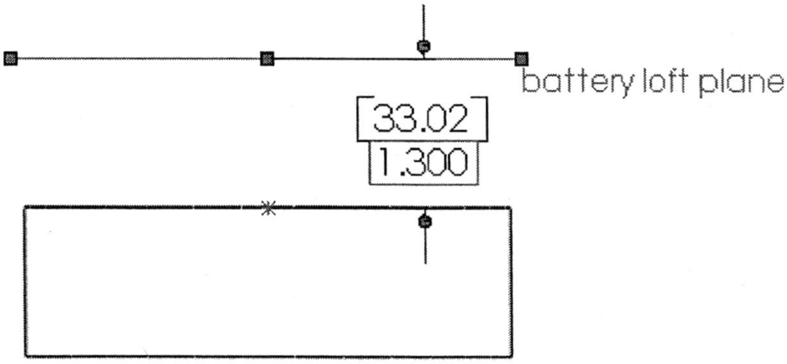

140) Rename **Plane4** to **battery loft plane**.

141) Click **Sketch**. Click **Back**. Extract the circular edge. Click the **circumference of the circle**. Click **Convert Entities**.

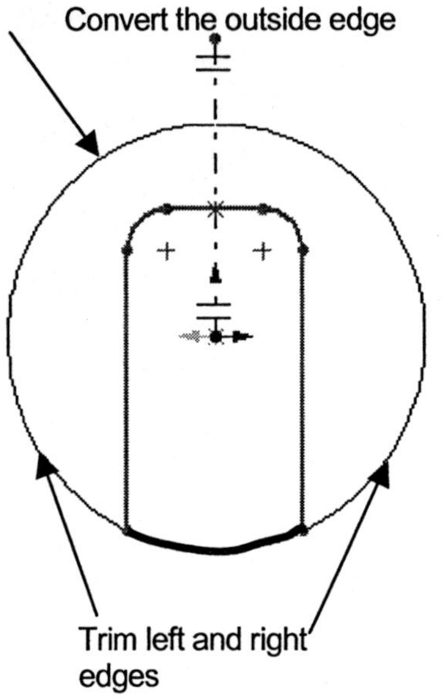

142) Click **Centerline**. Sketch a **centerline** collinear with the Right plane coincident to the Origin.

143) Click **Mirror**. Click **Line**. Sketch a **horizontal** line. Click **Tangent Arc**. Sketch a **90 degree arc**. Click **Line**. Sketch a **vertical** line. The endpoint of the vertical line is coincident with the edge of the circle. Turn the Mirror off. Click **Mirror**.

144) Trim unwanted geometry. Click **Trim**. Click the far **right edge of the circle**. Click the far **left edge of the circle**.

145) Add dimensions. Click **Dimension**. Create the **horizontal** dimension. Click the **left vertical line**. Click the **right vertical line**. Enter **3.100, [78.74]**. Create the **vertical** dimension from the Origin. Enter **1.600, [40.64]**. Create the **radial** dimension. Enter **.500, [12.7]**.

The FLASHLIGHT components must remain aligned to a common centerline.

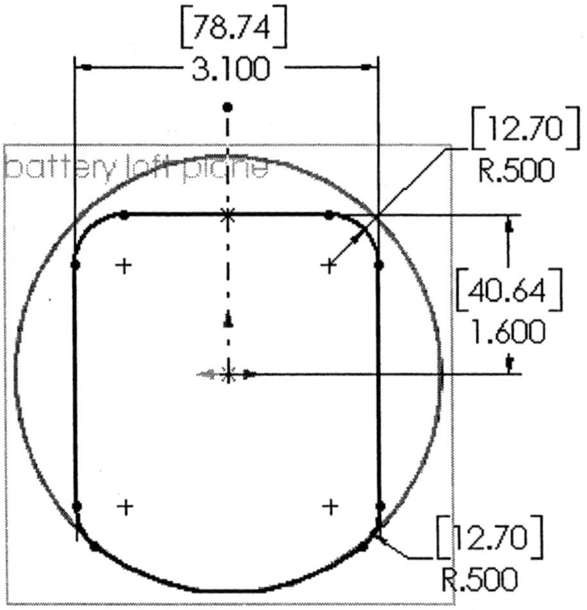

146) Remove all sharp edges. Add fillets at the lower corners. Click **Select**. Click the **lower left point**. Click **2D Fillet**. Enter **.500, [12.7]**. Click **Apply**. Click **the lower right point**.

147) Close Sketch3. Click **Sketch**.

148) Rename **Sketch3** to **Sketch Square**.

Create the Loft.

149) Display the Isometric view. Click **Isometric**. Click **Loft** from the Feature menu. Select the Profiles. Click the **top right corner of the Sketch Square**. Click the **upper right side of the Sketch Circle**. Display the Loft feature. Click **OK**.

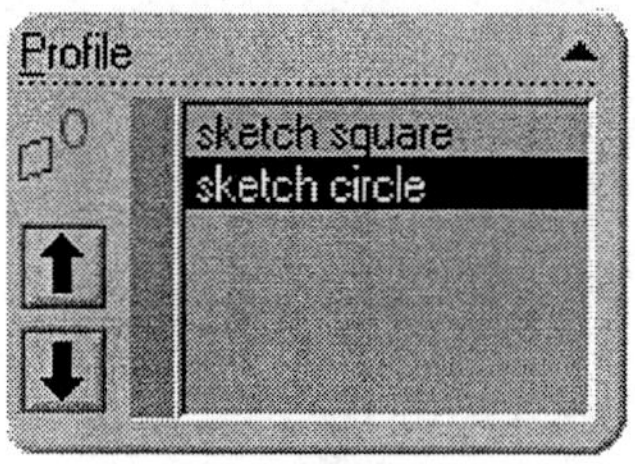

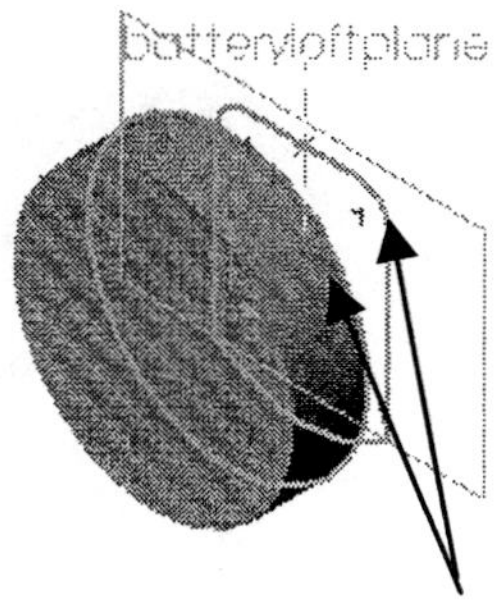

The Boss-Loft1 feature is composed of Sketch Square and Sketch Circle.

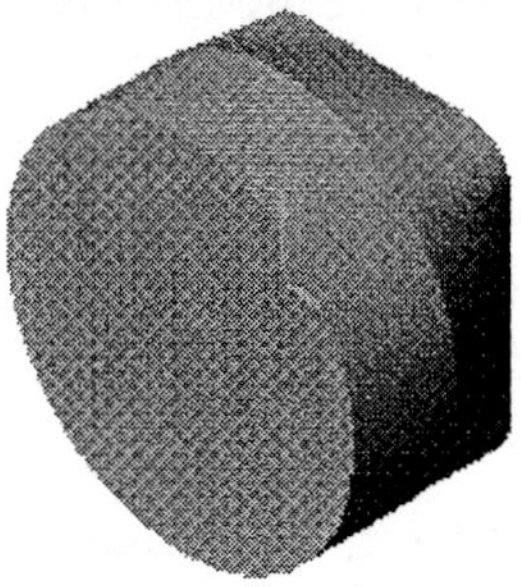

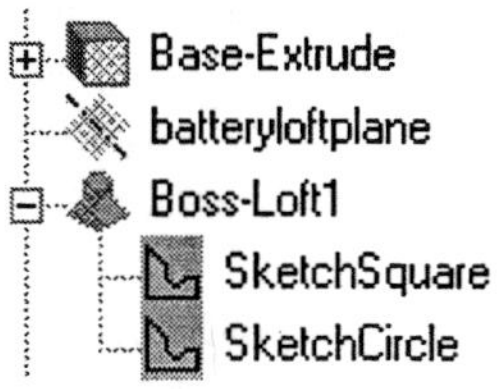

150) Save the HOUSING. Click **Save**.

Create the HOUSING – First Extruded Boss Feature

Create the first Extruded Boss feature from the square face of the Loft. How do you estimate the depth of this feature?

Answer: The Extruded Base feature of the BATTERY is 4.1inches, [104.14mm]. Ribs are required to support the BATTERY. Design for Rib construction. Use a 4.4inch, [111.76mm] depth as the first estimate. Adjust the estimated depth dimension later if required in the FLASHLIGHT assembly.

Create the first Extruded Boss feature.

151) Select the Sketch plane. Click the **back face** of the Loft.

152) Create the Sketch. Click **Sketch**. Extract the entire face. Click **Convert Entities**.

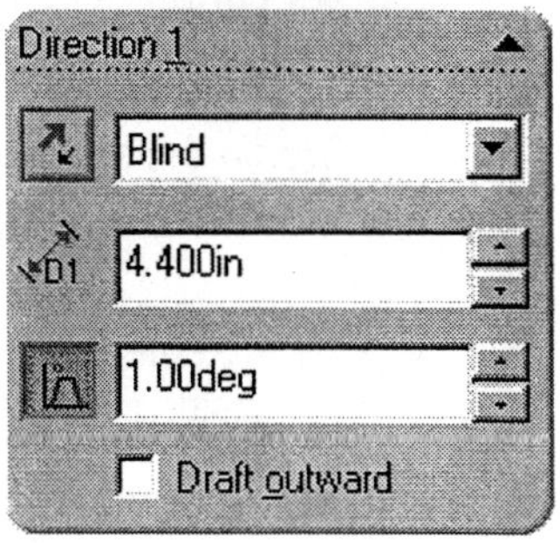

153) Extrude the Sketch. Click **Extruded-Boss/Base**. Enter **4.400**, **[111.76]** for Depth. Check **Draft While Extruding** check box. Enter **1** for Draft Angle.

154) Display the Boss feature. Click **OK**.

155) Rename **Boss-Extrude** to **Boss-Battery**.

156) Save the HOUSING. Click **Save**.

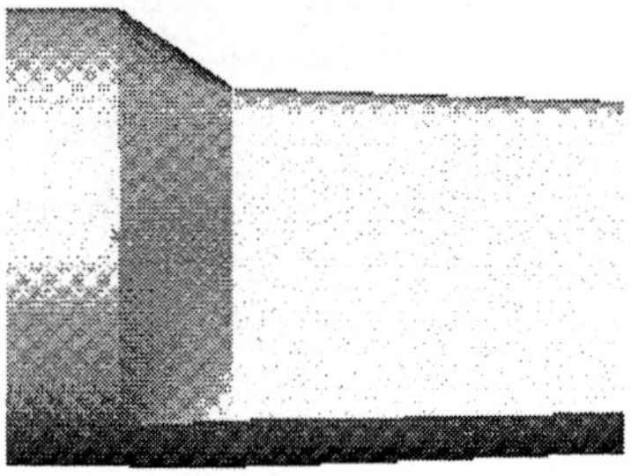

Create the HOUSING - Shell Feature

The Shell feature removes material. Use the Shell feature to remove the front face of the HOUSING.

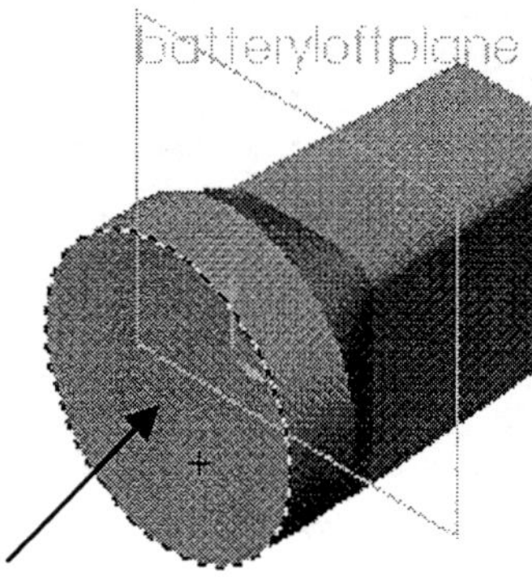

157) Create the Shell feature. Click **Shell** from the Feature toolbar. Click the **front face** of the Boss-Insert feature. The Shell Feature dialog box is displayed.

158) Enter **.100**, **[2.54]** for Thickness. Display the Shell. Click **OK**.

159) Save the HOUSING. Click **Save**.

Note: First create the Shell feature. Then create the Sweep feature for the handle. The handle is solid.

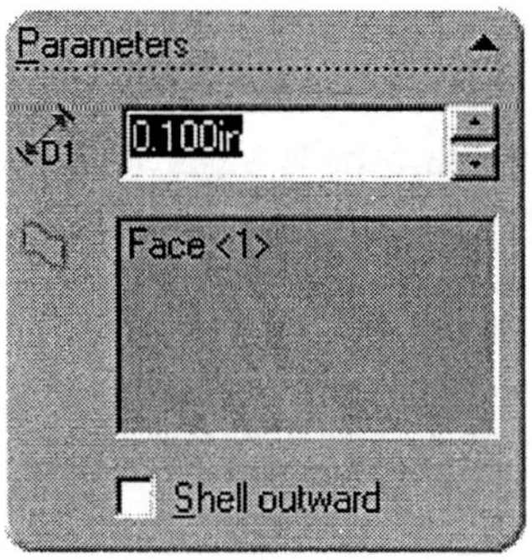

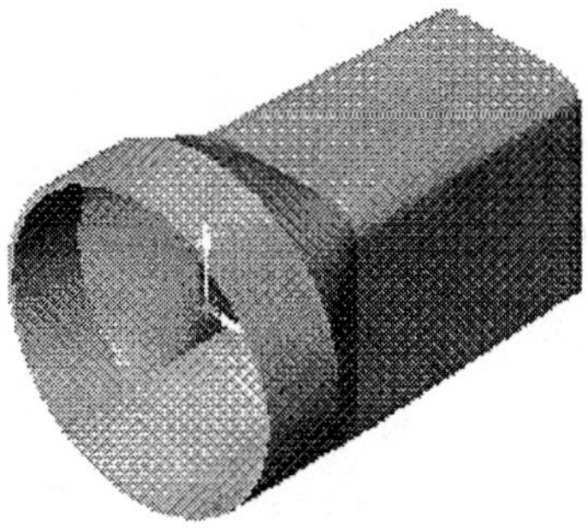

Create the HOUSING - Second Extruded Boss Feature

The second Extruded Boss feature creates a solid circular ring on the back circular face of the Extruded Base feature. The solid ring is a cosmetic stop for the LENSCAP and provides rigidity at the transition of the HOUSING.

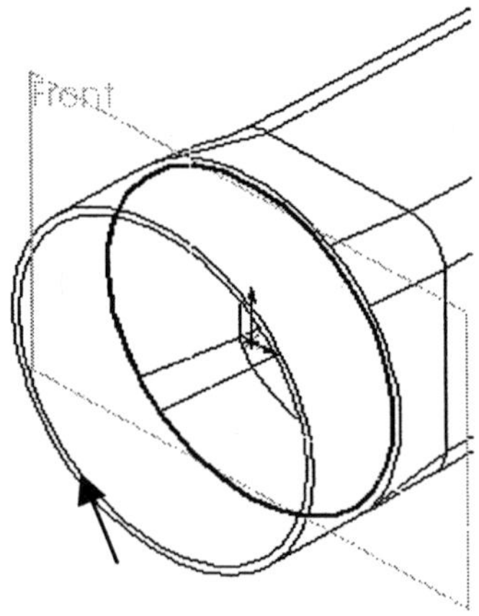

Create the second Extruded Boss.

160) Select the Sketch plane. Click the **Front** plane.

161) Create the **Sketch**. Click **Sketch** . Create the inside circle. Click the **front inside circular edge** of the Shell. Click **Convert Entities** .

162) Create the outside circle.

Click **Circle** . Create a **circle** centered at the Origin. Add a dimension. Click **Dimension** . Click the outside circle. Enter **5.125**, **[130.18]**.

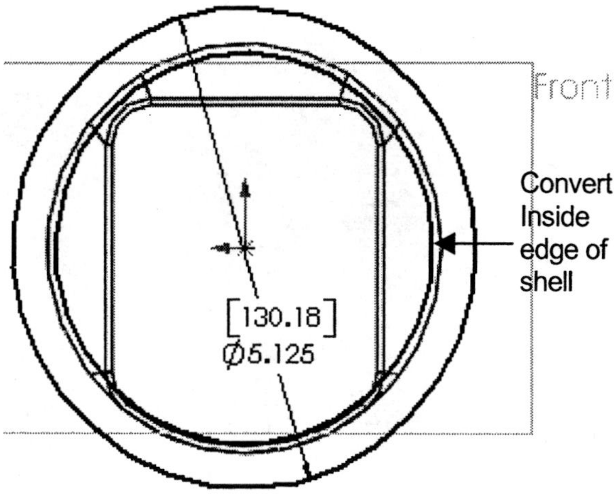

163) Extrude the Sketch. Click **Extruded Boss/Base** . Enter **.100**, **[2.54]** for Depth. Display the Boss feature. Click **OK**.

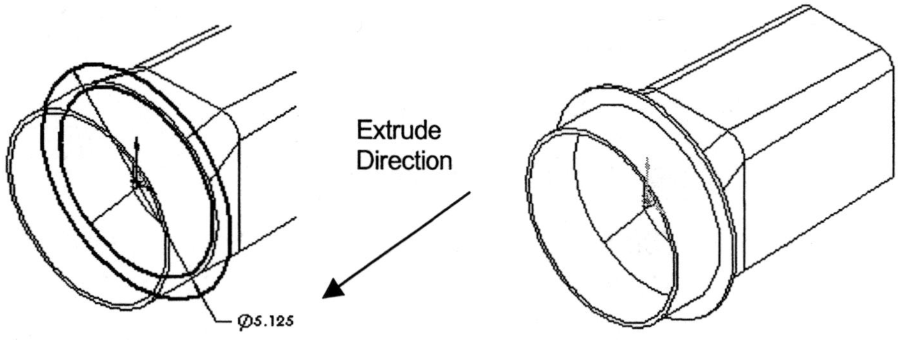

164) Rename **Boss-Extrude** to **Boss-Stop**.

165) Save the HOUSING. Click **Save**.

Create the HOUSING - Draft Feature

The LENSCAP Extruded Base feature has a 5 degree draft angle. The outside of the Extruded Base feature HOUSING requires a 5 degree draft angle. The 5 degree draft angle insures proper thread mating. The inside HOUSING wall does not contain a draft angle. Use the Draft feature to create a draft angle.

Create the Draft.

166) Click **Zoom In** on the front circular face. Click the thin **front circular face**. Click **Draft** . The Draft Feature dialog box is displayed.

167) The front circular face is displayed in the Neutral plane text box. Click the **Faces to draft** text box.

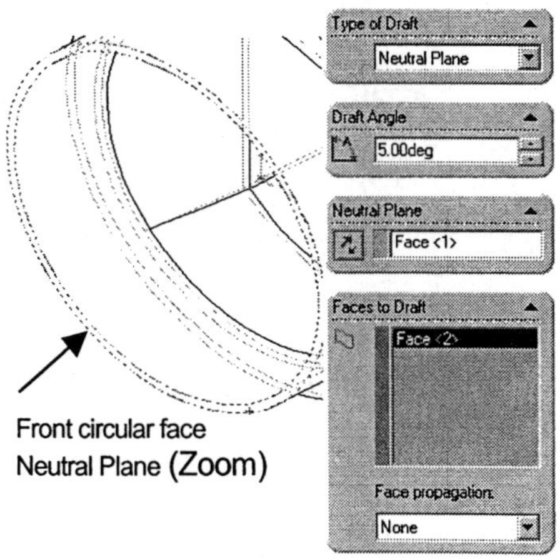

168) Click the **outside face**. Enter **5** for Draft angle. Display the Draft. Click **OK**.

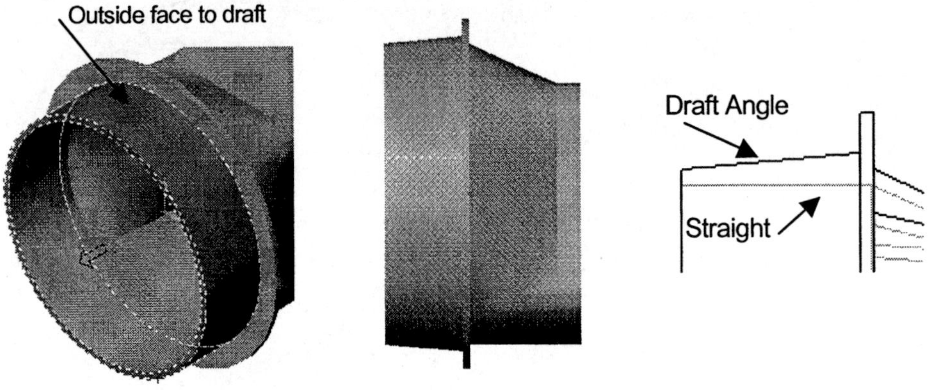

169) Display the Draft angle and the straight interior. Click the **Right** view. Click **Hidden Lines Removed**.

170) Save the HOUSING. Click **Save**.

Create the HOUSING – Thread with Sweep Feature

The HOUSING requires a thread. The threads for the HOUSING are created on the outside drafted face. Create a thread with the Sweep feature. The thread requires two sketches:

- Threadpath
- Threadsection

Create the Threadpath sketch.

171) Click the **front circular face** of the HOUSING. Create an Offset Reference plane. Click **Insert** from the Main menu. Click **Reference Geometry**. Click **Plane**. Click the **Reverse** check box. Enter **.125**, **[3.18]** for the Offset Distance. Click **OK**.

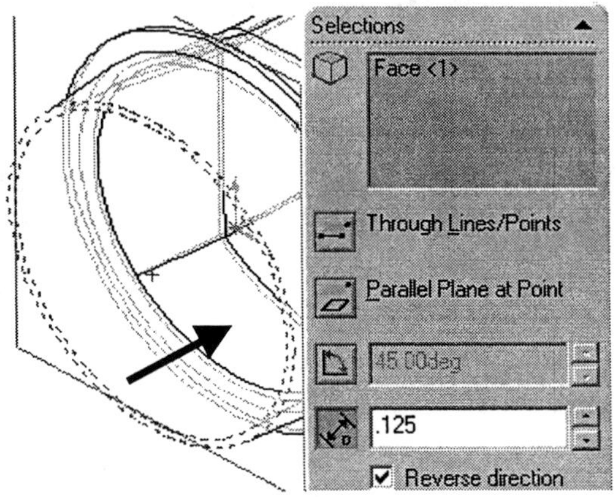

172) Rename **Plane** to **Threadplane**.

173) Verify the ThreadPlane position.

Click **Top**.

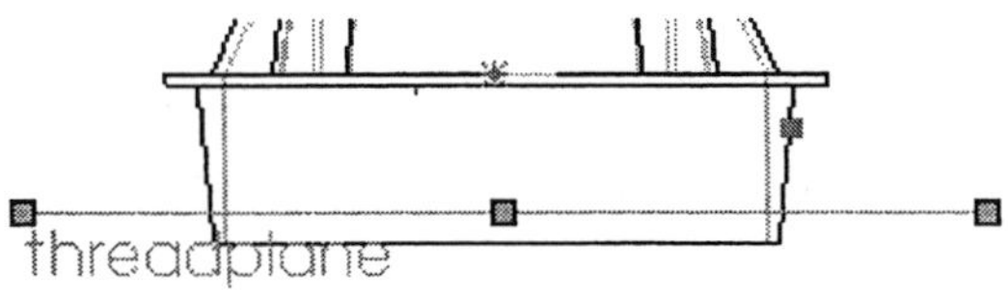

174) Select the Sketch plane. Click the **Threadplane**. Click **Isometric**.

175) Create the Sketch. Click **Sketch** . **Zoom in** on the front circular face. Select the **front** outside circular edge.

176) Click **Convert Entities** . The circular edge is displayed on the Threadplane.

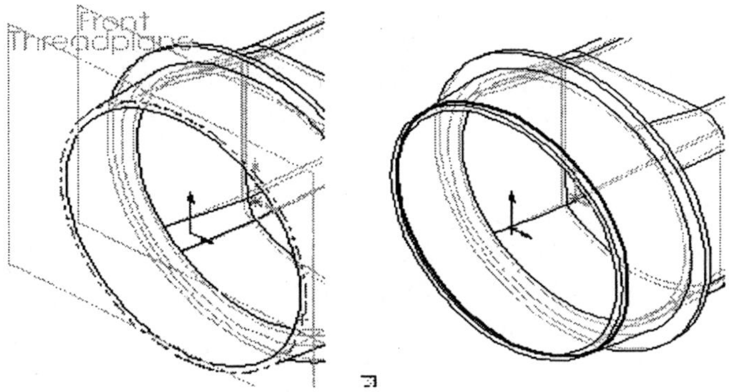

177) Create the Helix/Spiral curve from the circular edge. Click **Insert** from the Main menu. Click **Curve**. Click **Helix/Spiral**. The **Helix Curve** dialog box is displayed.

178) Enter **.250**, **[6.35]** for Pitch. Enter **2.5** for Revolutions. Click the Taper **Helix** check box. Enter **5** for Angle. Click the Taper **Outward** check box. Enter **180** in the Starting angle spin box. Click the **Reverse** direction checkbox. Display the Helix/Spiral curve. Click **OK**.

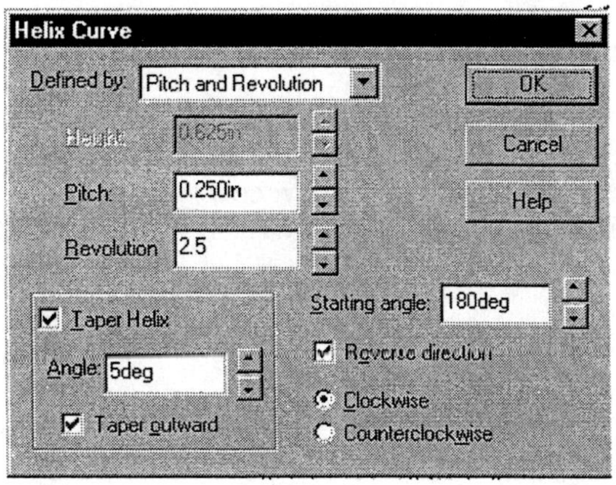

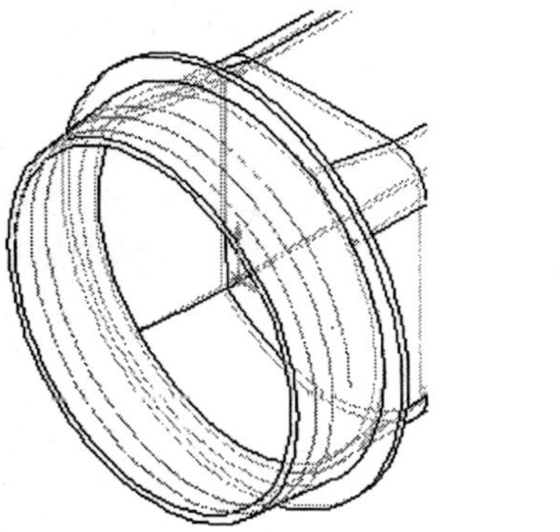

179) Rename **Helix1** to **Threadpath**.

180) Save the HOUSING. Click **Save**.

You have created the Threadsection for the LENSCAP.

Create the Threadsection sketch for the HOUSING. Copy the created LENSCAP sketched cross section.

LENSCAP

Copy the Threadsection from the LENSCAP.

181)Open the LENSCAP. **Expand** the Thread-Sweep feature from the FeatureManager. Click the **Threadsection** sketch. Click **Edit** from the Main menu. Click **Copy**. **Close** the LENSCAP.

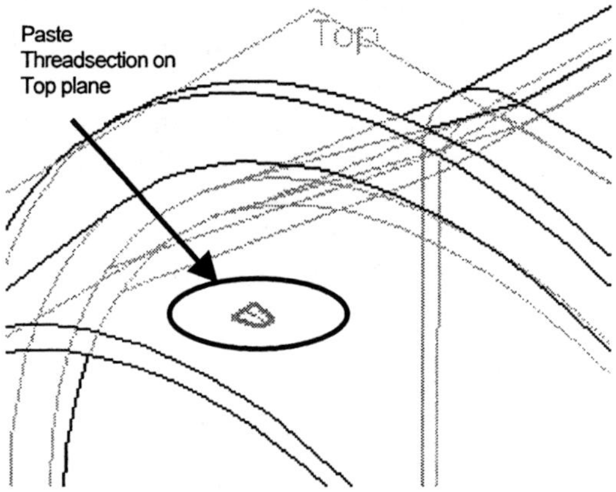

182) Click the **Top** plane in the HOUSING FeatureManager. Click **Edit** from the Main menu. Click **Paste**. The Threadsection is displayed on the Top plane. The new Sketch7 name is added to the FeatureManager.

183) Rename **Sketch7** to **Threadsection**.

184) Pierce Threadsection to Threadpath. Right-click on **Threadsection**. Click **Edit Sketch**. Enlarge the Sketch for improve visibility. Click the **ThreadSection**.

Click **Zoom to Selection** . Click the **Midpoint** of the Threadsection.

185) Display the Threadpath. Click **Zoom to Area** . Hold down the **Ctrl** key. Click the **right back edge of the Threadpath**. The Pierce relation positions the center of the cross section on the path. Click **Pierce**. Release the **Ctrl** key. Click **Apply**.

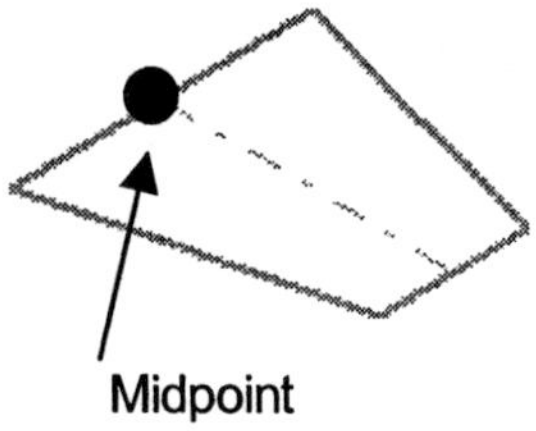

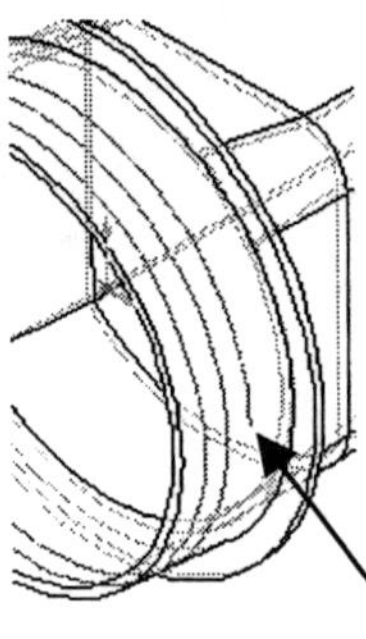

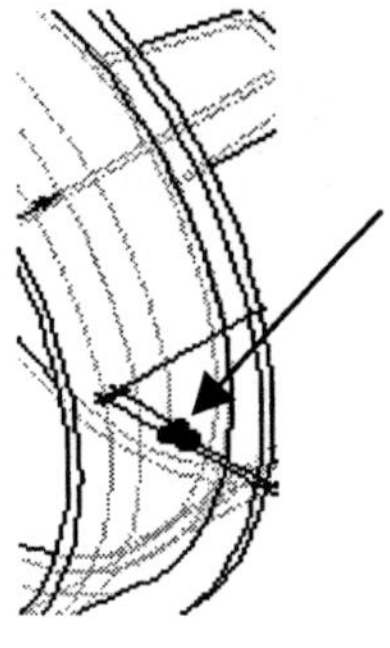

Caution: Do not click the front edge of the Thread path. The Thread is created out of the HOUSING.

186) Close the Sketch. Click **Sketch** .

Create the Sweep.

187) Click **Sweep** from the Feature toolbar. Click the **Boss-Sweep Property Manager tab** to display the FeatureManager. Select the cross section. Click the **Sweep section** text box. Click **Threadsection** from the FeatureManager. Click the **Sweep path** text box. Click the helical **Threadpath** from the FeatureManager. Display the Sweep. Click **OK**.

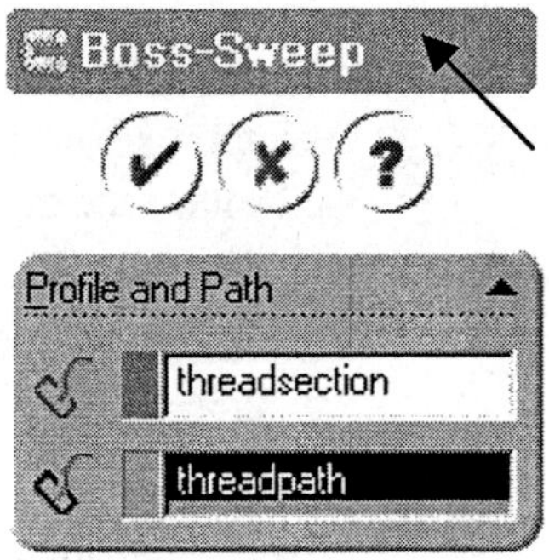

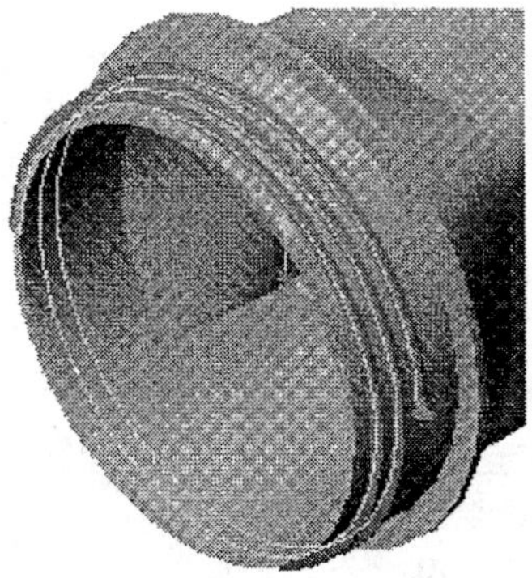

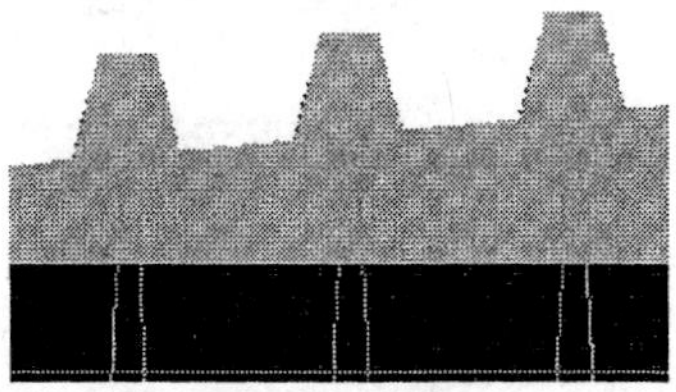

An external thread is created by piercing the midpoint of the Threadsection to the right side of the Threadpath. The Threadplane allows for a smooth lead. The Threadplane offset dimension is modified in the assembly by adjusting the mating threads of the LENSCAP and HOUSING. Creating a Threadplane provides flexibility to the design.

188) Rename **Boss-Sweep1** to **Thread**.

189) Save the HOUSING. Click **Save**.

Note: Conserve time. Use the Right-mouse button to invoke the Zoom and Pan commands. Selection Filters are helpful, however they require one or more clicks. View the mouse pointer feedback for entity confirmation.

Create the HOUSING – Handle with Sweep Feature

Create the handle with the Sweep feature. The Sweep feature consists of a sketched path and cross section profile. Sketch the path on the Right plane. The sketch uses edges from existing features. Sketch the cross section on the back circular face of the Boss-Stop feature.

Create the Sweep path.

190) Select the Sketch plane. Select the **Right** plane.

191) Create the Sketch. Click **Sketch**. Click the **top right small vertical edge**. Click **Convert Entities**.

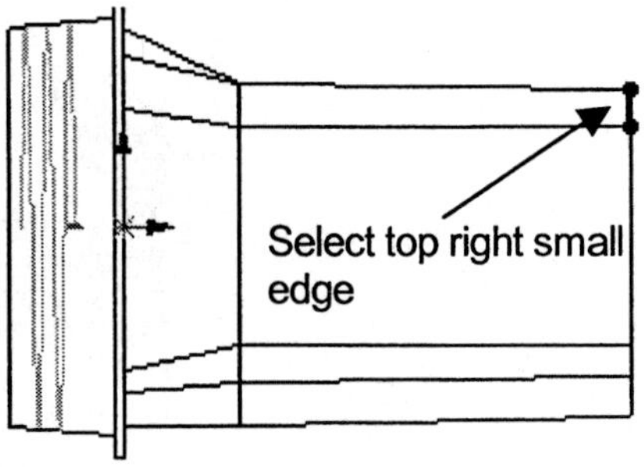

192) Drag the **bottom point** upward on the vertical line.

193) Click **Line**. Sketch a horizontal **line**. The left endpoint of the line is coincident with the back edge of the Boss-Stop.

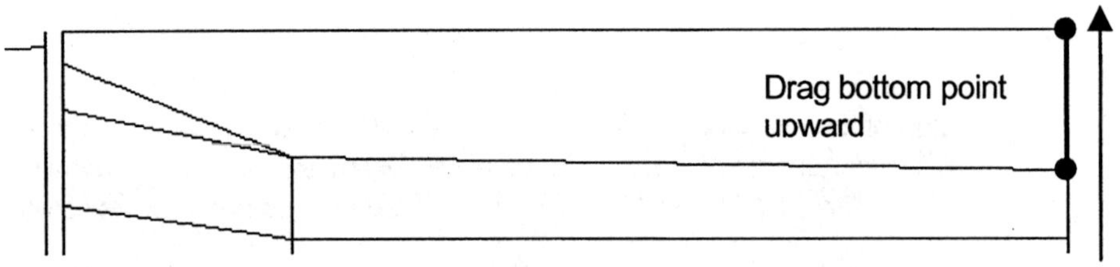

194) Create a 2D Fillet. Click **2D Fillet**. Click the **upper right intersection**. Enter **.500, [12.7]** for Radius. Click **Close**.

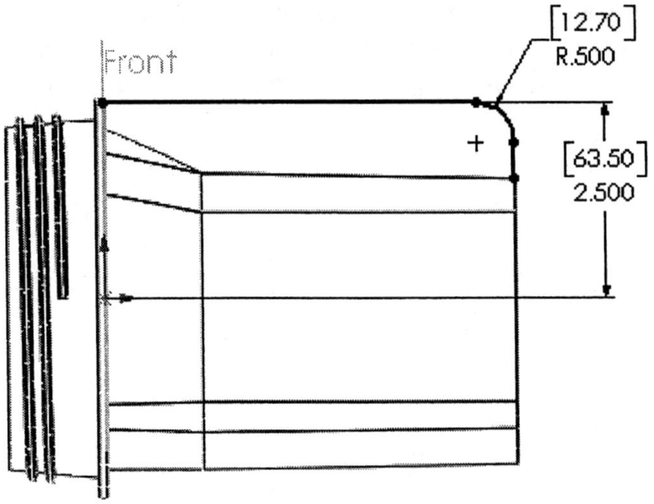

195) Add a dimension. Click **Dimension**. Add a vertical dimension from the **Origin** to the horizontal **line**. Enter **2.500, [63.5]**.

196) Close the **Sketch**. Click **Sketch**.

197) Rename **Sketch8** to **HandlePath**.

Create the Sweep Profile.

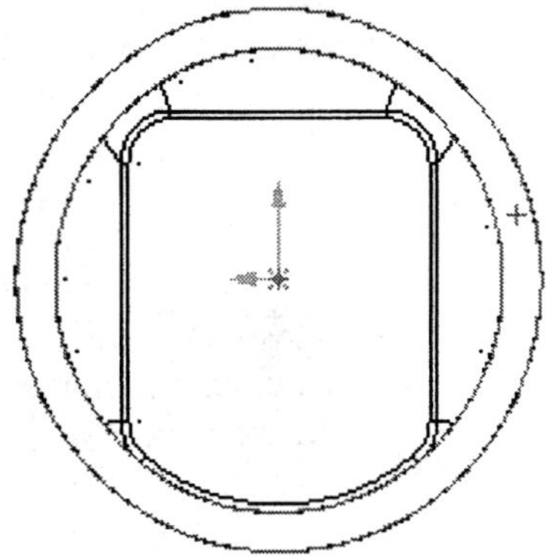

198) Select the Sketch plane. Click **Back**. Select the **back circular face** of the Boss-Stop feature.

199) Sketch the second profile. Click **Sketch**.

Click **Centerline**. Sketch a **centerline** collinear with the Right plane coincident to the Origin.

200) Click **Mirror**. Click **Line**. Sketch a **horizontal** line. Click **Tangent Arc**. Sketch a **180 degree arc**. Click **Line**. Sketch a **horizontal** line. Turn the Mirror off. Click **Mirror**.

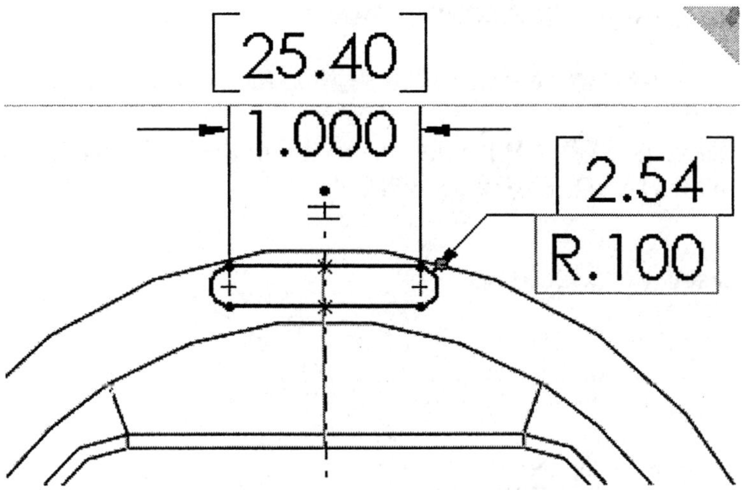

201) Add dimensions. Click **Dimension**. Create the **horizontal** and **radial dimensions.** Enter **1.000**, [**25.4**] between the arc center points. Enter **.100**, [**2.54**] for Radius.

202) Add a Pierce relation. Click the **top midpoint** of the Sketch profile. Hold down the **Ctrl** key. Click the **line** from the Handle Path. Click **Pierce**. Release the **Ctrl** key.

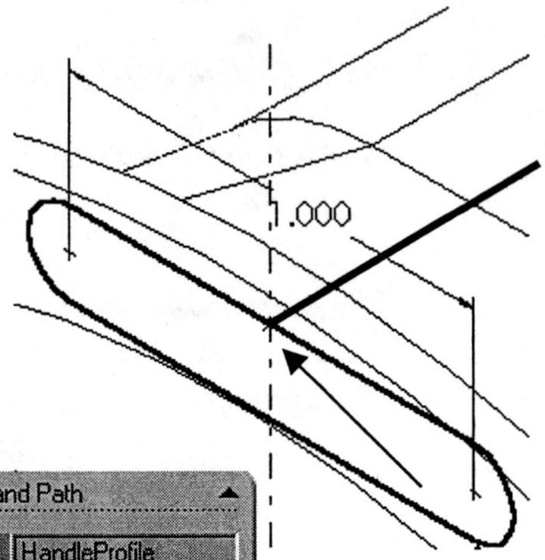

203)Close the Sketch.

204) Rename **Sketch9** to **HandleProfile**.

Sweep the Sketch.

205) Click **Sweep** from the Feature toolbar. Click the **Sweep tab** from the PropertyManager to display the FeatureManager. Click the **Sweep section** text box. Click the **HandleProfile** from the FeatureManager. Click the **Sweep path** text box. Click the **HandlePath** from the FeatureManager. Display the Sweep. Click **OK**.

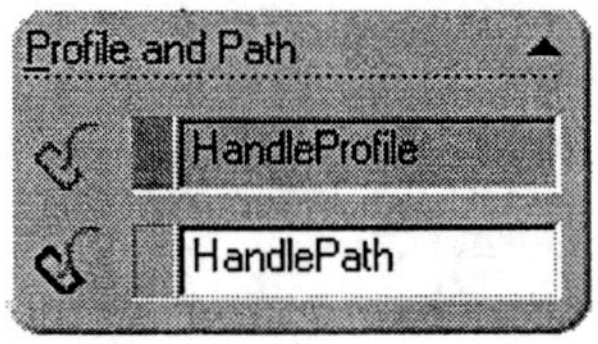

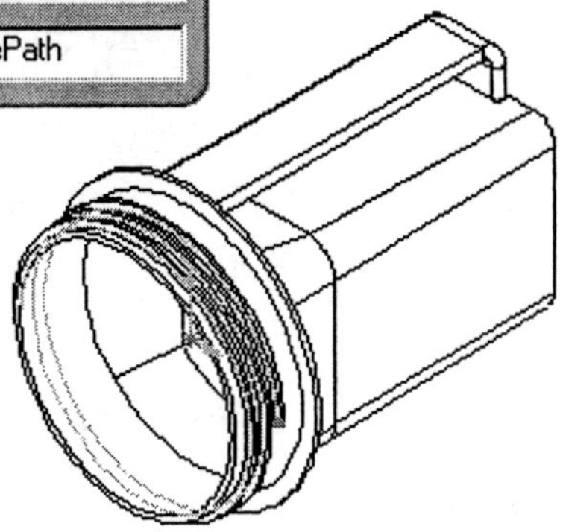

206) Rename **Sweep2** to **Handle**.

207) Save the HOUSING. Click **Save**.

Create the HOUSING – Extruded Cut Feature for SWITCH

Insert the SWITCH into the HANDLE of the HOUSING. Create an Extruded Cut in the HANDLE for the SWITCH.

Create the Extruded Cut feature.

208) Select the Sketch plane. Click the **top face** of the Handle.

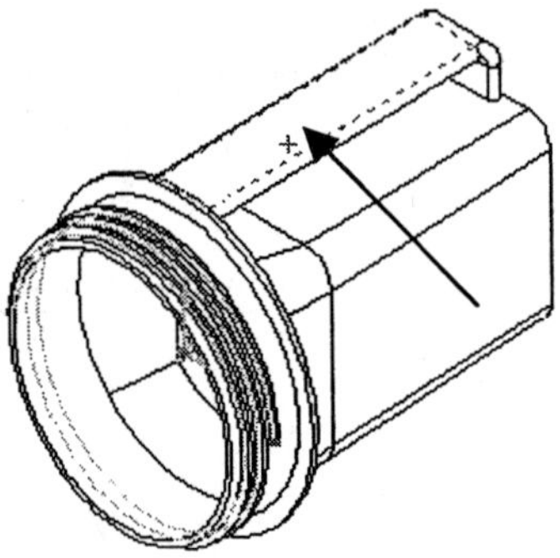

209) Create the Sketch.

Click **Sketch** .

Click **Normal To** .

Right click the **Right plane**. Click **Show**. Right click the **Front plane**. Click **Show**.

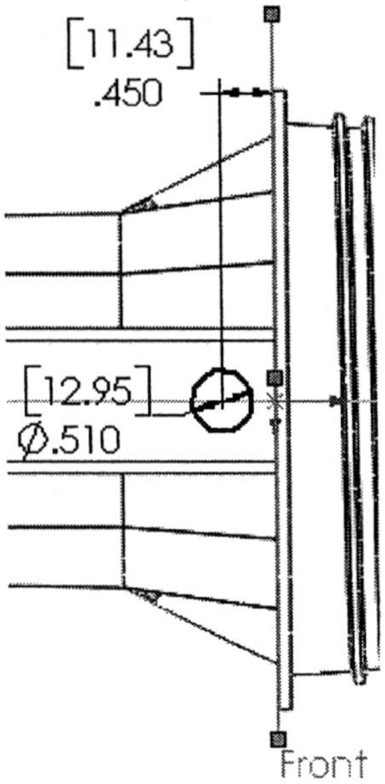

210) Create a circle. Click **Circle** . Center the **circle** collinear to the **Right** plane. Add dimensions. Click **Dimension** . Enter **.510, [12.95]** for diameter. Enter **.450, [11.43]** for the distance from the Front plane.

211) Extrude the Sketch. Click **Extruded-Cut** . Click the **UpTo Surface** option from the Type list box. Click the **top inside face** of the Shell. Click **OK**.

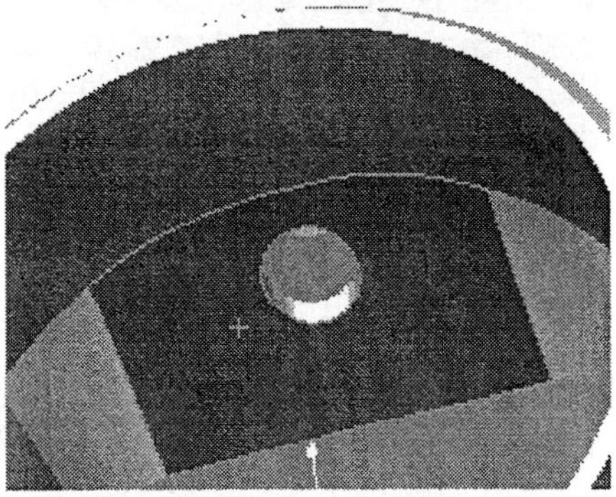

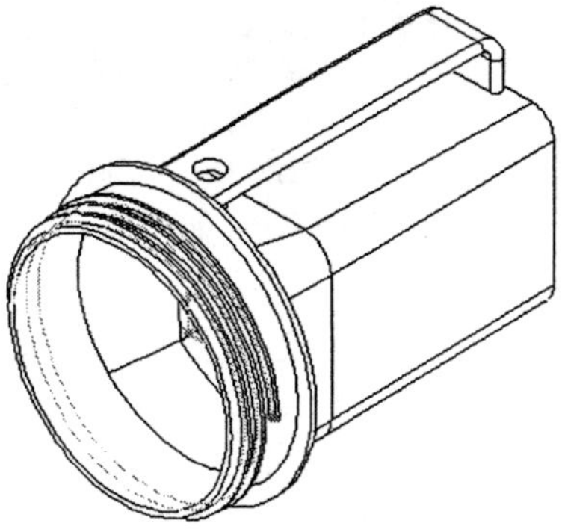

212) Rename **Extrude-Cut1** to **SwitchHole**.

213) Save the HOUSING. Click **Save**.

Create the HOUSING - First Rib Feature

The Rib feature adds material between contours of existing geometry. Use Ribs to add structural integrity to a part.

A Rib requires:

- A Sketch.
- Thickness.
- Extrusion direction.

Create the Rib.
214) Select the Sketch plane. Click the **Top** plane. Display all hidden lines to avoid unwanted relationships. Click **Hidden In Gray** .

215) Create the Sketch. Click **Sketch** . Display the Top view. Click **Top** . Click **Line** . The endpoints of the sketch are collinear with the inside wall of the Shell feature.

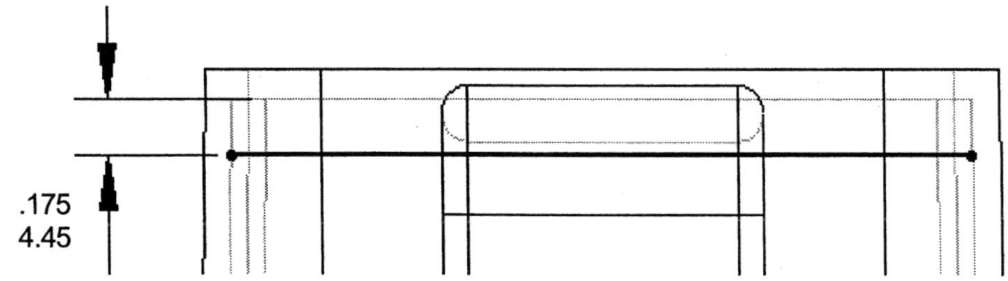

216) Add a linear dimension. Click **Dimension**. Click the **line**. Click the **inner back edge**. Enter **.175**, **[4.45]**.

217) Create the Rib. Click **Rib** from the Feature toolbar. Create the Rib on both sides of the Sketch plane. Click the **Both sides** button. Enter **.100**, **[2.54]** for Thickness. Click the **Parallel** button. The Rib direction arrow points to the back. Click the **Enable draft** check box. Click the **Draft outward** check box. Enter **1.00** for Draft Angle.

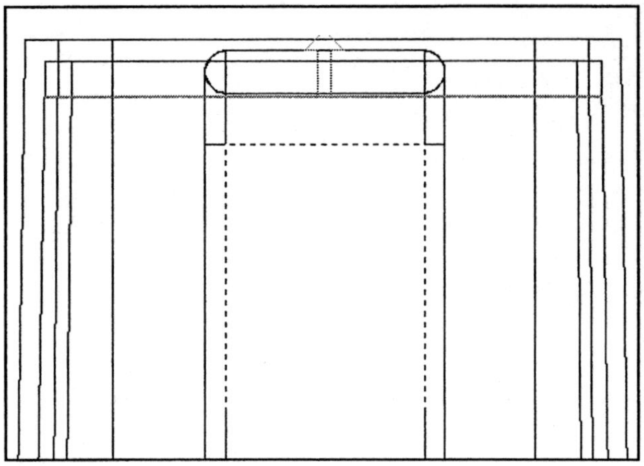

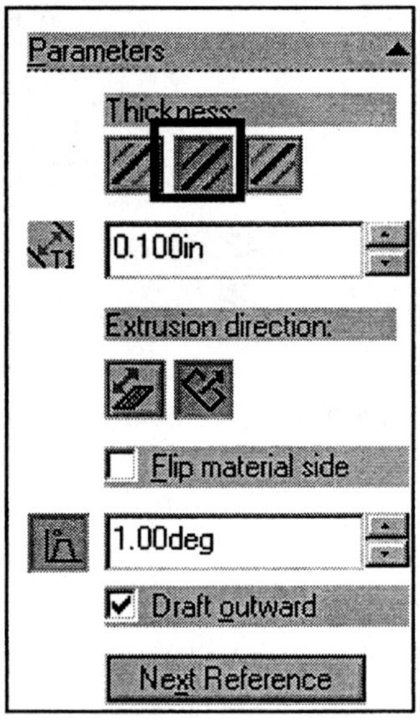

Note: Click Flip material side check box if the direction arrow does not point towards the back.

218) Display the Rib. Click **OK**. View the Rib. Click **Right**. Click **Isometric**.

Rib – Right view

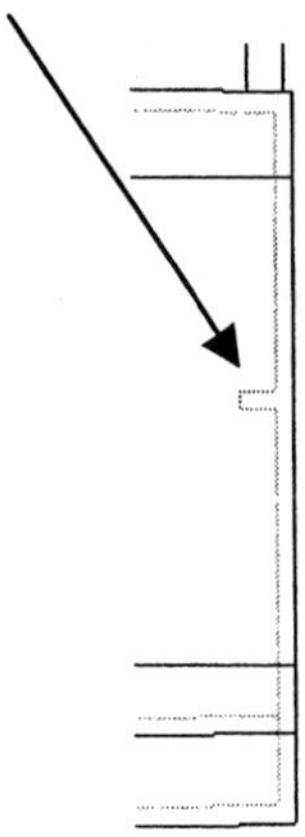

Rib

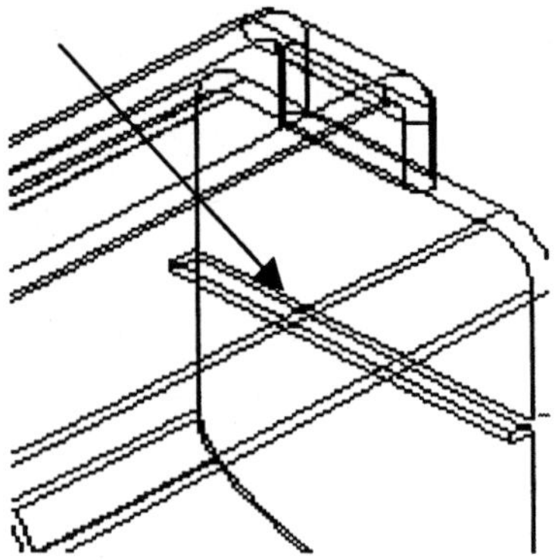

Existing geometry defines the Rib boundaries. The Rib does not penetrate through the wall.

Create the HOUSING - Linear Pattern of Ribs

The HOUSING requires multiple Ribs to support the BATTERY. A Linear Pattern creates multiple Instances of a feature along a straight line. The Linear Pattern is created in two directions along the same edge.

Create a Linear Pattern of the Rib feature.

219) Click **Linear Pattern** . The Linear Pattern PropertyManager is displayed.

220) Enter Direction 1 properties. Click **Rib1** in the Items to copy text box. Click the **Direction 1 Pattern Direction** text box. Click the **hidden upper back vertical edge**. Click the **Reverse direction** button. The direction arrow points upward.

221) Instances are created inside the HOUSING. Enter **.500**, **[12.7]** for Spacing. Enter **3** for Total Instances.

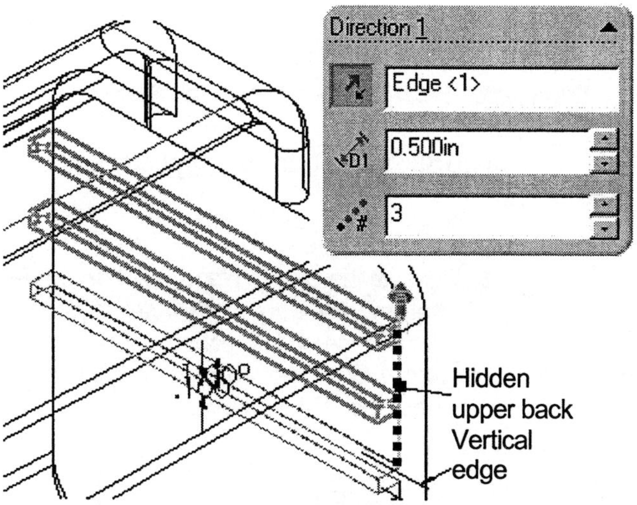

222) Expand the Direction 2 properties. Click the hidden **lower back vertical edge**. The direction arrow points downward. Enter **.500**, **[12.7]** for Spacing. Enter **3** for Total instances. Select the **Pattern seed only** check box.

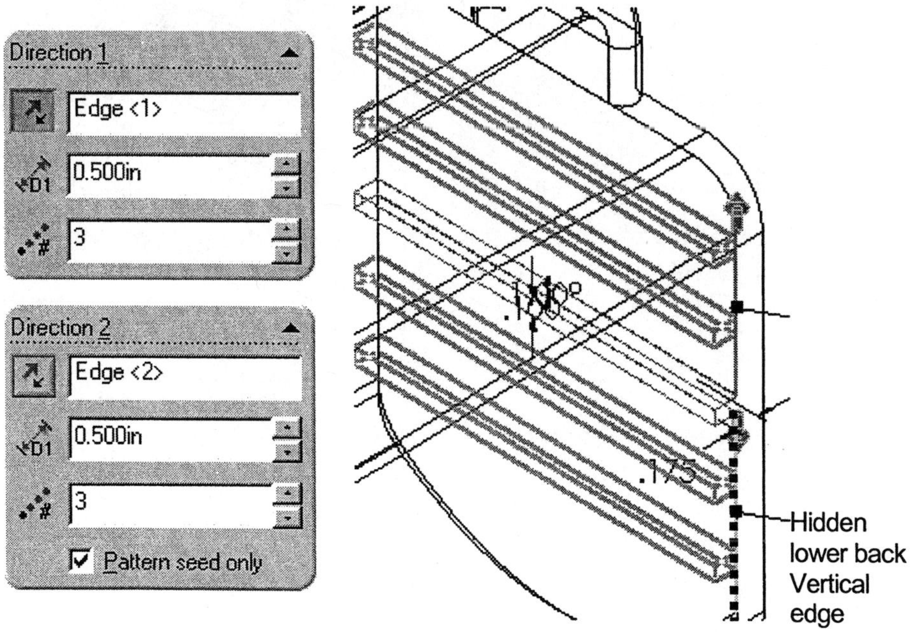

223) Scroll down the Linear Pattern Property Manager. Click the **Geometry Pattern** check box from the Options section.

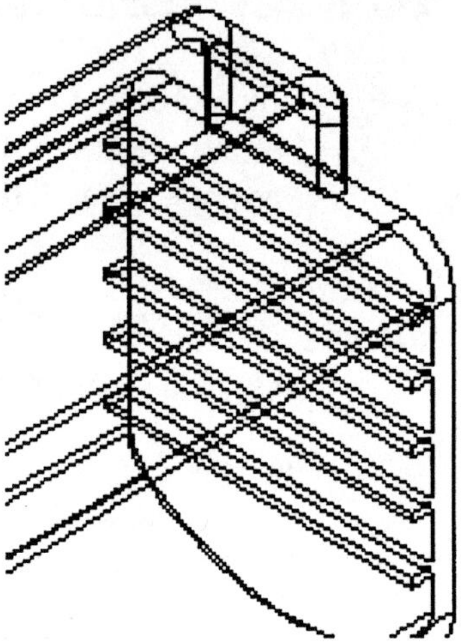

224) Display the Linear Pattern of Ribs. Click **OK**.

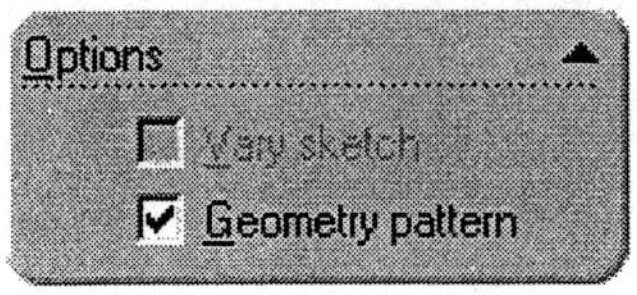

225) Save the HOUSING. Click **Save**.

Create the HOUSING - Second Rib Feature

The Second Rib feature supports and centers the battery.

Create the Plane.

226) Create the Sketch plane. Click **Insert** from the Main menu. Click **Reference Geometry, Plane**. Click **Parallel Plane at Point**. The Point-Plane option requires a point and a plane. Click the **Right** plane. Click the **vertex** at the back right of the handle. Display the plane. Click **OK**.

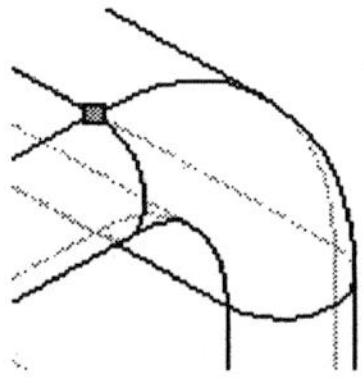

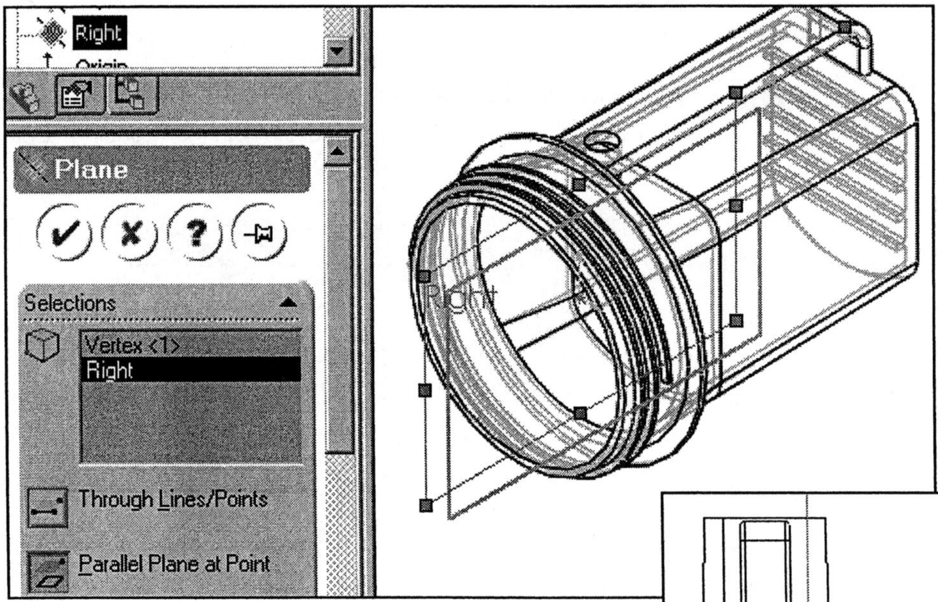

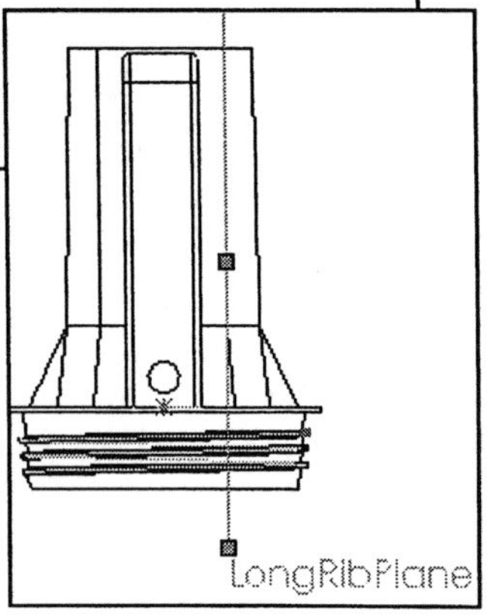

227) Rename **Plane1** to **LongRibPlane**.

Create the second Rib.

228) Select the Sketch plane. Click the **LongRibPlane**. Display the Right view. Click **Right** .

229) Create the Sketch. Click **Sketch** . Click **Line** . Sketch a **horizontal line.** The right endpoint of the line is coincident with the inside back wall of the Shell feature.

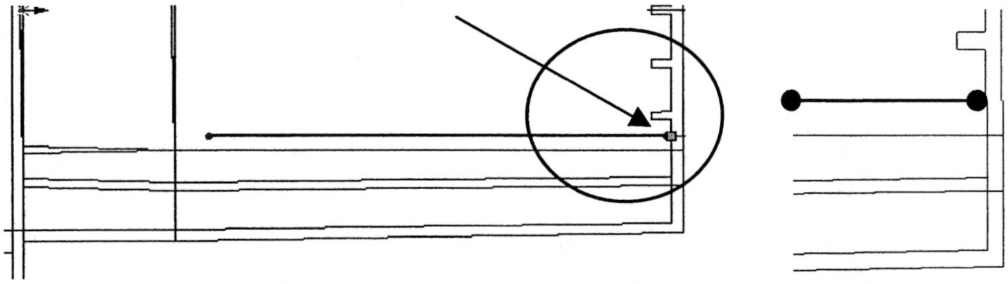

Note: When sketch and reference geometry become complex, create dimensions by selecting reference planes in the FeatureManager. For design flexibility, dimension the Rib from the Origin rather than from the bottom HOUSING surface.

Do not sketch the Rib line collinear to the horizontal lines of the HOUSING or coincident with the Linear Pattern of Ribs. Split the FeatureManager to select Planes.

230) Click **Dimension** . Click the **horizontal line**. Click **Origin**. Click the location for the vertical linear dimension **text**. Enter **1.300**, **[33.02]**.

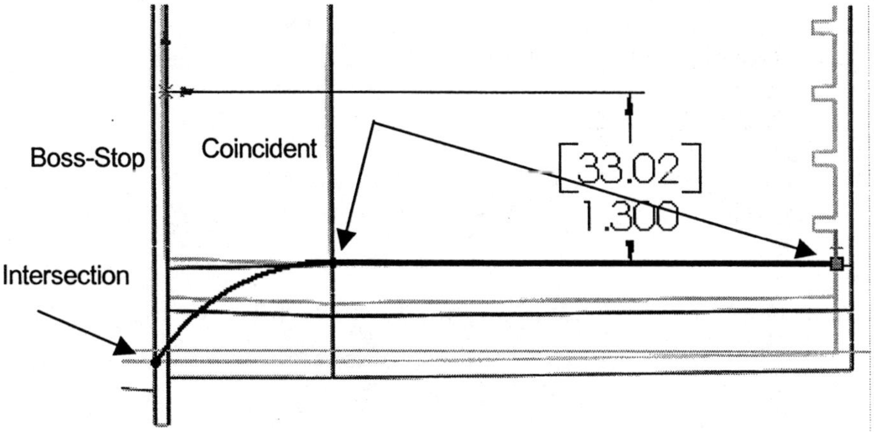

231) Click **Tangent Arc** . Click the **left end** of the horizontal line. Click the **intersection of the inside wall**. Click the **Boss-Stop** feature.

232) Add an intersection relation. Hold down the **Ctrl** key. Click the **end point** of the arc. Click the intersection of the **horizontal inside wall**. Click the **vertical Boss-Stop edge**. Release the **Ctrl** key. Click **Apply**.

233) Add a coincident relation between the left end point of the line and the batteryloftplane. Hold down the **Ctrl** key. Click the **left end point**. Click the **batteryloftplane** from the FeatureManager. Release the **Ctrl** key. Click **Apply**.

234) Create the Rib. Click **Rib** from the Feature toolbar. The Rib Property Manager is displayed. Create the Rib on both sides of the Sketch plane. Click the **Both sides** button. Enter **.075**, [**1.91**] for Thickness.

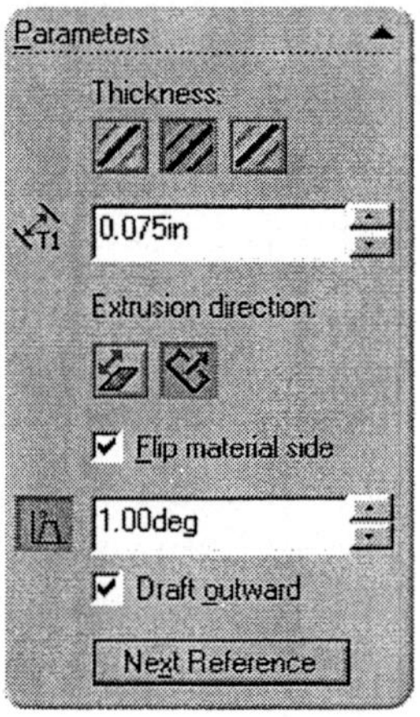

235) Create the draft. Click the **Enable draft** check box. Enter **1** for Angle. Click the **Draft outward** check box. The direction arrow for Rib creation points towards the bottom.

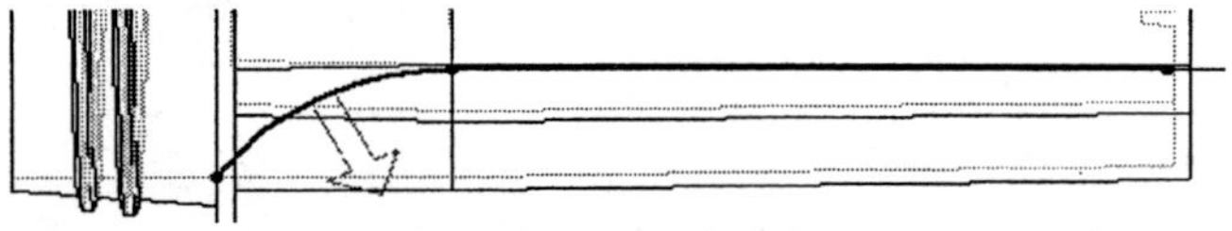

236) Display the Rib. Click **OK**. Display the Isometric. Click **Isometric**.

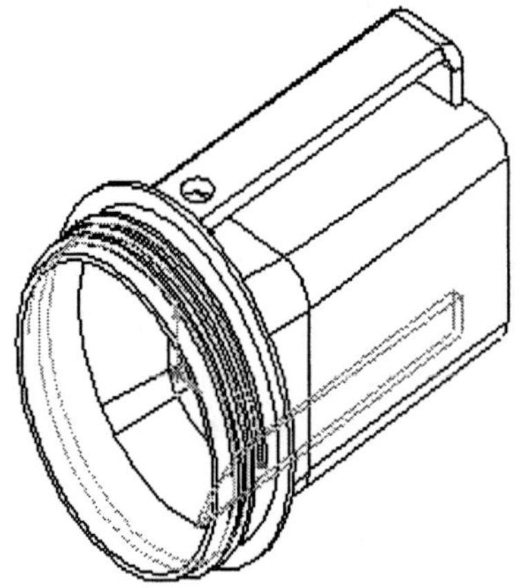

Mirror the Second Rib

An additional Rib is required to support the BATTERY.

Mirror the second Rib feature about the Right plane.

237) Click **Insert** from the Main menu. Click **Pattern/Mirror**, **Mirror Feature**. The Mirror Pattern Feature dialog box is displayed. Click the **Mirror plane** text box. Click the **Right** plane. The Rib2 feature is displayed in the Feature to mirror text box. Display the Mirror Pattern feature. Click **OK**.

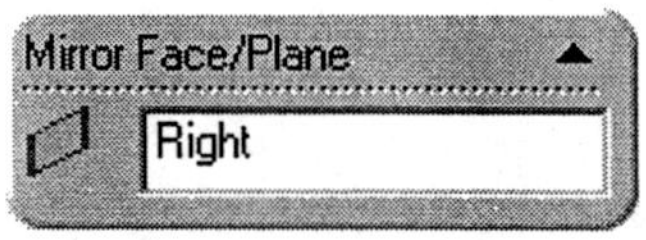

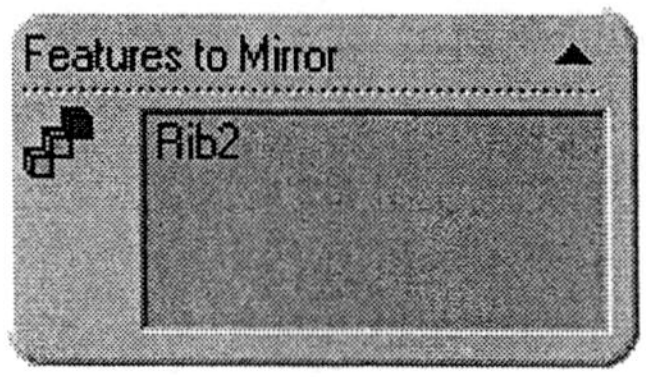

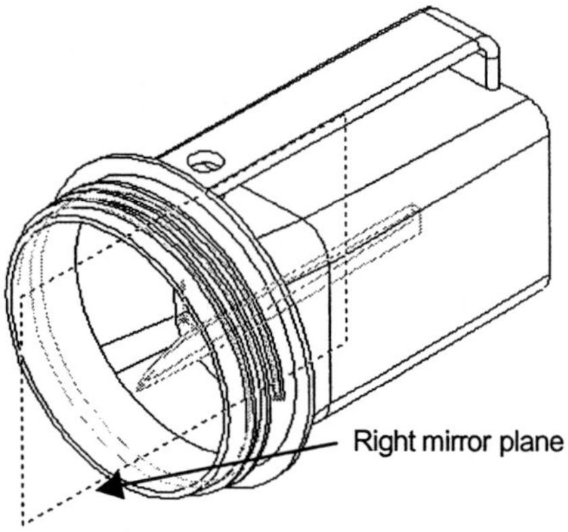

238) Save the HOUSING. Click **Save**.

239) Close all parts before you begin the next assembly exercise. Click **Window, Close All.**

Note: Adjust Ribs in the assembly to improve fit.

The parts for the FLASHLIGHT are complete!

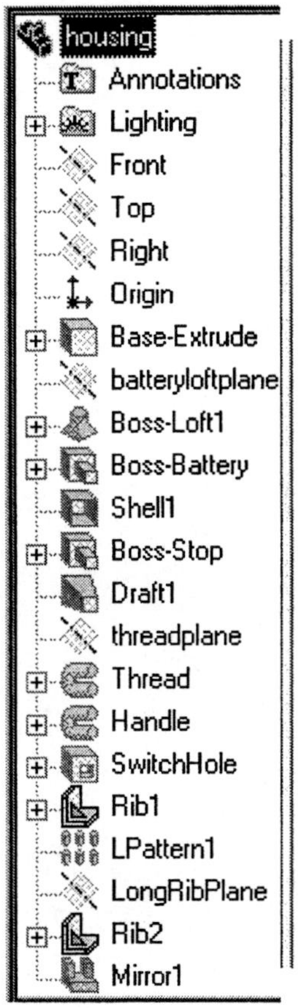

Questions

1. Identify the function of the following features:
   - Sweep
   - Shell
   - Revolved Cut Thin
   - Loft
   - Rib
2. Define a Split Curve.
3. What are suppressed features?
4. Why would you suppress a feature?
5. The Rib features require a Sketch, thickness and a ___________ direction.
6. What is a Pierce relation?
7. How do you create a thread using the Sweep feature?
8. How do you create a Linear Pattern?
9. How do you create a Circular Pattern?
10. What is the purpose of the Thread Plane for both the LENSCAP and the HOUSING?

Exercises

Create O-RINGs with various cross sections using the Base-Sweep feature.

Exercise 3.1: Create C-SHAPE-O-RING.

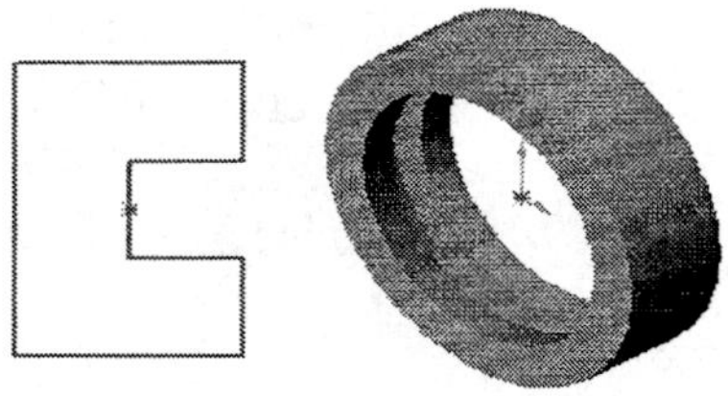

Exercise 3.1 C-Shaped O-RING

Exercise 3.2: Create QUATTRO-SEAL-O-RING.

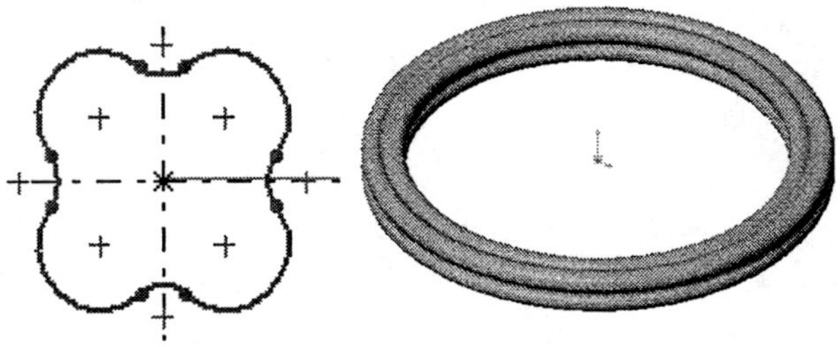

Cross Section Quattro Seal O-RING

Exercise 3.2

Exercise 3.3: Create the OFFSET-LOFT.

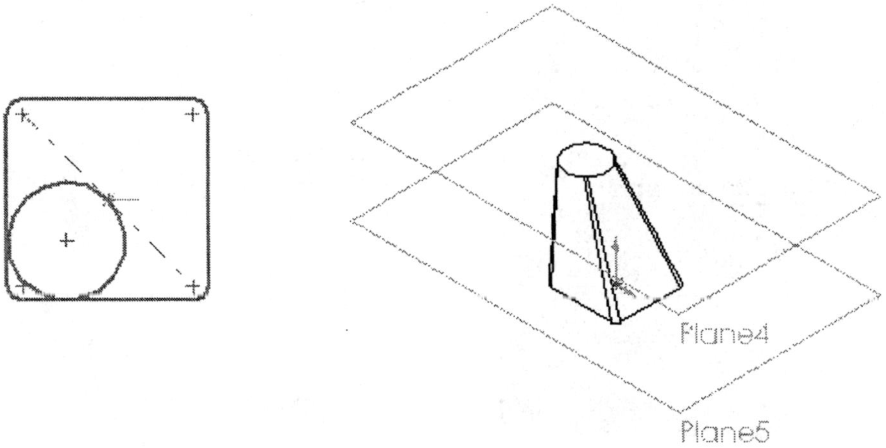

Exercise 3.4a: Review the HOUSING. The mold for the HOUSING requires that all sharp edges are removed. Add Fillet features throughout the FeatureManager to remove sharp edges from the inside and outside of the HOUSING. The Long Ribs are too thick. Divide the Rib thickness by 2. Add three Ribs to support the BATTERY.

Exercise 3.4b: Design Issue. The HOUSING Handle feature is too close to the Boss Battery feature. How would you redesign the Handle feature to allow for ample room for a human hand?

A plastic BOTTLE is created from a variety of SolidWorks features. Create the shoulder of the BOTTLE with the Loft Base feature.

Exercise 3.5: Create the Triangular Shaped Bottle.

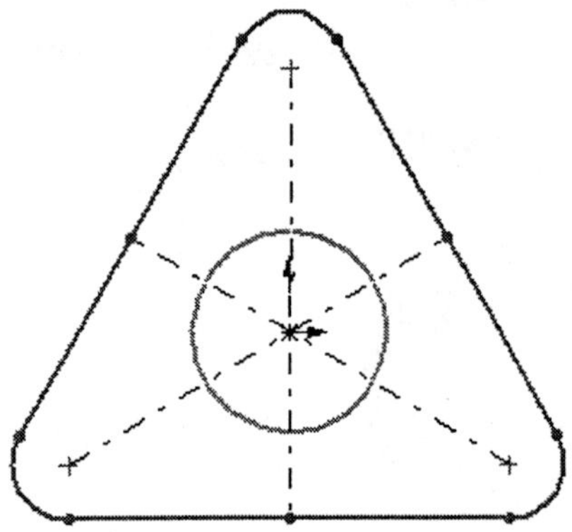

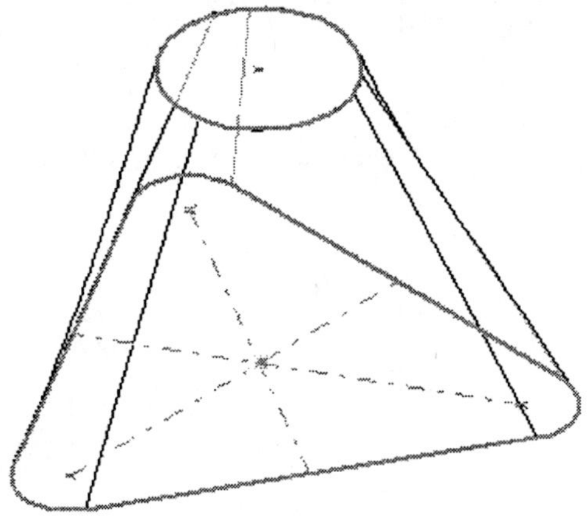

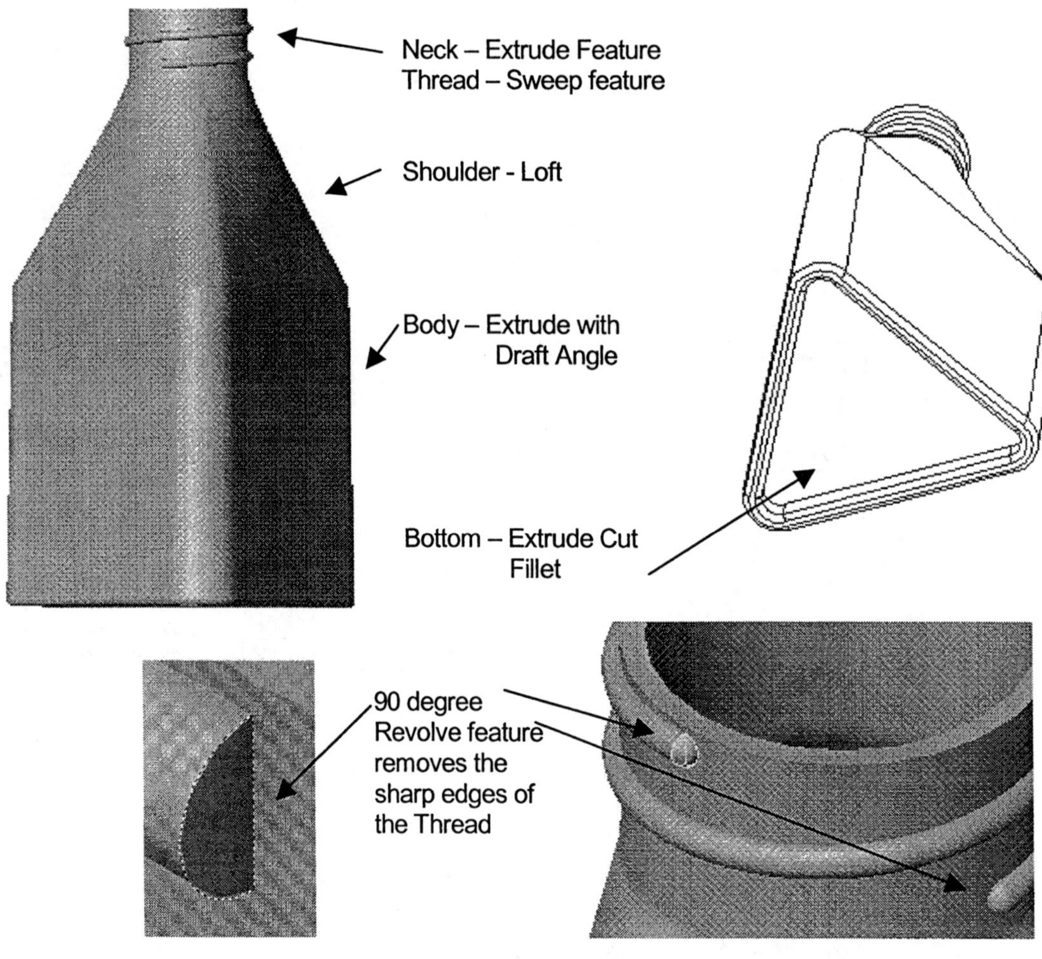

Exercise 3.6: Create the Oval Shaped Bottle.

Exercise 3.7: Create the Rectangular Shaped Bottle.

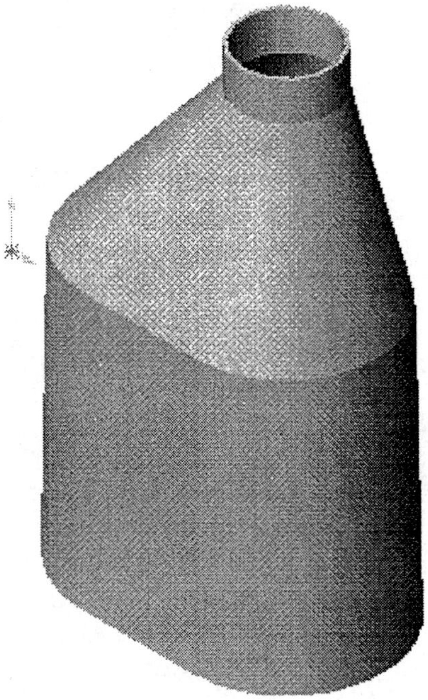

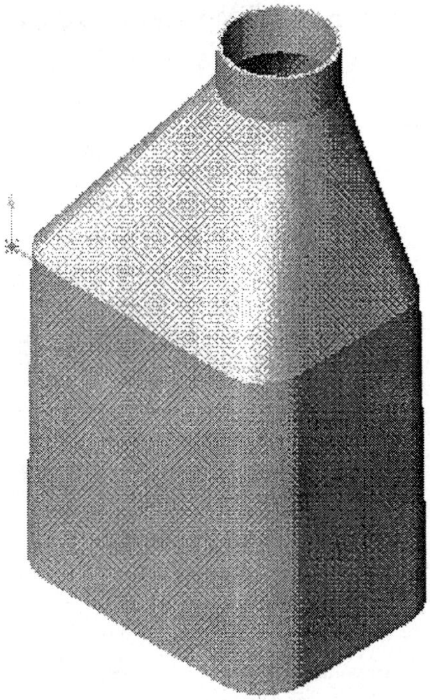

Exercise 3.8: Create the Rectangular Shaped Bottle with Handle.

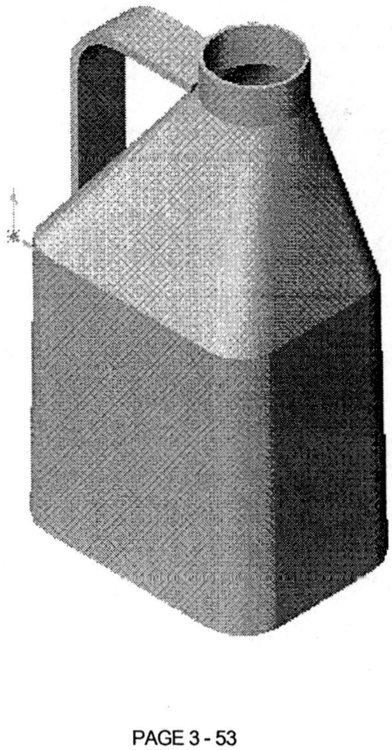

Exercise 3.9: Industry Collaborative Exercise.

Companies incorporate parts and assemblies from SolidWorks to create web pages.

Example: The company, 3-D Design Solutions creates a Technical Web page that is written as follows:

- Clear message
- Concise text
- Accurate and up to date product information.

Models and images courtesy of
3-D Design Solutions
Carlsbad, CA USA

a) Create an image of the O-RING. Save the O-RING in a Tiff format.

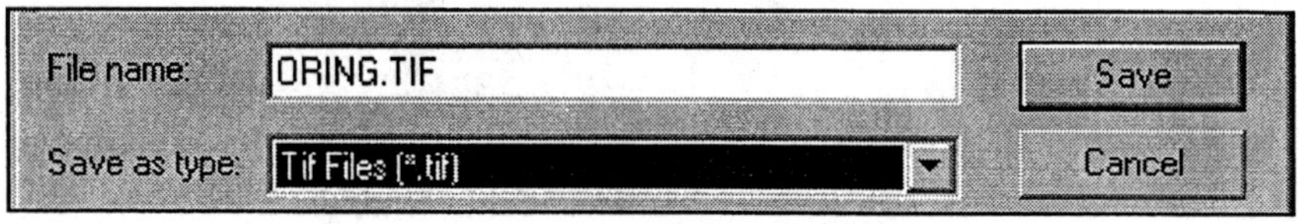

b) Research the world wide web to find O-RING manufactures (Hint:globalspec.com). What key words would you utilize to describe an O-RING on a technical web page?

c) Create a new Web Page document using Microsoft Word 2000. Add text and a Tiff image file of the O-RING to the document. View the web page using File, Web Page Preview in your web browser.

Note: Other web creation software tools can be utilized to create this web page.

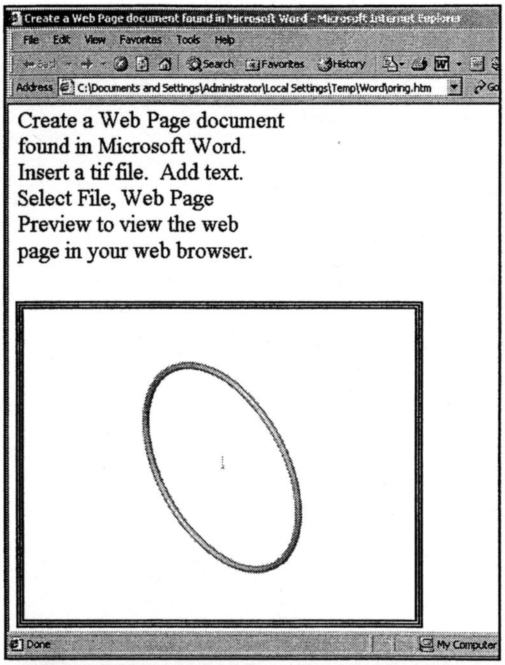

d) Create a JPEG file of the O-RING using PhotoWorks.

Note: PhotoWorks Add in is required for this exercise.

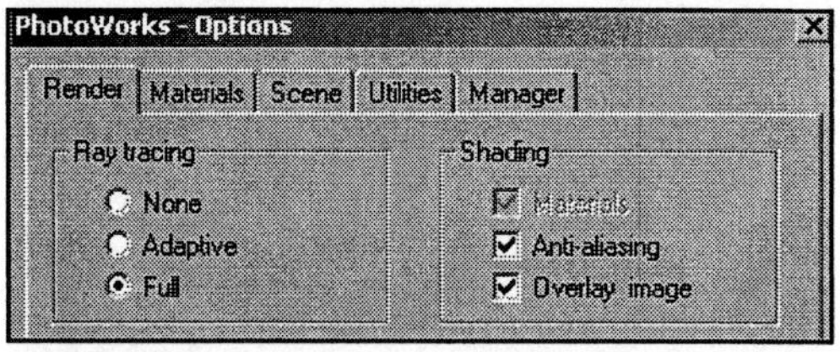

Select PhotoWorks, Options. Select Full for Ray tracing and Anti-aliasing for Shading to create a smooth image.

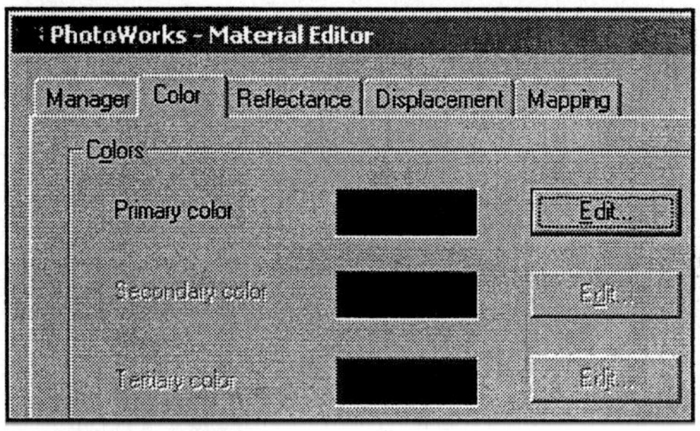

Select PhotoWorks, Material, Stock Procedural, Plastics, Rubber. Select Black as Primary Color from the Color tab.

Select PhotoWorks, Render Image to File. Select 320 x 240.

Insert the JPEG file into your web page.

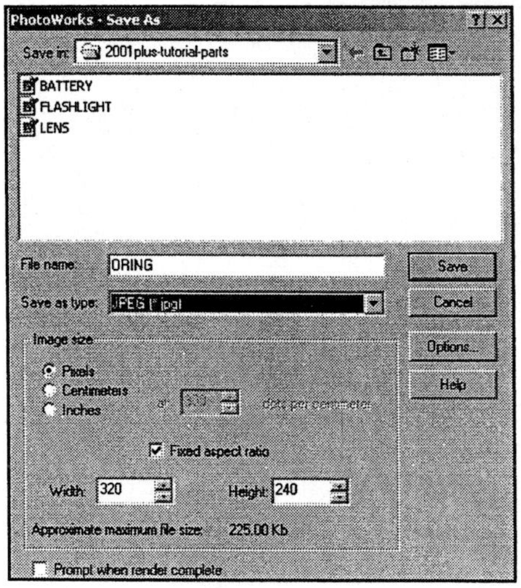

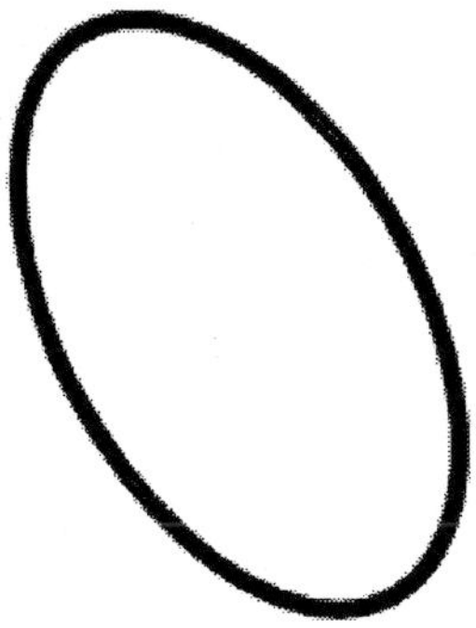

Notes:

Project 4

Assembly Fundamentals

Below are the desired outcomes and usage competencies based upon the completion of Project 4.

Project Desired Outcomes	Usage Competencies
Knowledge and experience in assembly creation.	Ability to create new assemblies and combine assemblies.
Create three sub-assemblies: • LENSANDBULB • CAPANDLENS • BATTERYANDPLATE	An understanding of the Mate options and assembly mating functionality.
Create the FLASHLIGHT assembly	

Notes:

Project 4 – Assembly Fundamentals

Project Objective

Create three sub-assemblies: LENSANDBULB, CAPANDLENS & BATTERYANDPLATE. Create the final FLASHLIGHT assembly.

Project Situation

Design constraints directly influence the assembly design process. Other considerations indirectly impact the assembly design, namely: cost, manufacturability and serviceability.

In Project 1 through Project 3, you created eight parts: BATTERY, BATTERYPLATE, LENS, BULB, O-RING, SWITCH, LENSCAP and HOUSING. In Project 4, create the final FLASHLIGHT assembly, Figure 4.1.

An assembly combines two or more parts. In an assembly, parts are referred to as components. A sub-assembly is contained within an assembly.

Figure 4.1

Project Overview

How do you combine the created parts into the FLASHLIGHT? The action of assembling components in SolidWorks is defined as Mates. Mates are relationships between components that simulate the construction of the assembly in a manufacturing environment, Figure 4.2.

Do not assemble all of the components at the top level. Create sub-assemblies that represent major components of the assembly.

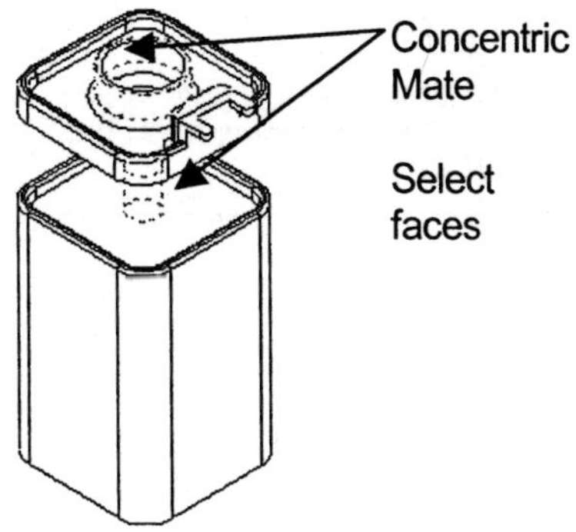

Figure 4.2

Identify the base component for each sub-assembly. The base component is the first component is each sub-assembly. The first component is fixed to the Origin. A fixed component is stationary.

Plan the FLASHLIGHT assembly component layout, Figure 4.3.

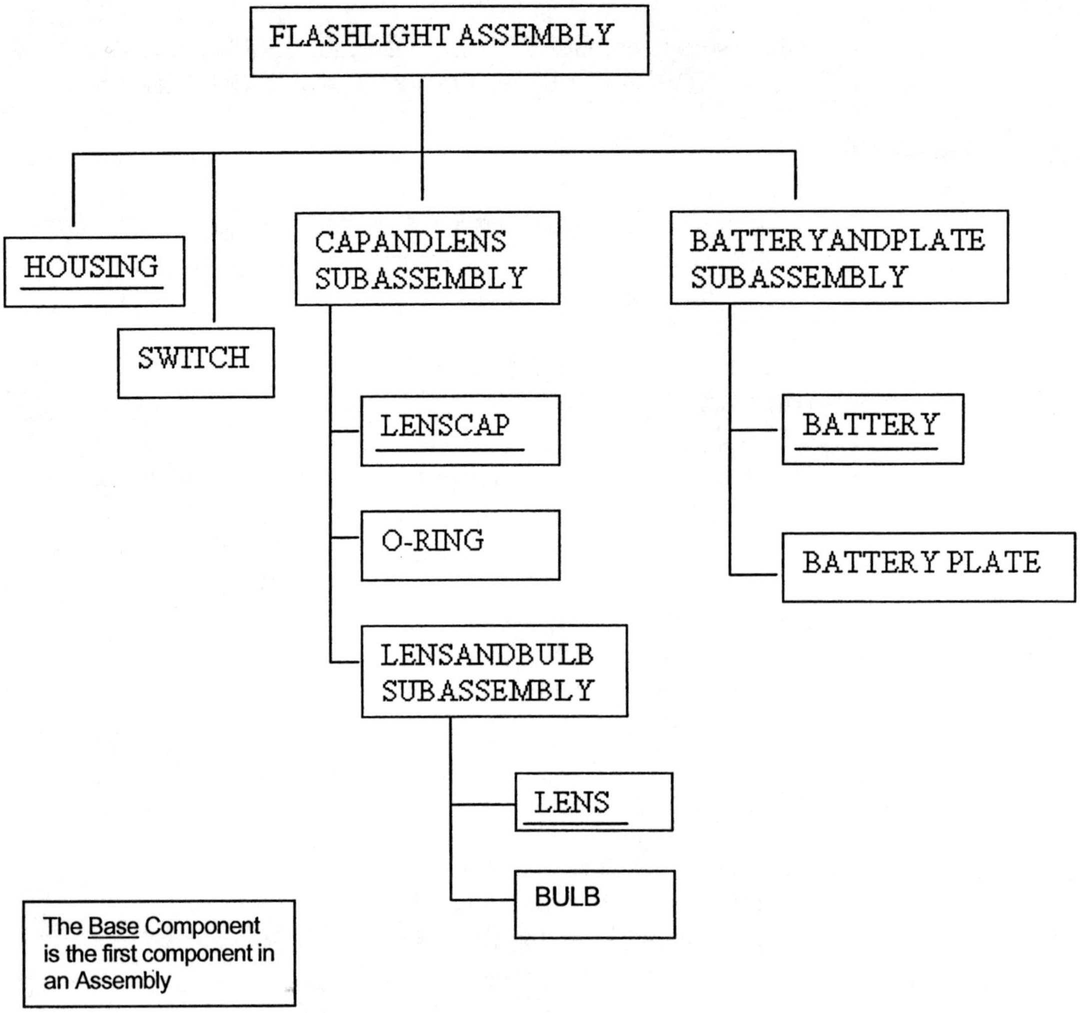

FLASHLIGHT Layout Structure
Figure 4.3

FLASHLIGHT Assembly Overview

The FLASHLIGHT assembly steps are as follows:

- Create the LENSANDBULB sub-assembly from the LENS and BULB, Figure 4.4a. The LENS is the Base component.
- Create the BATTERYANDPLATE sub-assembly from the BATTERY and BATTERYPLATE, Figure 4.4b. The BATTERY is the Base component.
- Create the CAPANDLENS sub-assembly from the LENSCAP, O-RING and LENSANDBULB sub-assembly, Figure 4.4c. The LENSCAP is the Base component.

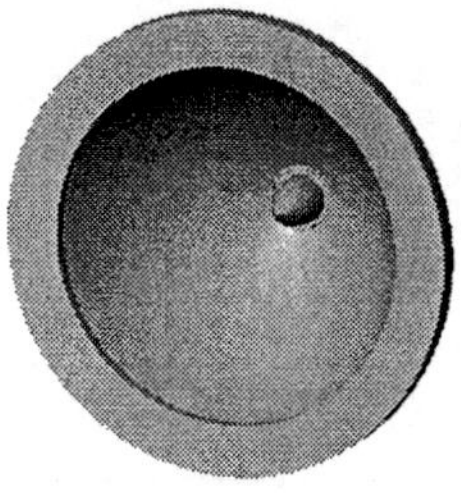

Figure 4.4a

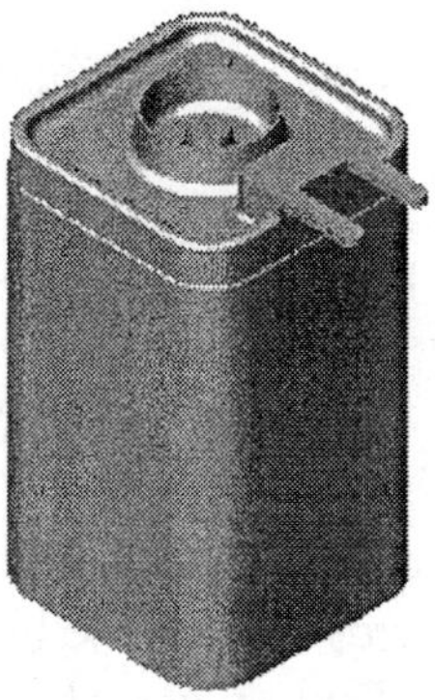

Figure 4.4b

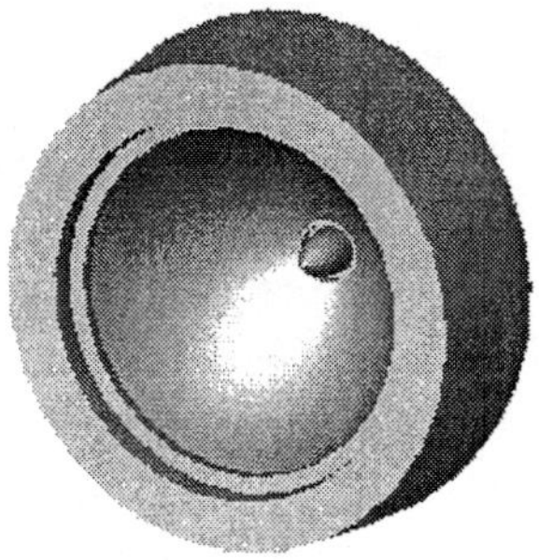

Figure 4.4c

- Create the FLASHLIGHT assembly. The HOUSING is the Base component, Figure 4.4d. Add the SWITCH, CAPANDLENS and BATTERYANDPLATE, Figure 4.4e.

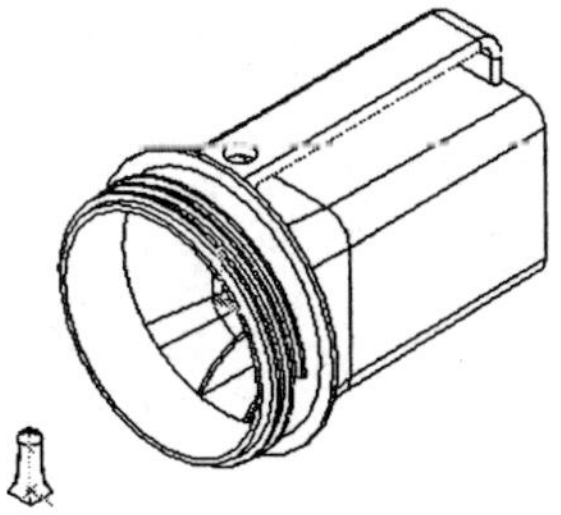

Figure 4.4d

Figure 4.4e

Figure 4.4f

- Modify the dimensions to complete the FLASHLIGHT assembly, Figure 4.4f.

Assembly Techniques

Assembly modeling requires practice and time. Below are a few helpful hints and techniques to address the Bottom up design modeling approach.

- Create an assembly layout structure. This will organize the sub-assemblies and components.
- Insert sub-assemblies and components as lightweight components. Lightweight components save on file size, rebuild time and complexity. Set Lightweight components in the Tools, Options command.
- Use the Zoom and Rotate commands to select the geometry in the mate process. Zoom to select the correct face. Use filters to select geometry.
- Apply various colors to features and components to improve display.
- Mate with reference planes when addressing complex geometry. Example: The O-RING does not contain a flat surface or edge.
- Activate Temporary axis and Planes from the View menu.
- Select reference planes from the FeatureManager. Expand the component to view the planes. Example: Select the Right plane of the LENS and the Right plane of the BULB to be collinear. Do not select the Right plane of the HOUSING if you want to create a mate between the LENS and the BULB, Figure 4.5.

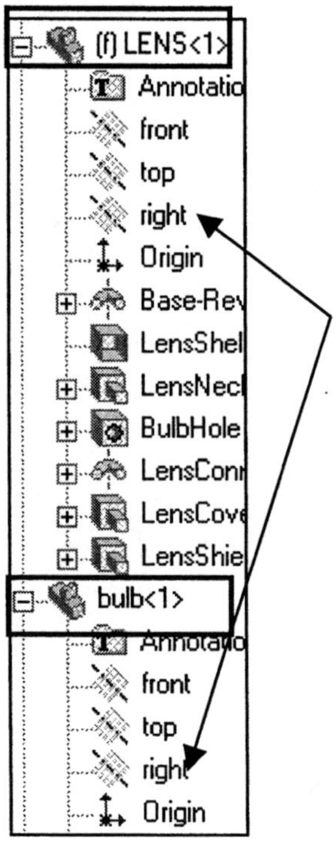

Figure 4.5

- Remove display complexity. Hide components and Suppress features when not required.
- Use the Move Component and Rotate Component commands before mating. Position the component in the correct orientation.
- Conserve time. Use Preview during the mate operation. Suppress unwanted features. Create additional flexibility into a mate. Use a distance mate with a zero value. Flip the direction if required.
- Remove unwanted entries. Use the Delete key from the Assembly Mating Items Selected text box.

- Verify the position of the mated components. Use Top, Front, Right and Section views.
- Use caution when you view the color red in an assembly. Red indicates that you are editing a part in the context of the assembly.
- Avoid unwanted references. Verify your selections with the PropertyManager.

Linear Motion and Rotational Motion

In dynamics, motion of an object is described in linear and rotational terms. Components possess linear motion along the x, y and z-axes and rotational motion around the x, y, and z-axes. In an assembly, each component has 6 degrees of freedom: 3 translational (linear) and 3 rotational. Mates remove degrees of freedom. All components are rigid bodies. The components do not flex or deform.

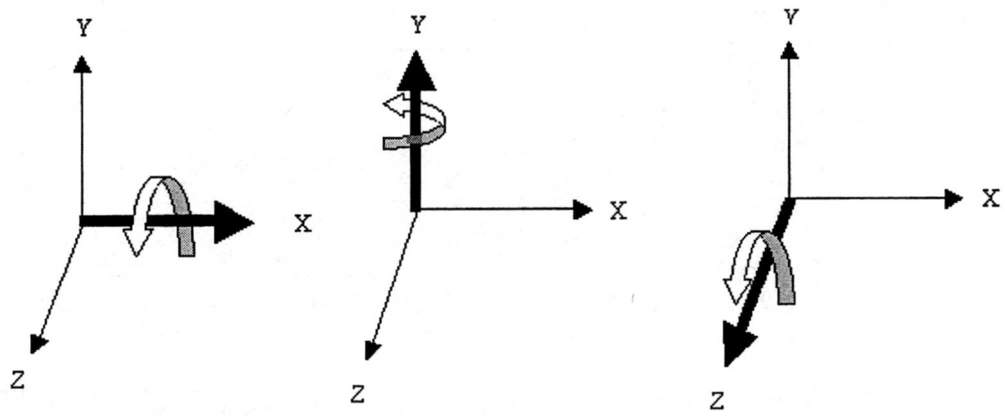

Create an Assembly Template

An Assembly Document Template is required to create the Flashlight Assembly and its sub-assemblies.

For English Components: Create an Assembly Document Template using Inch units.

For Metric Components: Create an Assembly Document Template using Millimeter units.

Create an assembly template.

1) Click **New** . Click **Assembly** Assembly from the Template dialog box. Click **OK**.

2) Set the Assembly Document Template options. Click **Tools, Options, Document Properties**. Select **ANSI** from the Detailing option.

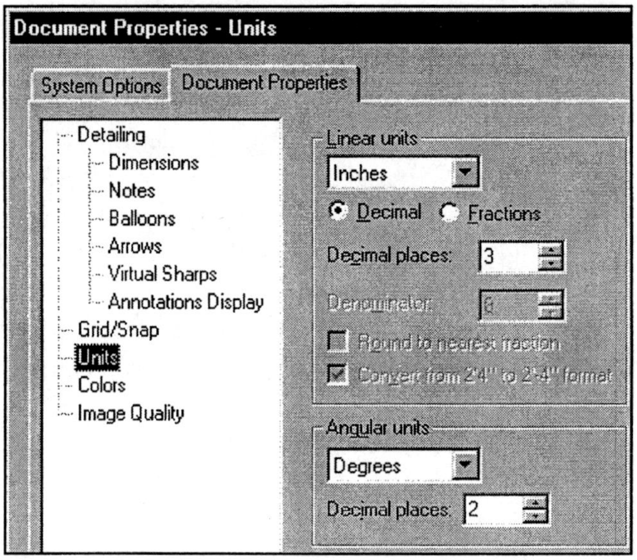

3) For English components: Click **Units**. Enter **Inches** from the Linear units list box. Enter **3** in the Decimal places spin box. Click **OK**.

4) For English components: Save the assembly template. Click **File, SaveAs**. Click the ***.asmdot** from the Save As type list box. Enter **ASSEMBLY-INCH** in the File name text box. Click the **Save** button. Use the ASSEMBLY-INCH for all FLASHLIGHT assemblies and sub-assemblies built with inch components.

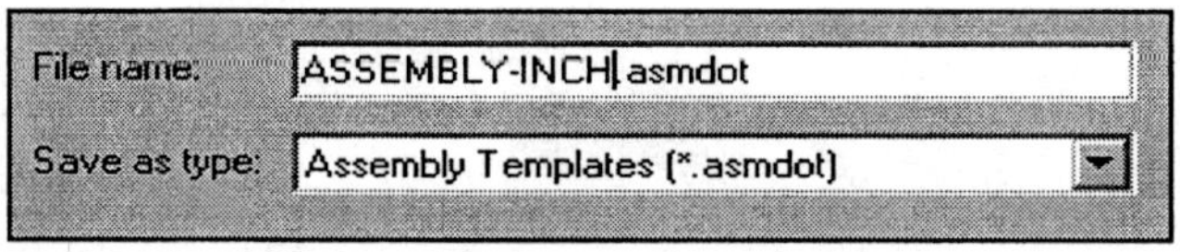

5) For Metric components: Click **Units**. Enter **Millimeters** from the Linear units list box. Enter **2** in the Decimal places spin box. Click **OK**.

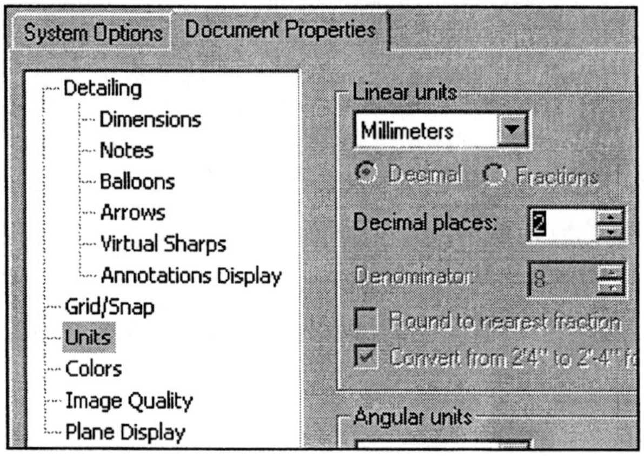

6) For Metric components: Save the assembly template. Click **File, SaveAs**. Click the ***.asmdot** from the Save As type list box. Enter **ASSEMBLY-MM** in the File name text box. Click the **Save** button. Use the ASSEMBLY-MM for all FLASHLIGHT assemblies and sub-assemblies built with millimeter components.

7) Close all documents. Click **Windows, Close All**.

LENSANDBULB Sub-assembly

Create the LENSANDBULB sub-assembly. The LENS is the Base component. LENSANDBULB sub-assembly mates the BULB component to the LENS component.

Create the LENSANDBULB sub-assembly.

8) Click **New** from the Standard toolbar. The New SolidWorks Documents dialog box is displayed. Click **ASSEMBLY-INCH**, [**ASSEMBLY-MM**] template. Click **OK**.

9) Click **Save**. Enter the name of the sub-assembly. Select **My Documents** folder in the Save In text box. Enter **LENSANDBULB**. Select **Assembly Files** (*.asm, *sldasm) for Save As Type. Click the **Save** button. The LENSANDBULB assembly

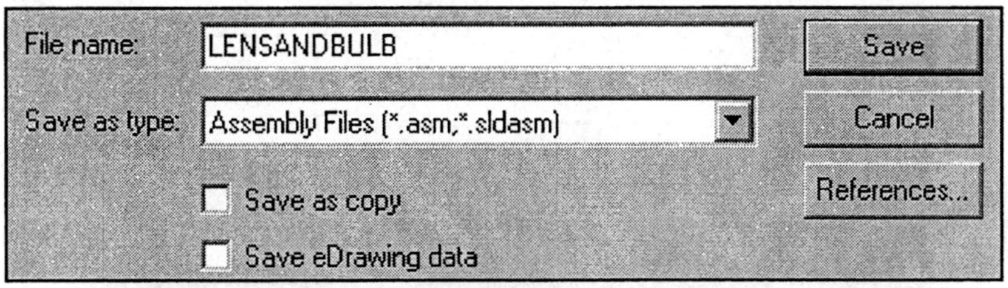

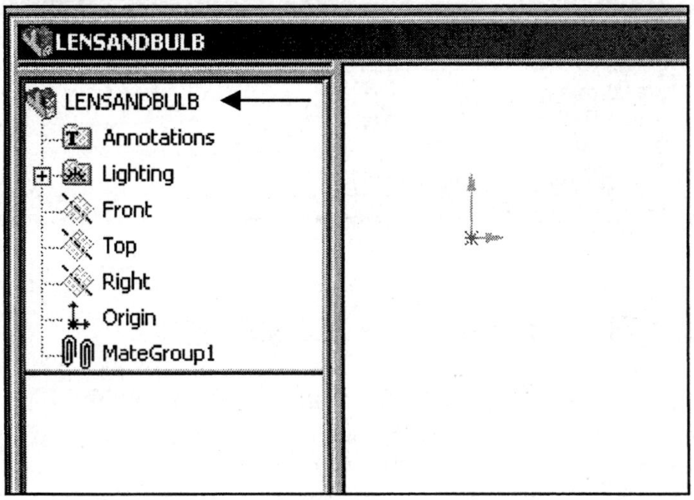

icon is displayed in the FeatureManager.

10) Open the LENS. Click **Open**. Enter the name of the part. Enter **LENS**.

11) Open the BULB. Click **Open**. Enter the name of the part. Enter **BULB**.

12) Click **Window** from the Main menu. The following documents are displayed: BULB, LENS and LENSANDBULB. Display the assembly and components. Click **Tile Horizontally** from the Window menu. Review the windows. The Origin feature of the LENSANDBULB assembly is displayed in the Graphics window. The FeatureManager is displayed on the left side of the three Graphics windows.

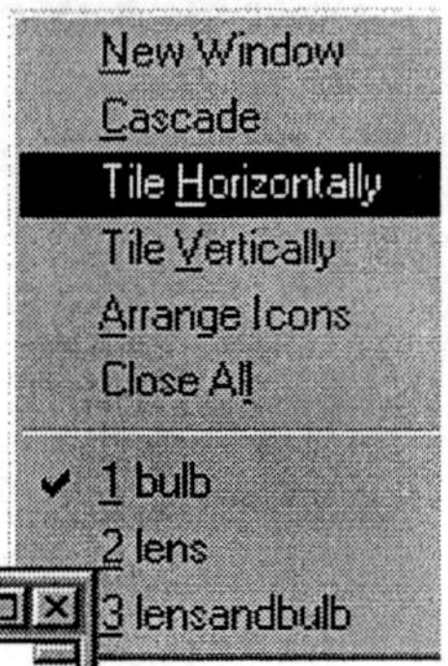

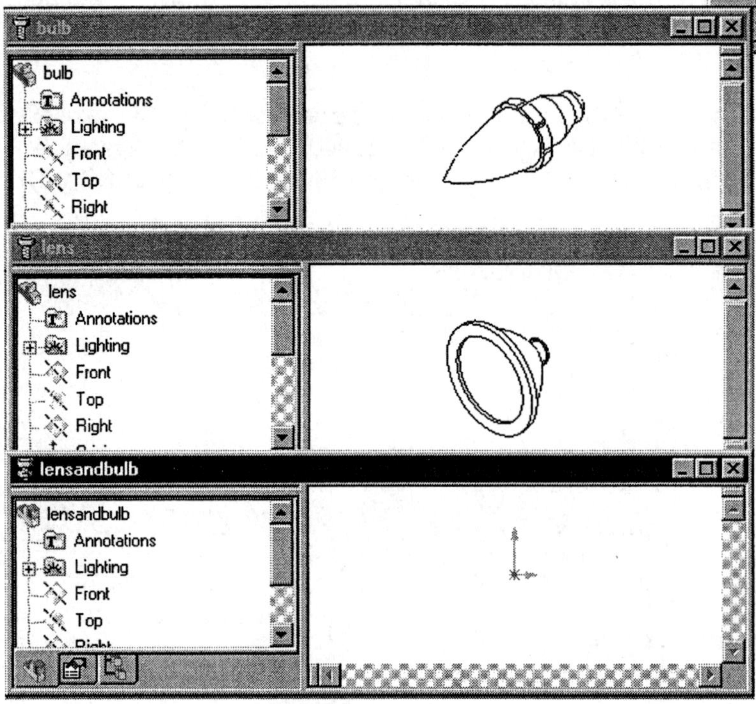

LENSANDBULB Assembly – Insert Components

The first component is the foundation of the assembly. The LENS is the first component. Components are added to the assemblies utilizing the following techniques:

- Click Insert from the Main menu. Click Component. Click From File.
- Drag components from Windows Explorer.
- Drag components from the Feature Palette window.
- Drag components from the Open part files.

Add the first component.

13) Position the LENS inside the assembly Graphics window. Click the **LENS** lens icon from the top of the FeatureManager. Drag the **LENS** to the Origin of the **LENSANDBULB** assembly window. The mouse pointer displays when position on the Origin. Display the LENS component in the assembly Graphics window. Release the **mouse button**.

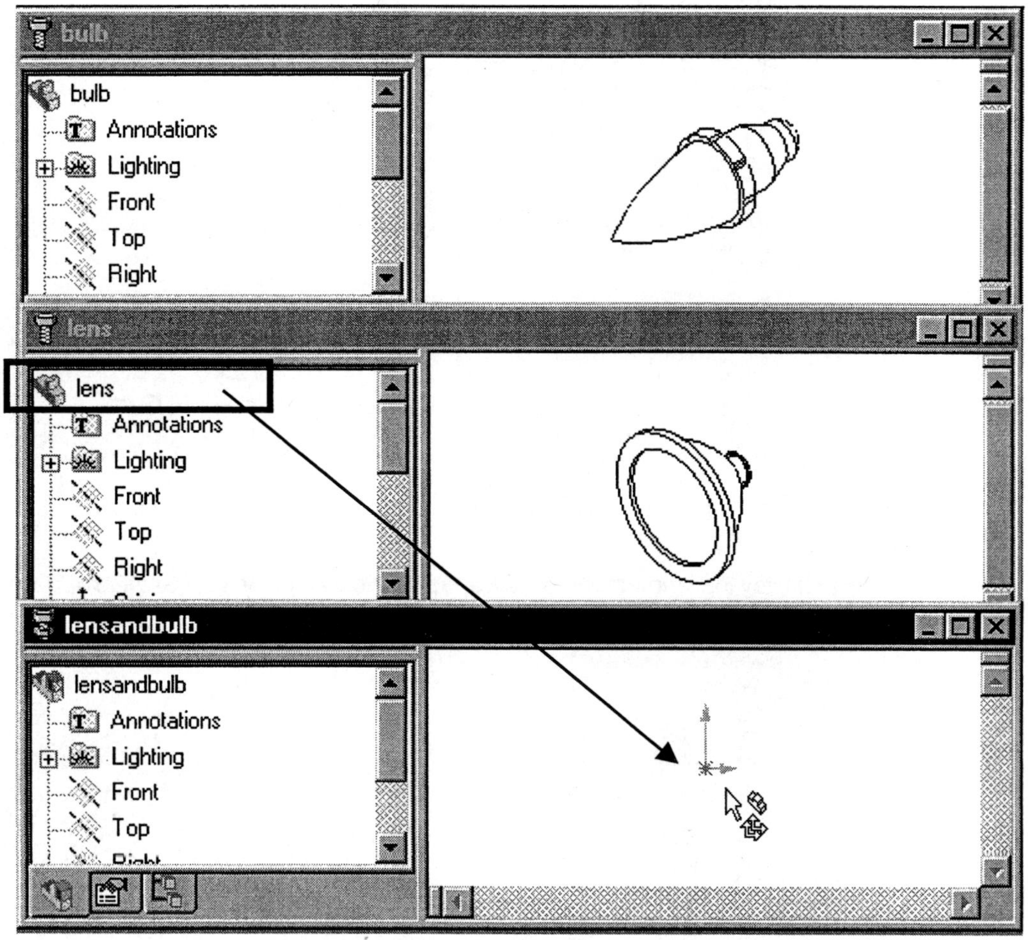

Note: The front view of the LENS is displayed in the LENSANDBULB assembly. The LENS name is added to the LENSANDBULB assembly FeatureManager with the symbol (f). The symbol (f) represents a fixed component. A fixed component cannot move and is locked to the assembly Origin. To remove the fixed state, Right-click a component name in the FeatureManager. Click Float.

14) Numerous views are used in an assembly. **Pin** the View Orientation dialog box to the right corner of the Graphics window.

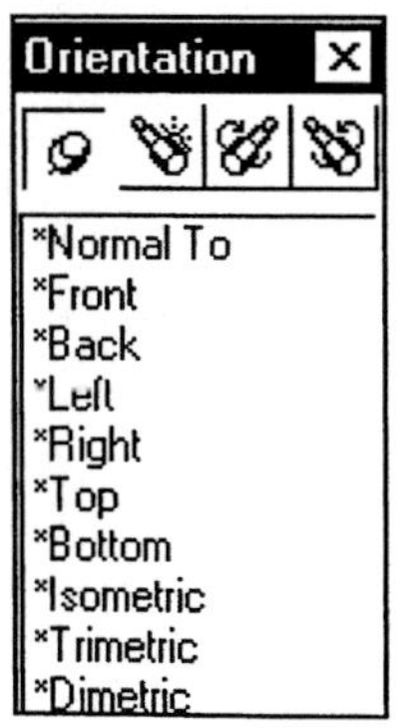

Add the second component.

15) Click the **BULB** bulb icon from the top of the FeatureManager. Drag the **BULB** component to the assembly window, left of the LENS. The mouse pointer displays when positioned inside the LENSANDBULB Graphics window. Display the BULB component. Release the **mouse button**.

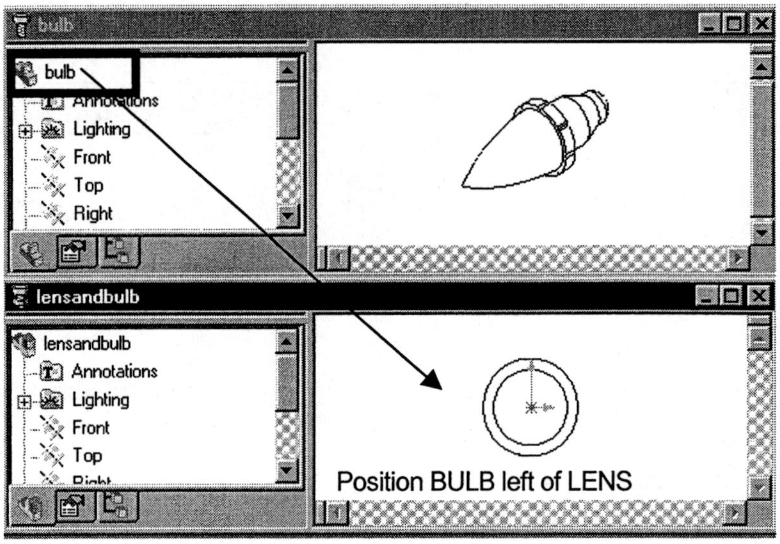

16) Close the LENS. Click **Close**. Close the BULB. Click **Close**.

17) Display the Isometric view. Click **Isometric** in the LENSANDBULB window.

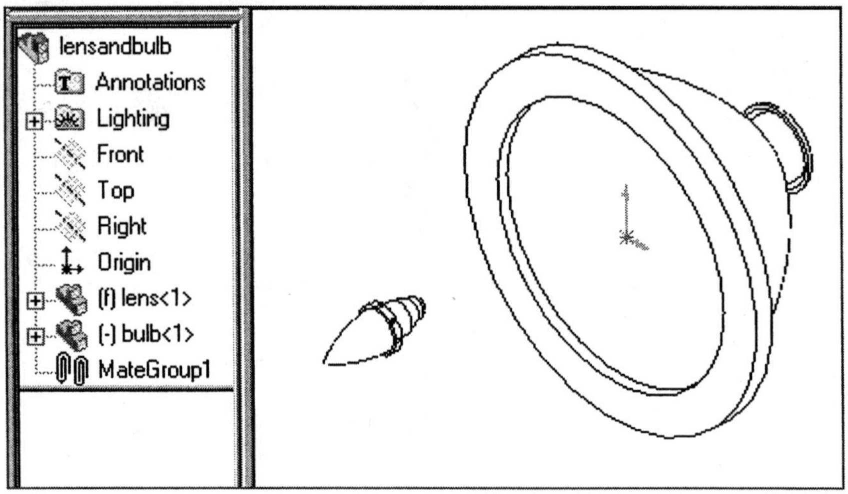

18) Enlarge the assembly window. Click **Maximize** in the upper right hand corner of the LENSANDBULB window. Fit all components in the Graphics window. Press the **f** key.

19) Save LENSANDBULB. Click **Save**.

Review the FeatureManager Syntax.

20) Double click the **LENS** component inside the FeatureManager. The LENS component lists features. Example: Base-Revolve, Shell, Lensneck. Sketches are contained within the LENS features that display the Plus sign ⊞ icon.

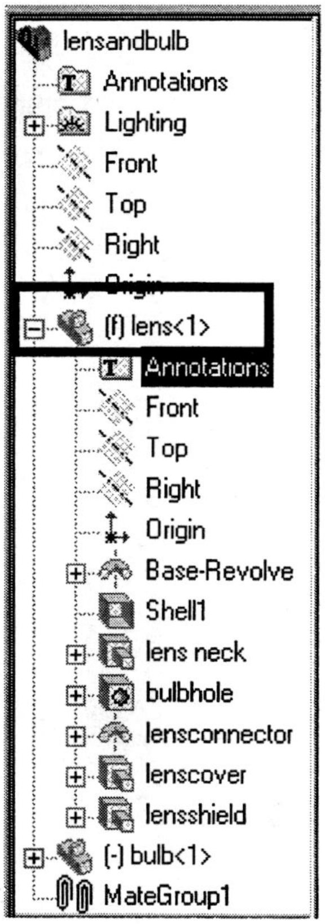

Review the BULB component syntax in the FeatureManager.

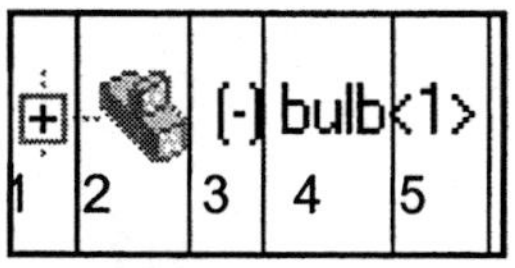

1. A Plus sign ⊞ icon indicates that additional feature information is available. A minus sign indicates that the feature list is fully expanded.

2. A component icon indicates that the BULB is a part.

   The assembly icon indicates that the LENSANDBULB is an assembly. Sub-assemblies display the same icon as an assembly.

3. Column 3 identifies the Component State:

   - A minus sign (–) indicates that the component is under defined and requires additional information.
   - A plus sign (+) indicates that the component is over defined.
   - A fixed symbol (f) indicates that the component does not move.
   - A question mark (?) indicates that additional information is required.

4. BULB - Name of the component.

5. The symbol <#> indicates the number of copies in the assembly. The symbol <1> indicates the original component, "BULB" in the assembly.

Move the BULB Component.

21) Components located in an assembly are free to move. Click the **BULB**. Click **Move Component**  from the Assembly toolbar. The PropertyManager is displayed on the left side of the Graphics window. The mouse pointer displays 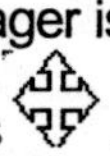. Position the **BULB** in front of the LENS. Click **OK**.

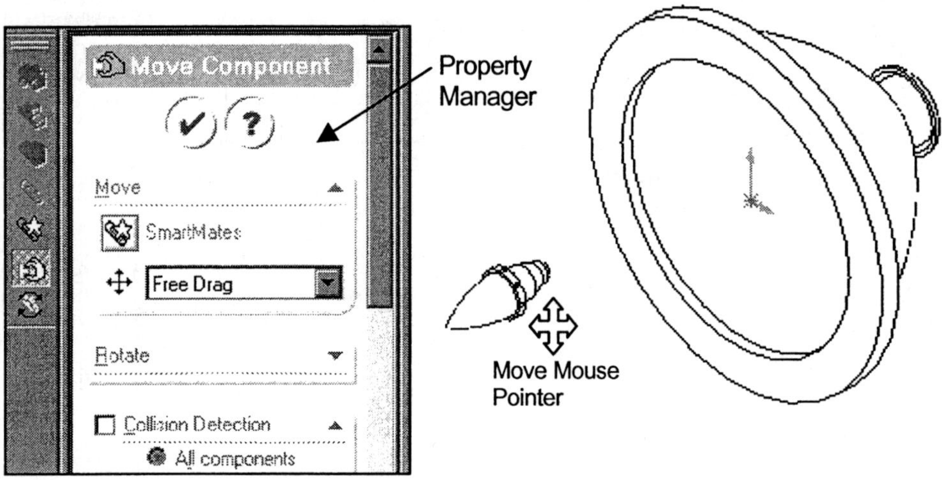

Mates are relationships that align and fit components in an assembly. You will create three Mates in this section.

Components are inserted into an assembly with various intuitive options: Coincident, Parallel, Tangent, Concentric, Distance, Angle and Perpendicular.

Establishing the correct component relationship in an assembly requires forethought on component interaction.

Suppress the Lens shield feature to view all surfaces during the mate process.

Suppress the Lens shield feature.

22) Right-click **Lens Shield** in the LENS FeatureManager. Click **Properties**. Check the **Suppressed** box. Click **OK**.

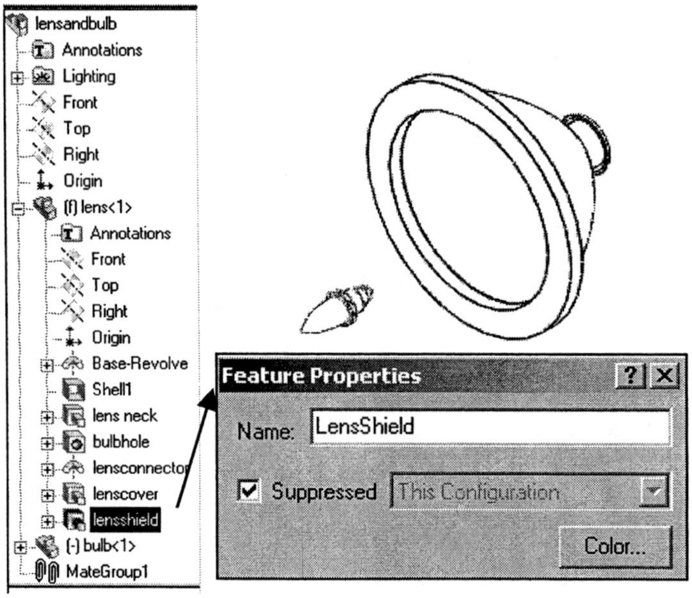

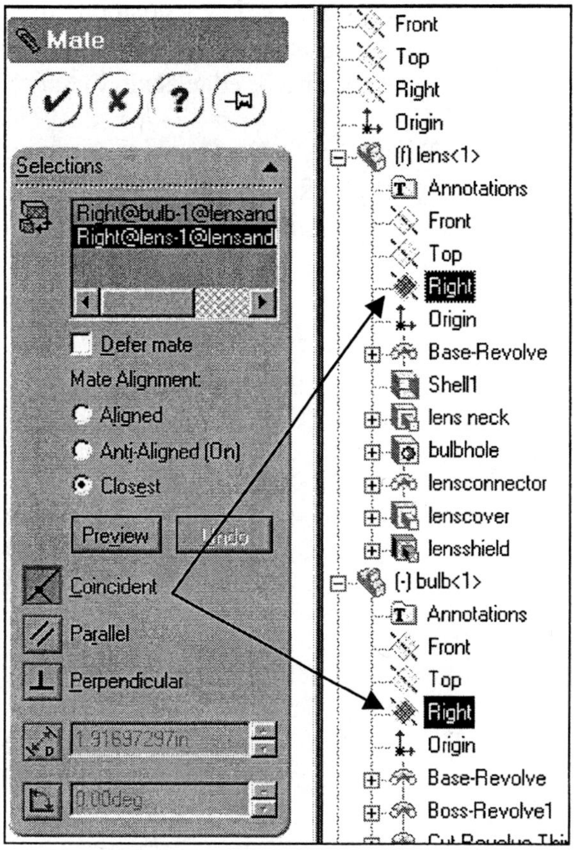

Create the first Mate.

23) Click **Mate** from the Assembly toolbar. The Assembly Mating PropertyManager is displayed. Click the **Mate PropertyManager** Mate icon to display the FeatureManager.

24) Create a Coincident Mate. Click the **Right** plane of the **LENS** from the FeatureManager. Click the **Right** plane of the **BULB** from the FeatureManager. Click **Preview**. Click **OK**.

The Right plane of the LENS and the Right plane of the BULB are coincident. Mates remove degrees of freedom.

Move and Rotate the BULB.

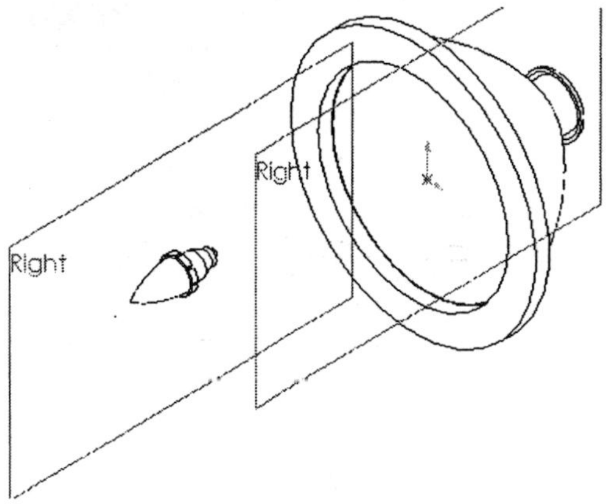

25) Click **Move Component** from the Assembly toolbar. Drag the **BULB** in a horizontal direction. The BULB travels linearly towards/away from the LENS. Display the Top view. Click **Top**. Drag the **BULB** and view the movement constrained to the Right plane. Click **Isometric**.

26) Click the **Rotate** button from the Move PropertyManager. Drag the **BULB** in a vertical direction. The BULB rotates about its Origin. Position the **BULB** in front of the LENS. Click **OK** from the Move PropertyManager.

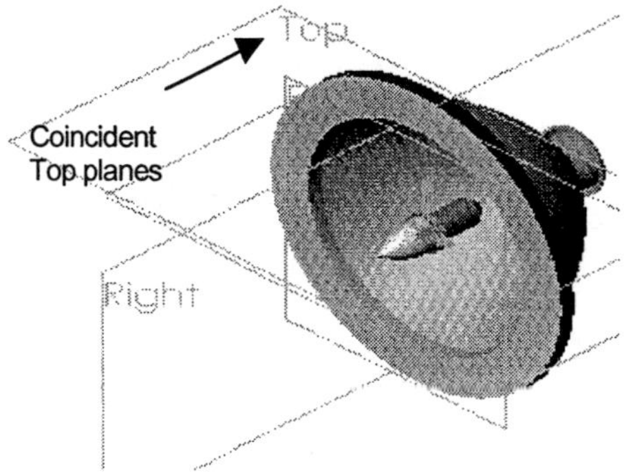

Create the second Coincident Mate.

27) Click **Mate** from the Assembly toolbar. Click the **Top** plane of the LENS from the FeatureManager. Click the **Top** plane of the BULB from the FeatureManager. Click **Preview**. Click **OK**.

When selecting faces, position the mouse pointer in the middle of the face. Do not position the pointer near the edge of the face. If the wrong face or edge is selected, click the face or edge again to remove it from the Items Selected text box.

Right-click in the Graphics window. Click Clear Selections to remove all geometry from the Items Selected text box. The next step is an example of the Select Other option.

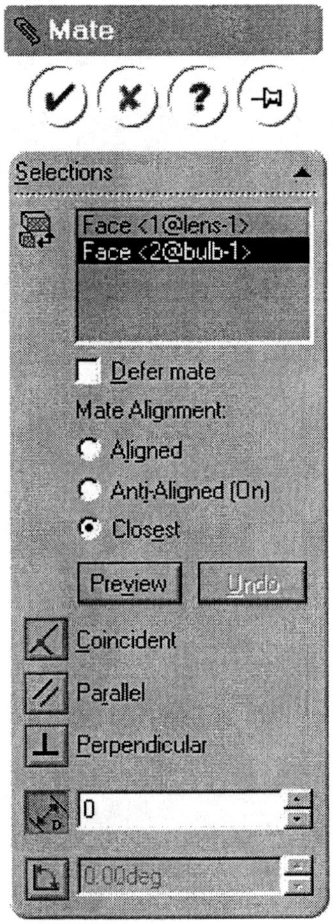

Select hidden geometry.

28) Create the third Mate. **Zoom in** on the counter bore hole. Click **Mate**. Select the **Counterbore face** of the LENS. Right-click in the **Graphics** window.

Click **Select**. The Select icon is displayed from the Pop-up menu. Right-click **N** until the correct geometry is selected. The system highlights the inside counter bore face of the LENS. Accept the counter bore face. Click **Y**.

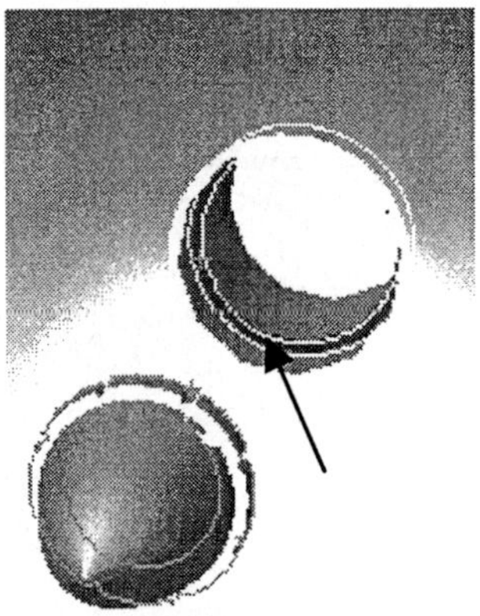

29) Create a Distance Mate. Rotate the LENS and BULB. Right-click **Rotate View** in the Graphics window. Click the bottom center **face** of the BULB. Click the **Distance** button. Enter **0** for Distance. Click **Preview**. Click **OK**. The LENSANDBULB sub-assembly is fully defined.

30) Display the mate types. Double-click on **MateGroup1** in the FeatureManager. Display the full mate names. Drag the **vertical FeatureManager border** to the right.

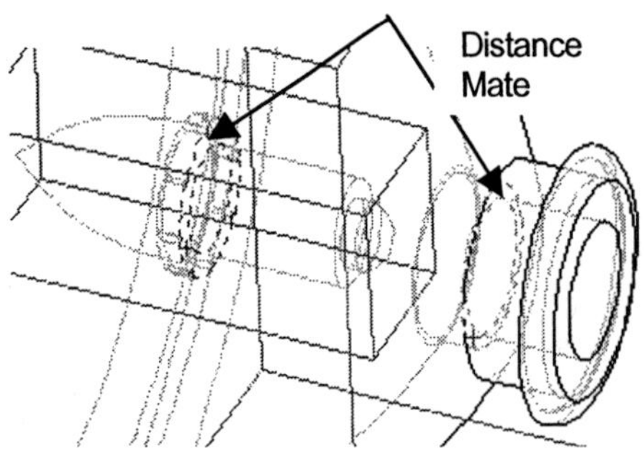

Rotate and Zoom in on view

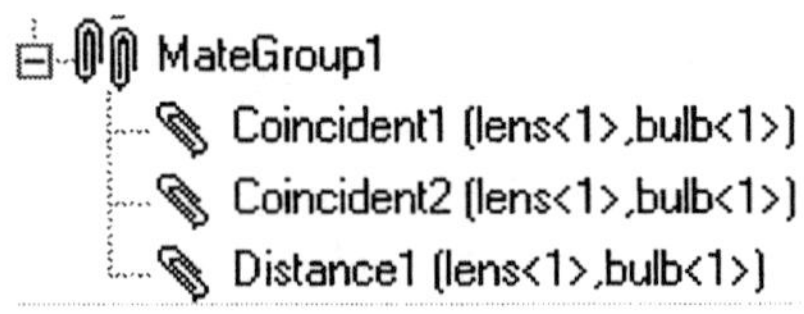

31) Verify the position of the BULB in the Right view. Click **Right**.

32) Save the LENSANDBULB sub-assembly. Click **Save.**

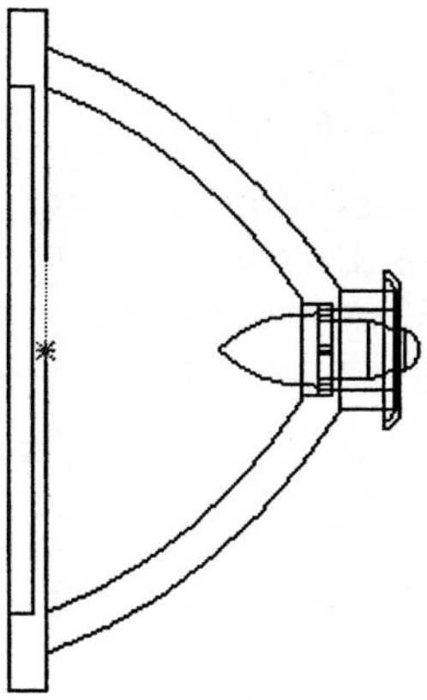

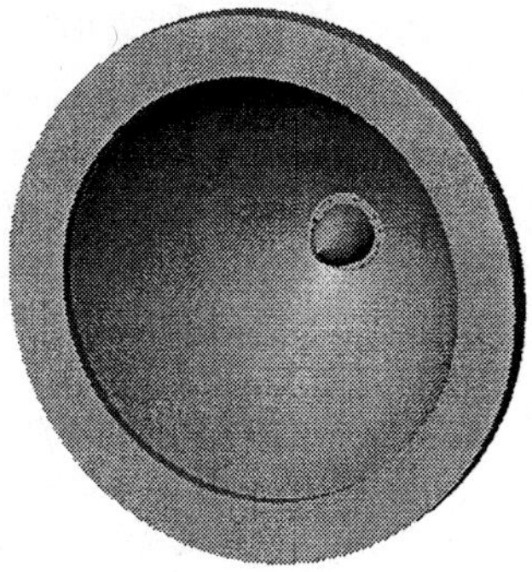

BATTERYANDPLATE

Create the BATTERYANDPLATE sub-assembly. The BATTERY is the Base component. BATTERYANDPLATE sub-assembly mates the BATTERYPLATE component to the BATTERY component.

Create the BATTERYANDPLATE sub-assembly.

33) Click **New**. Click **ASSEMBLY-INCH, [ASSEMBLY-MM]** template. Click **OK.**

34) Click **Save**. Select **My Documents** folder in the Save In text box. Enter the name of the sub-assembly. Enter **BATTERYANDPLATE**. Click **Save.**

35) Open the BATTERY. Click **Open**. Enter the name of the part. Enter **BATTERY.** Open BATTERYPLATE. Click **Open**. Enter the name of the part. Enter **BATTERYPLATE.**

36) Click **Window** from the Main menu. The following documents are displayed: BATTERY, BATTERYPLATE and BATTERYANDPLATE. Display the assembly and components. Click **Tile Horizontally** from the Window menu.

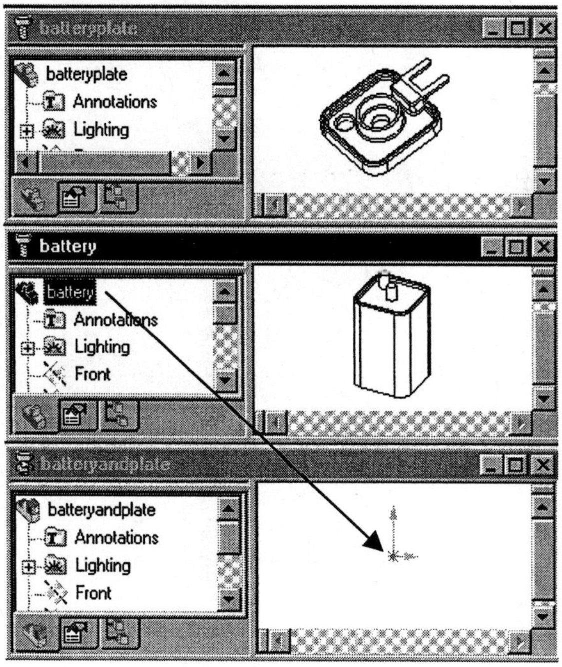

Add the first component.

37) Position the BATTERY inside the assembly Graphics window. Click the **BATTERY** icon from the top of the FeatureManager. Drag the **BATTERY** icon to the Origin of the **BATTERYANDPLATE** assembly window. The mouse pointer displays when position on the Origin.

Add the second component.

38) Click the **BATTERYPLATE** icon from the top of the FeatureManager. Drag the **BATTERYPLATE** component to the assembly window, left of the BATTERY. The mouse pointer displays when positioned inside the BATTERYANDPLATE Graphics window.

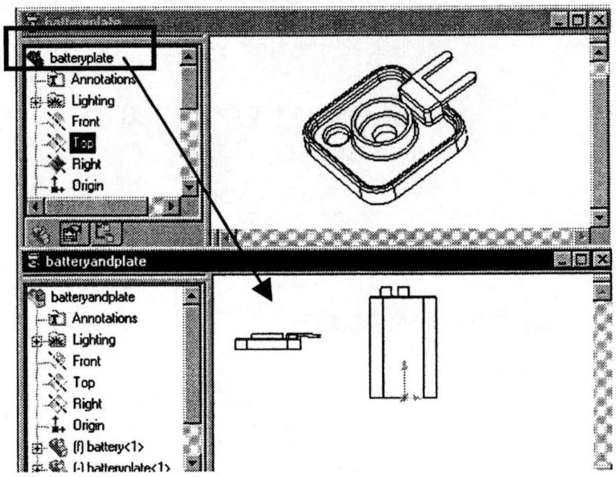

39) Display the Isometric view. Click **Isometric** in the BATTERYANDPLATE window.

40) Close the BATTERYPLATE. Click **Close**. Close the BATTERY. Click **Close**.

Create the first Mate between the BATTERYPLATE and BATTERY components.

41) Create a Distance Mate. Click **Mate** from the Assembly toolbar. Click the **bottom face** of the BATTERYPLATE. Click the **top narrow flat face** of the BATTERY. Click the **Distance** button. Enter **0** for Distance. Click **Preview.** Click **OK.**

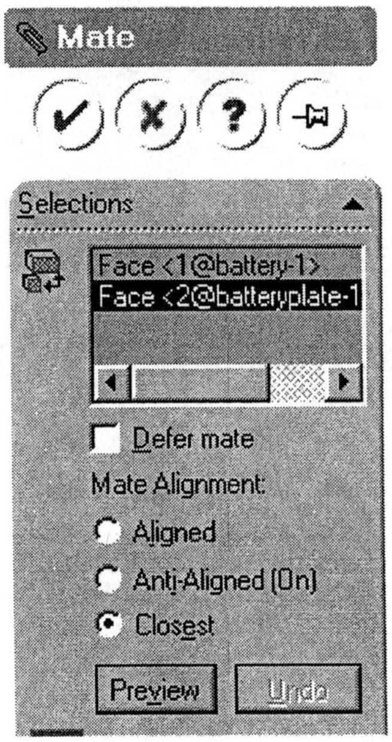

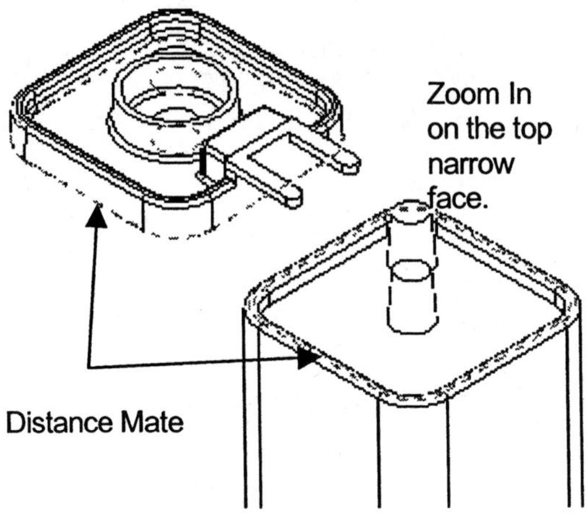

Create the second Mate.

42) Click **Mate** from the Assembly toolbar. Click the **Mate PropertyManager** icon to display the FeatureManager. Create a Coincident Mate. Click the **Right** plane of the BATTERY. Click the **Right** plane of the BATTERYPLATE. Click **Preview**. Click **OK**.

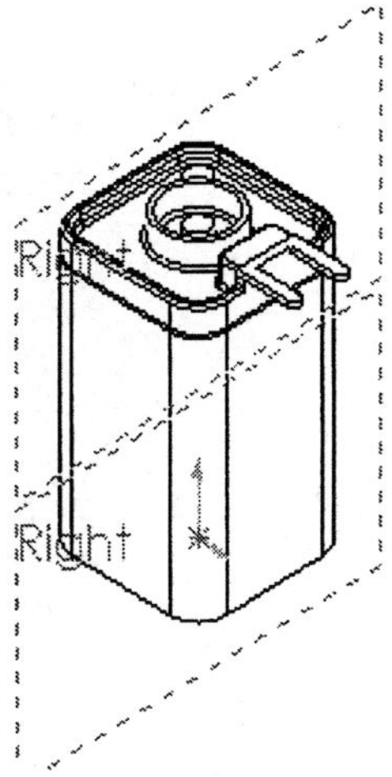

Create the third Mate.

43) Click **Mate** from the Assembly toolbar. Create a Concentric Mate. Click the **cylindrical face** Terminal feature of the BATTERY. Click the **cylindrical face** Holder feature of the BATTERYPLATE. Click **Preview**. Click **OK**. The BATTERYANDPLATE sub-assembly is fully defined.

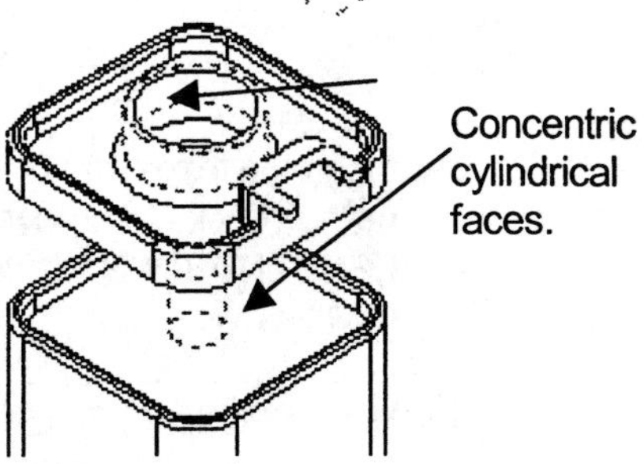

44) Save the BATTERYANDPLATE. Click **Save**.

45) Close all documents. Click **Windows, Close All**.

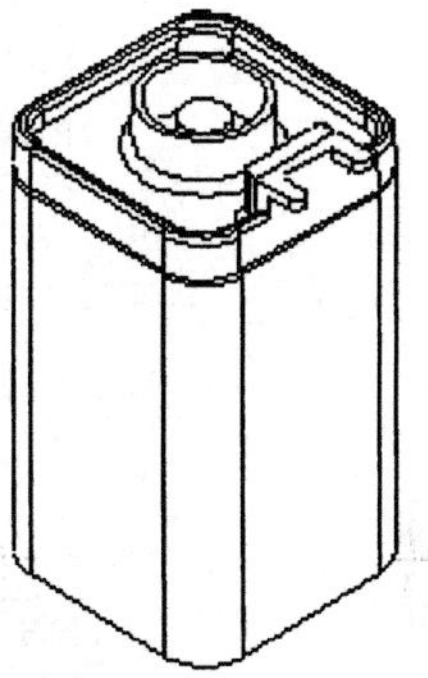

CAPANDLENS Sub-assembly

Create the CAPANDLENS sub-assembly. The LENSCAP is the Base component. The CAPANDLENS sub-assembly mates the O-RING component to the LENSCAP component. The LENSANDBULB sub-assembly is mated to the LENSCAP component. Components are directly inserted into the assembly with the Insert, Component, From File command.

Create the CAPANDLENS sub-assembly.

46) Click **New**. Click **ASSEMBLY-INCH**, [**ASSEMBLY-MM**] template. Click **OK**. Click **Save**. Select **My Documents** folder in the Save In text box. Enter the sub-assembly name. Enter **CAPANDLENS**. Click **Save**.

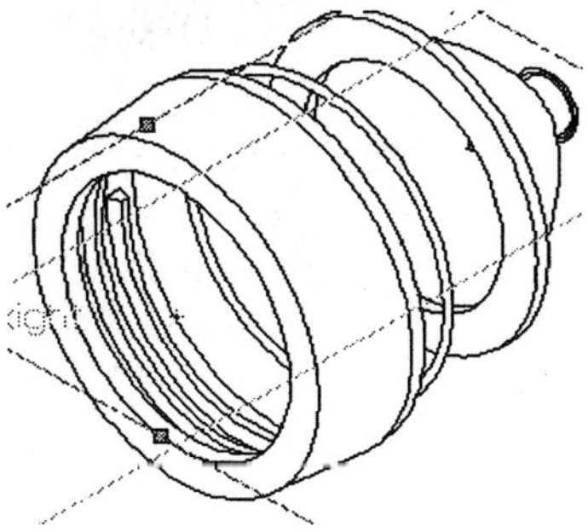

47) Insert the Base component. Click **Insert** from the Main menu. Click **Component, From File**. Enter **LENSCAP**. Click **Open**. Click the **Origin** in the CAPANDLENS Graphics window.

48) Insert the second component. Click **Insert** from the Main menu. Click **Component, From File**. Enter **O-RING**. Click **Open**. Click **behind** the LENSCAP.

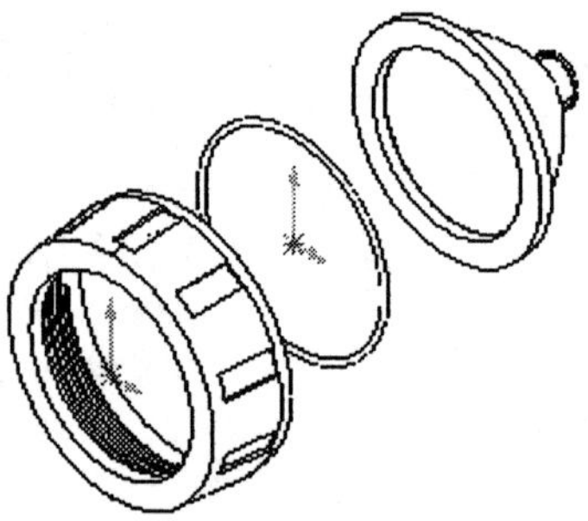

49) Insert the third component. Click **Insert** from the Main menu. Click **Component, From File**. Enter **LENSANDBULB**. Click **Open**. Click **behind** the O-RING.

Note: LENSANDBULB is an assembly. Select File type ".sldasm" from the Open dialog box.

Caution: Select the correct reference. Expand the LENSCAP and O-RING. Click the Right plane within the LENSCAP. Click the Right plane within the O-RING. Do not select the Right plane from the top level FLASHLIGHT assembly. You will create unwanted references.

50) Hide Components when not required for Mates. Right-click on the **LENSANDBULB** sub-assembly in the FeatureManager. Click **Hide**.

Create the Mates between the LENSCAP and O-RING component.

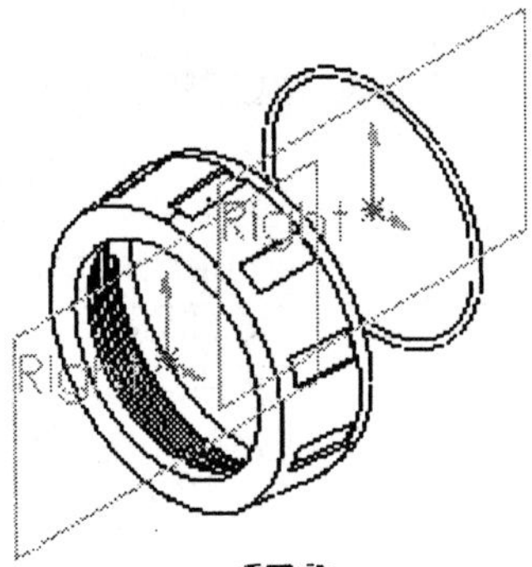

51) Click **Mate** from the Assembly toolbar. Create a **Coincident Mate**. Click the **Right** plane of the LENSCAP. Click the **Mate** button (Mate) to display the FeatureManager. Click the **Right** plane of the O-RING. Click **Preview**. Click **OK**.

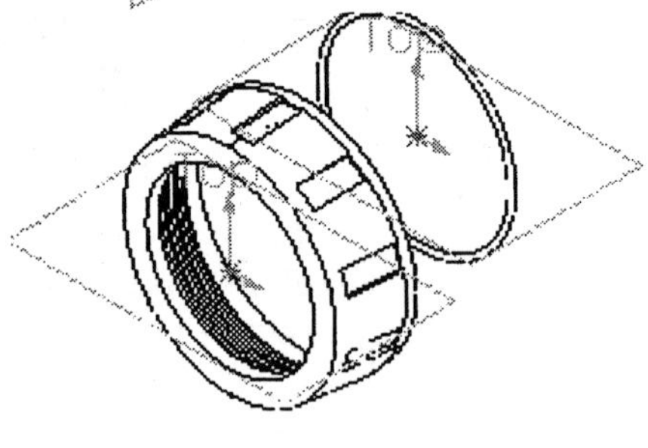

52) Click **Mate**. Create a **Coincident Mate**. Click the **Top** plane of the LENSCAP. Click the **Top** plane of the O-RING. Click **Preview**. Click **OK**.

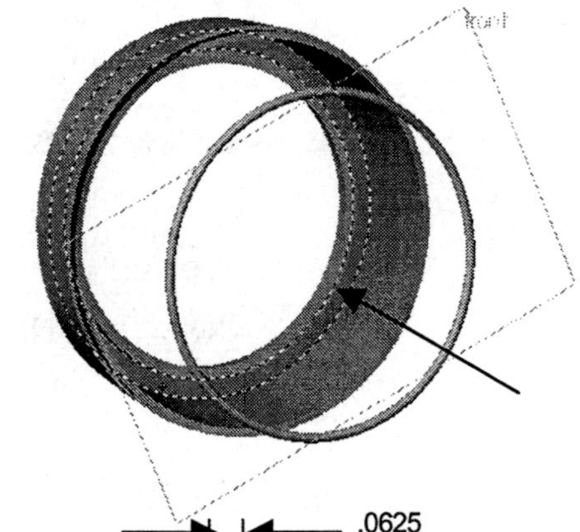

53) Click **Mate**. Create a **Distance Mate**. Click the **back inside face** of the LENSCAP. Click the **Front** plane of the O-RING. Click **Distance**. Enter **.125/2**, **[3.18/2]** for Distance.

54) Click the **Right** view. Click **Preview.**

55) If required, click the **Flip the dimension to the other side** check box. Click **OK.**

How is the Distance, .0625, [1.588] calculated?

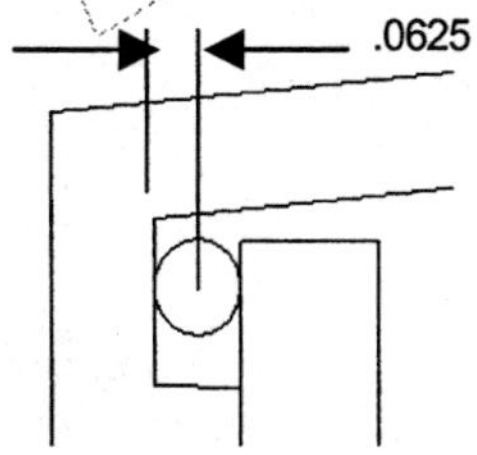

Answer: O-RING Radius (.1250in/2) = .0625in.

O-RING Radius [3.18mm/2] = [1.589].

Note: The Distance Mate option offers additional flexibility over the Coincident Mate option. A Distance Mate value can be modified.

56) Display the LENSANDBULB. Right-click **LENSANDBULB** in the FeatureManager. Click **Show.**

Create the Mates between the LENSCAP and the LENSANDBULB sub-assembly.

57) Click **Mate**. Create a Coincident Mate. Click the **Right** plane of the LENSCAP. Click the **Right** plane of the LENSANDBULB.

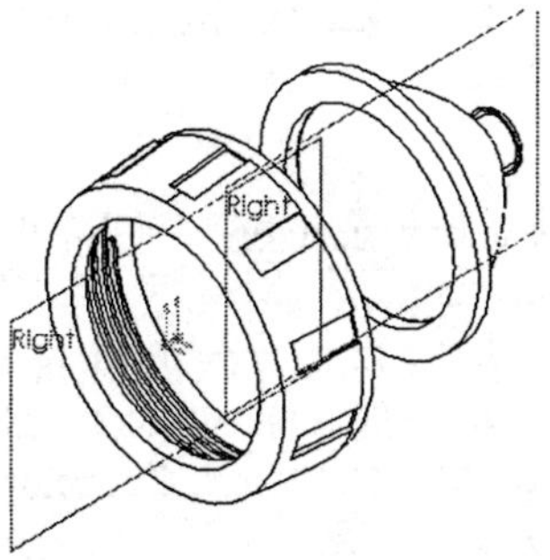

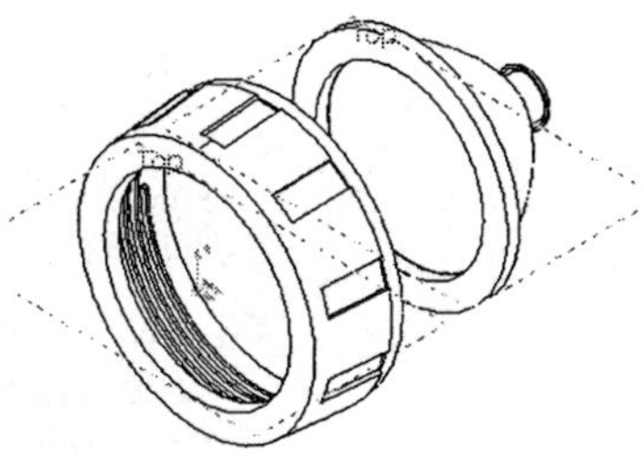

58) Click **Mate**. Create a Coincident Mate. Click the **Top** plane of the LENSCAP. Click the **Top** plane of the LENSANDBULB.

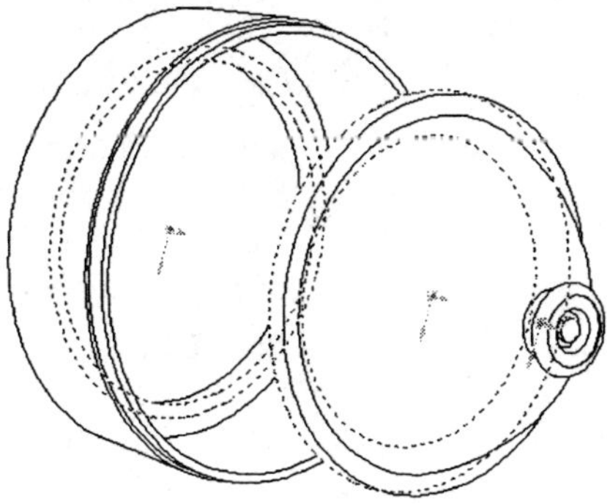

59) Click **Mate**. **Create a Distance Mate.** Click the **narrow inside face** of the LENSCAP. Click the **front face** of the LENSANDBULB. Click **Distance**. Enter **0** for Distance. Click **Preview.** Click **OK.**

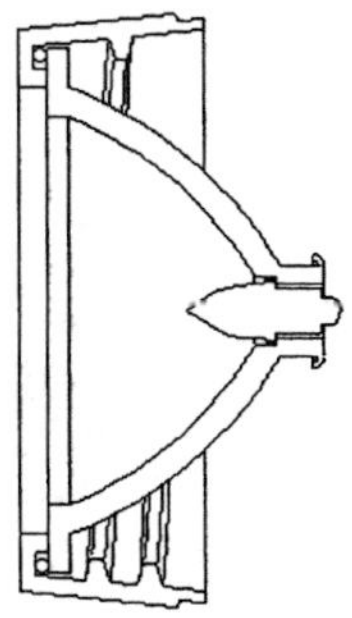

60) The CAPANDLENS sub-assembly is fully defined. Confirm the location of the O-RING. Click the **Right** plane. Click **View** from the Main menu. Click **Display.** Click **Section.**

61) Remove the Section view. Click **View, Display, Section.**

62) **Save** the CAPANDLENS.

Complete the FLASHLIGHT Assembly

Create the FLASHLIGHT assembly. The HOUSING is the Base component. The FLASHLIGHT assembly mates the HOUSING to the SWITCH component. The FLASHLIGHT assembly mates the CAPANDLENS and BATTERYANDPLATE.

Create the FLASHLIGHT assembly.

63) Click **New**. Click the **ASSEMBLY-INCH**, [**ASSEMBLY-MM**] template. Click **OK**.

Click **Save**. Select **My Documents** folder in the Save In text box. Enter **FLASHLIGHT**. Click **Save**.

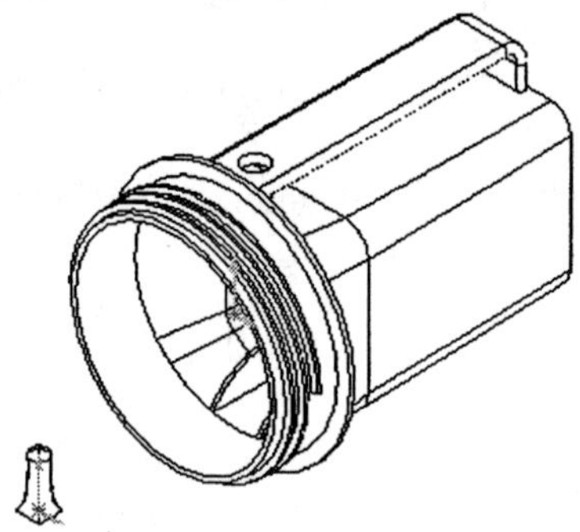

64) Insert the Base component. Click **Insert** from the Main menu. Click **Component**, **From File**. Select **All Files** from Files of Type list box. Enter **HOUSING**. Click **Open**. Click the **Origin** of the FLASHLIGHT.

65) Insert the second component. Click **Insert** from the Main menu. Click **Component**, **From File**. Enter **SWITCH**. Click **Open**. Click **in front** of the HOUSING.

Create the Mates between the HOUSING and the SWITCH component.

66) Click **Mate**. Create a Coincident Mate. Click the **Right** plane of the HOUSING. Click the **Right** plane of the SWITCH. Click **Coincident**. Click **Preview**. Click **OK**.

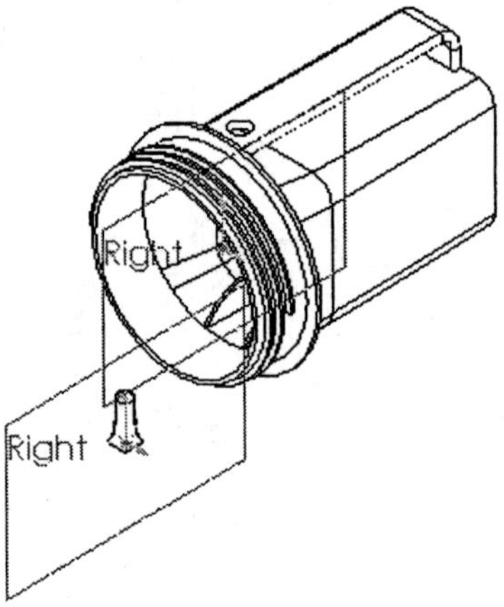

67) Click **View** from the Main menu. Click **Temporary axis**. Click **Mate**. Create a Coincident Mate. Click the **temporary axis** inside the Switch Hole of the HOUSING. Click the **Front** plane of the SWITCH. Click **Coincident**. Click **Preview**.

Click **OK**.

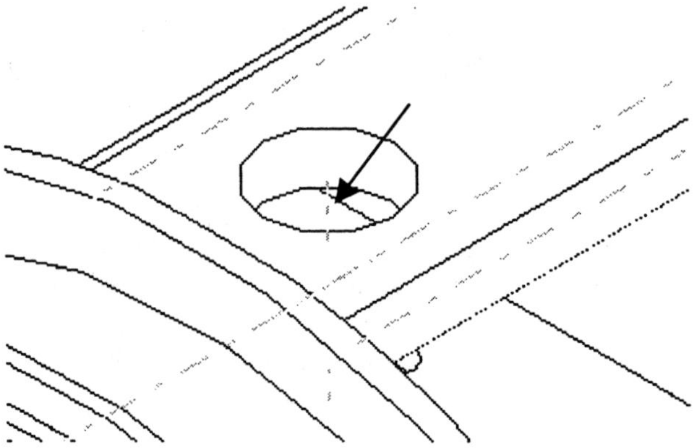

68) Click **Mate**. Create a Distance Mate. Click the **top face** of the Handle. Click the **Vertex** on the Dome of the SWITCH. Click **Distance**. Enter **.200**, [**5.08**] for Distance. Click **Preview**. Click **OK**.

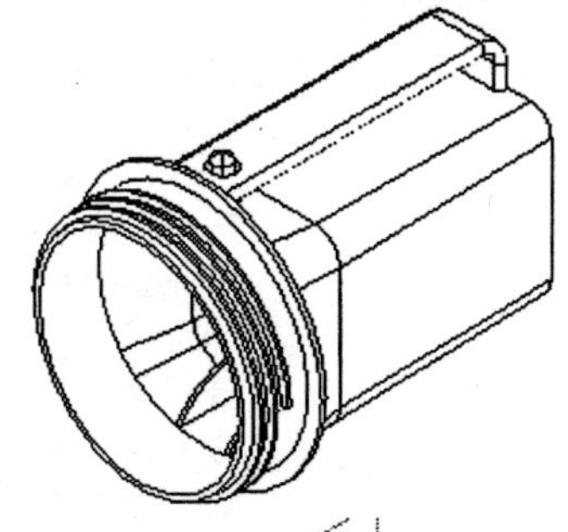

69) Save the FLASHLIGHT assembly. Click **Save**.

70) Insert the CAPANDLENS sub-assembly. Click **Insert** from the Main menu. Click **Component, From File**. Enter **CAPANDLENS**. Click **Open**. Click **in front** of the HOUSING.

Create the Mates between the HOUSING component and the CAPANDLENS sub-assembly.

71) Create a Coincident Mate. Click the **Right** plane of the HOUSING. Click the **Right** plane of the CAPANDLENS. Click **Preview**. Click **OK**.

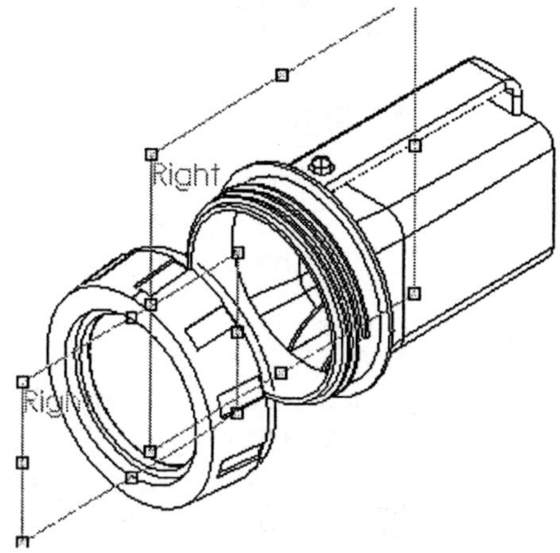

72) Create a Coincident Mate. Click the **Top** plane of the HOUSING. Click the **Top** plane of the CAPANDLENS. Click **Preview**. Click **OK**.

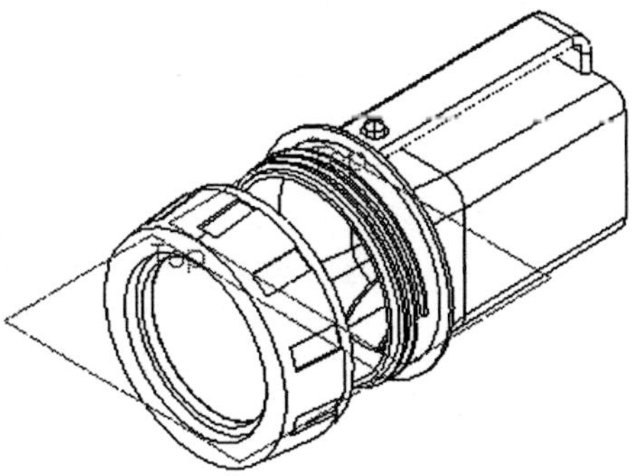

73) Create a Distance Mate. Click the **front face of the Boss-Stop** on the HOUSING. Click the **back face** of the CAPANDLENS.

Click **Distance**. Enter **0** for Distance.

74) Save the FLASHLIGHT assembly. Click **Save**.

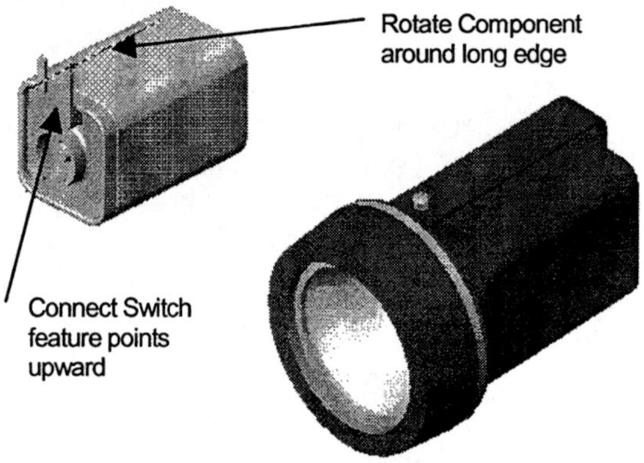

75) Insert the BATTERYANDPLATE sub-assembly. Click **Insert** from the Main menu. Click **Component, From File.** Enter **BATTERYANDPLATE.** Click **Open.** Click the FLASHLIGHT **Graphics window** to the left of the HOUSING.

76) Rotate the part. Click **Rotate Component**. Rotate the **BATTERYANDPLATE.** until the ConnectorSwitch feature points vertical. Click **OK.**

Create the Mates between the HOUSING component and the BATTERYANDPLATE.

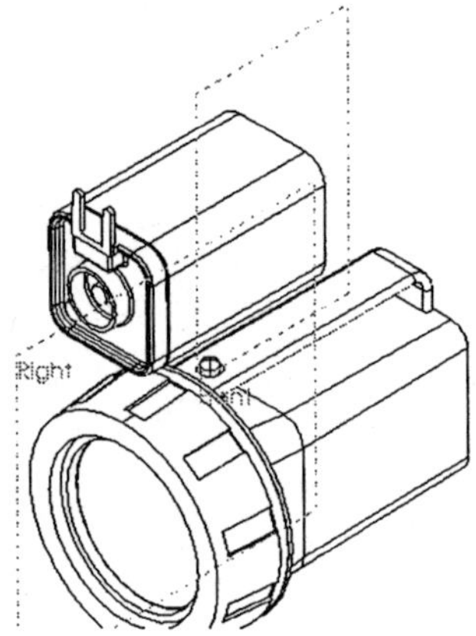

77) Create a Coincident Mate. Click the **Right** plane of the HOUSING. Click the **Front** plane of the BATTERYANDPLATE. Click **Preview.** Click **OK.**

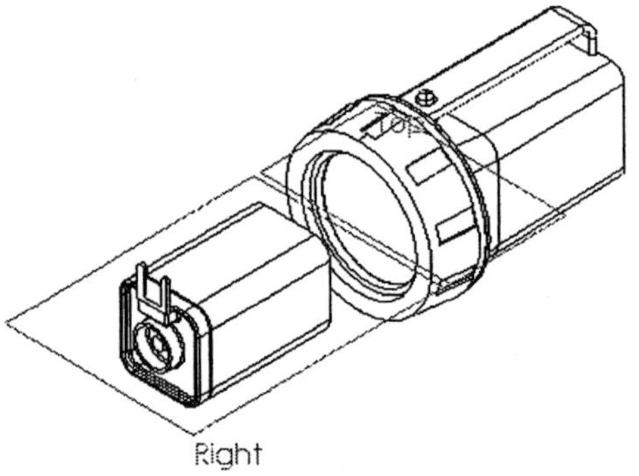

78) Create a Coincident Mate. Click the **Top** plane of the HOUSING. Click the **Right** plane of the BATTERYANDPLATE. Click **Preview.** Click **OK.**

79) Create a Distance Mate. Click the **back face** of the HOUSING. Click the **bottom face** of the BATTERYANDPLATE. Click **Distance** . Enter **.275**, **[6.99]** for Distance. Click **Flip** if required. Click **Preview**. Click **OK**.

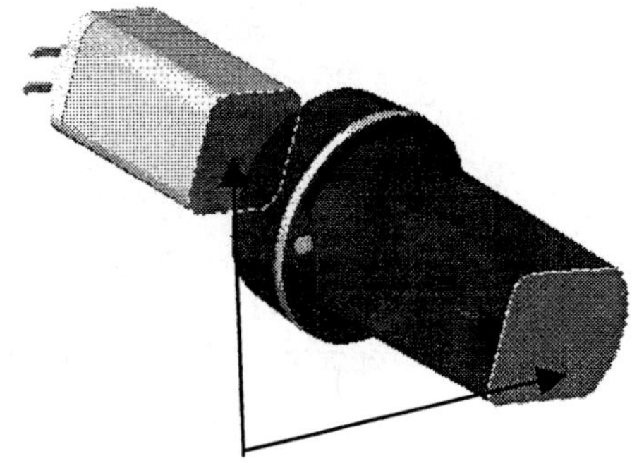

80) Rebuild the FLASHLIGHT.

81) Save the FLASHLIGHT. Click **Save**. Click **Yes** to update all components.

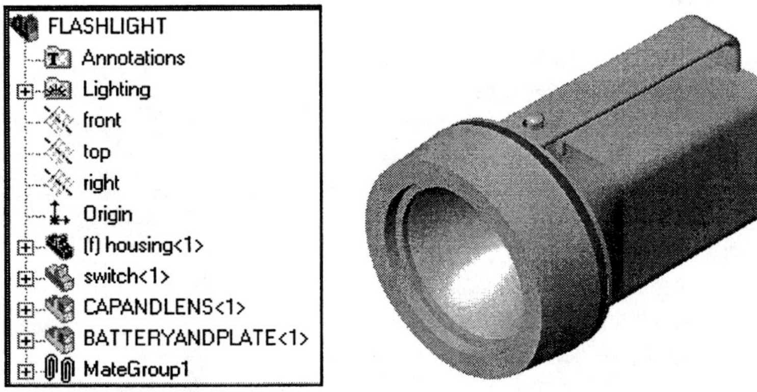

This FLASHLIGHT is displayed in PhotoWorks and is saved in TIF format. marie ?TIF?

Addressing Design Issues

There are interference issues with the assembly. Address these design issues:

- Reduce the SwitchConnector feature size on the BATTERYPLATE.
- Adjust Rib2 on the HOUSING. Test with the Interference Check command.

The ConnectorSwitch feature of the BATTERYPLATE was created too long. Contain the ConnectorSwitch within the HOUSING. Use the Measure function to determine an acceptable distance between the ConnectorSwitch feature of the BATTERYPLATE.

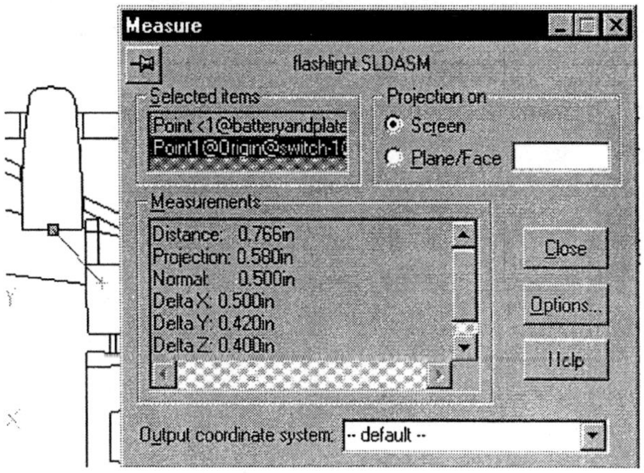

Check for Interference.

82) Hide the CAPANDLENS. Right click on **CAPANDLENS** in the FeatureManager. Click **Hide Component**.

83) Expand the BATTERYANDPLATE sub-assembly. Click **Plus** ⊞. Expand the BATTERYPLATE. Resolve Lightweight Components. Double-click the **ConnectorSwtich** feature. Double click **1.000**, **[2.54]** dimension text. Enter **.500**, **[12.7]**. Click **Rebuild**.

Displayed with Section view

There is an interference fit issue between the HOUSING and the BATTERY.

84) Click **Tools** from the Main menu. Click **Interference Detection**. The Interference Volumes dialog box is displayed. Click the **BATTERY**. Click the **HOUSING**. Click the **Check** button. The interference is displayed in the Graphics window. Rib2 overlaps the BATTERY by .050, [1.27].

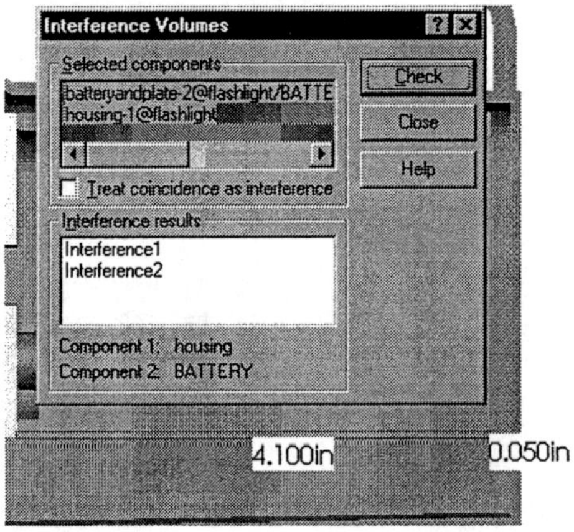

85) Expand the **HOUSING** in the FeatureManager. Double-click on the **Rib2** feature. Double click **1.300, [33.02]**. Enter **1.350, [34.29]**. Click **Rebuild**.

Note: Interference must exist between the BULB and the BATTERY to create an electrical connection.

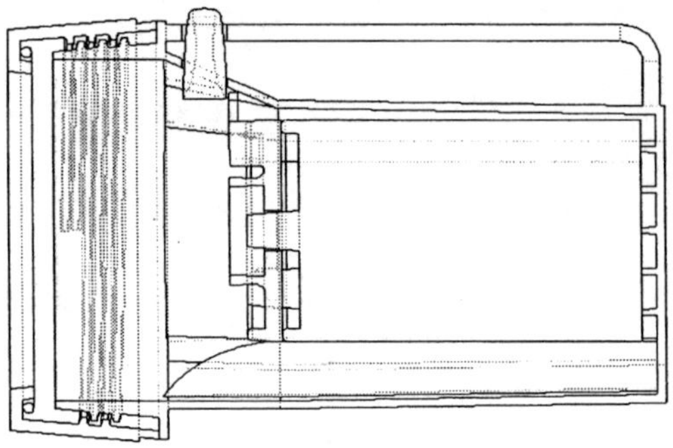

86) Show the CAPANDLENS. Right-click on **CAPANDLENS**. Click **Show**. **Show** all hidden components and features. The FLASHLIGHT design is complete.

87) **Save** the FLASHLIGHT. Click **YES** to the question, "Rebuild the assembly and update the components".

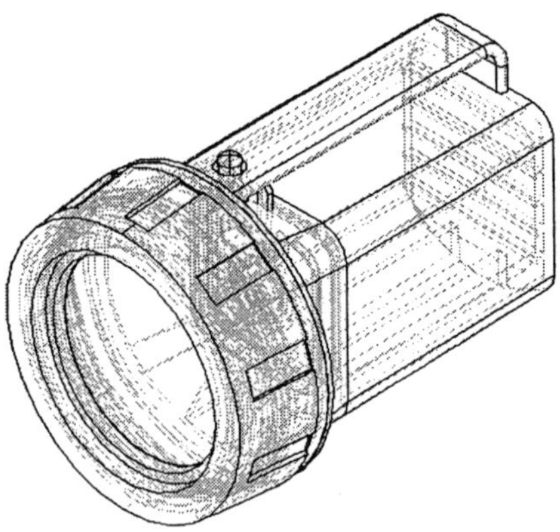

Exploded View

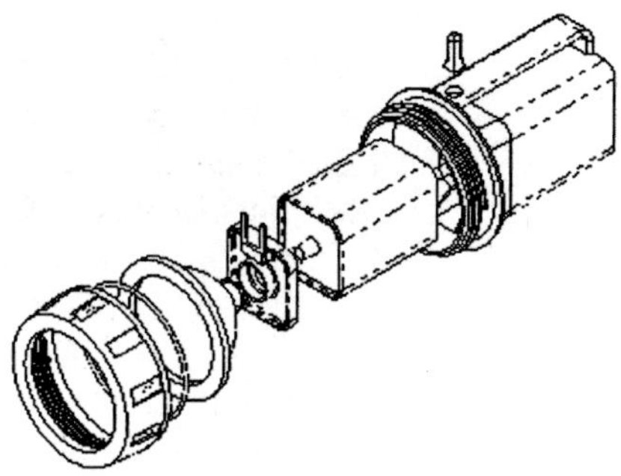

Exploded views assist the designer in the viewing of the design creation. You can fully explode an assembly in a single or multi step procedure. AutoExplode explodes an assembly in a single step procedure. In a multi step procedure, create individual explode steps. Explode the assembly in a multi step approach.

Edit steps in the Exploded view with the following Step Editing tools. From left to right the following icons represents: 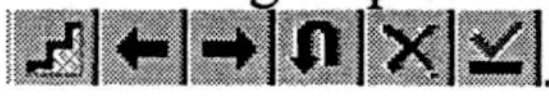.

- New Step.
- Edit Previous Step.
- Edit Next Step.
- Undo Changes to Current Step.
- Delete Current Step.

Apply the Current Steps.

Explode Step 1.

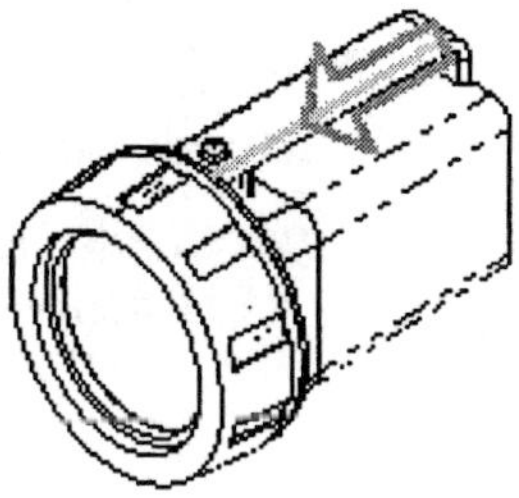

88) Click **Insert** from the Main menu. Click **Exploded View**. Create the first explode step. Click **New Step**  from the Step Editing Tools. Click the handle **top right** edge of the HOUSING. Enter **14**, **[355.6]** for Distance. Click the **Component to explode** text box. Click **FeatureManager** . Click **LENSCAP** from the FeatureManager. Click **Component part only**. View the first explode step, named Explode Step1. Click the **Apply Check Mark** .

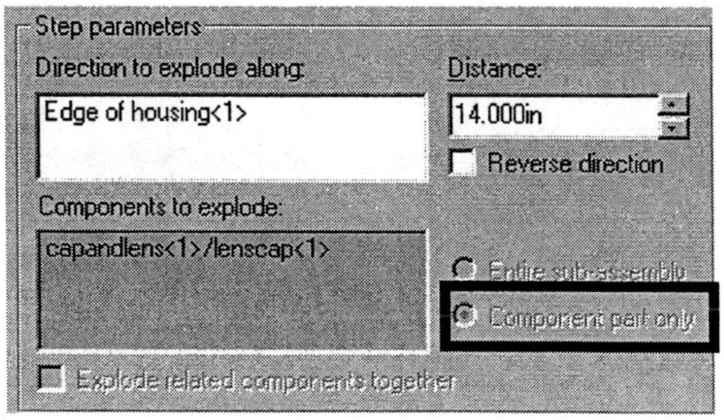

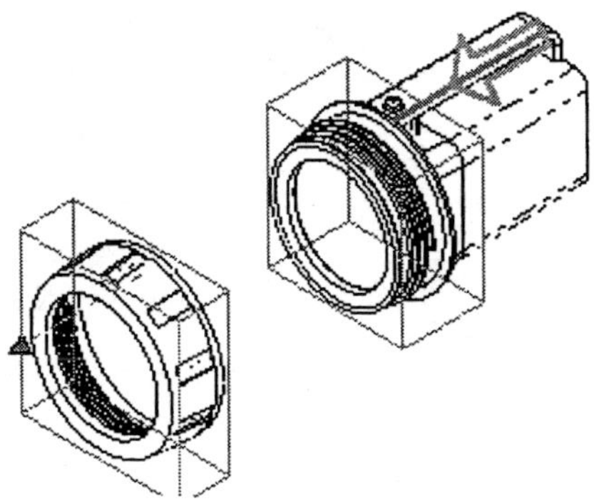

Explode Step 2.

89) Create the second step. Click **New Step** . Click the handle **top right** edge of the HOUSING. Enter **11.5, [292.1]** for Distance. Click the **Component to explode** text box. Click **O-RING** from the FeatureManager. Click **Component part only**. View Explode Step 2. Click the **Apply Check Mark** .

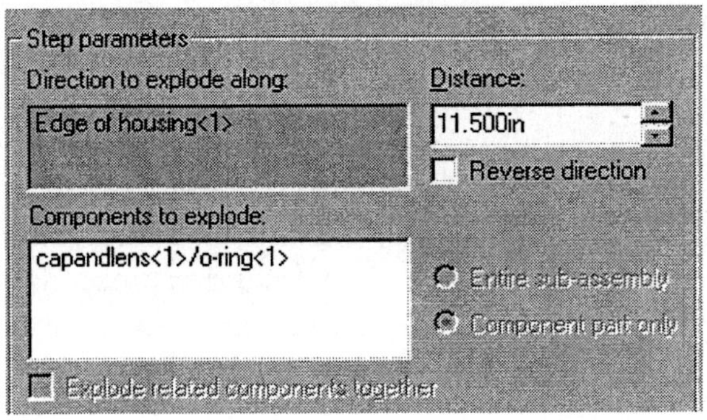

Explode Step 3.

90) Create the third step. Click **New Step** . Click the handle **top right** edge of the HOUSING. Enter **10, [254]** for Distance. Click the **Component to explode** text box. Click **LENSANDBULB** from the FeatureManager. Click **Component part only.**

View Explode Step 3. Click the **Apply Check Mark** .

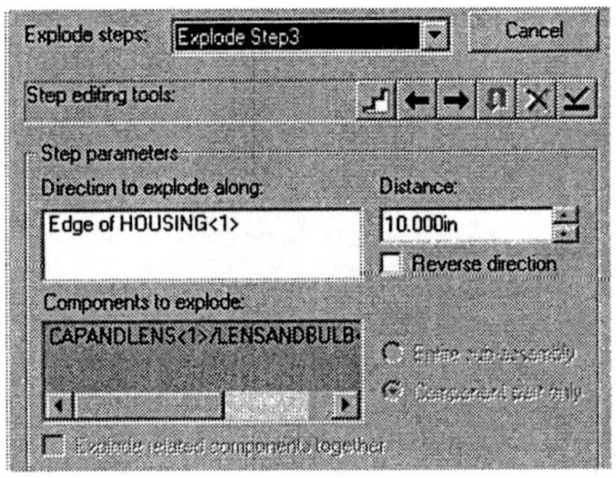

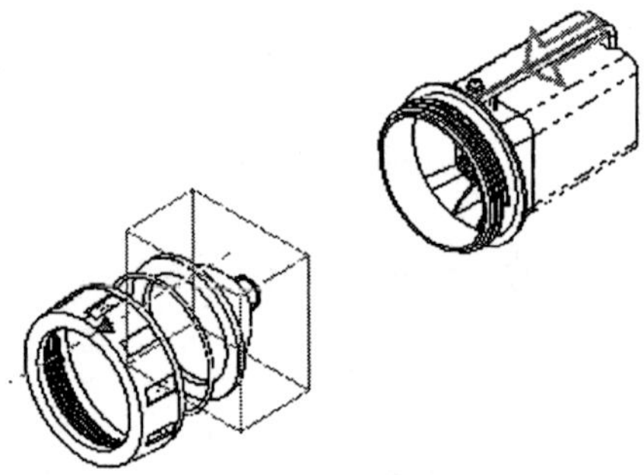

Explode Step 4.

91) Create the forth step. Click **New Step** . Click the handle **top right** edge of the HOUSING. Enter **8.5, [215.9]** for Distance. Click the **Component to explode** text box. Click **BATTERYPLATE** from the FeatureManager. Click **Component part only**. View Explode Step 4. Click the **Apply Check Mark** .

Explode Step 5.

92) Create the fifth step. Click **New Step** . Click the handle **top right** edge of the HOUSING. Enter **7, [177.8]** for Distance. Click the **Component to explode** text box. Click **BATTERY** from the FeatureManager. Click **Component part only**. View Explode Step 5. Click the **Apply Check Mark** .

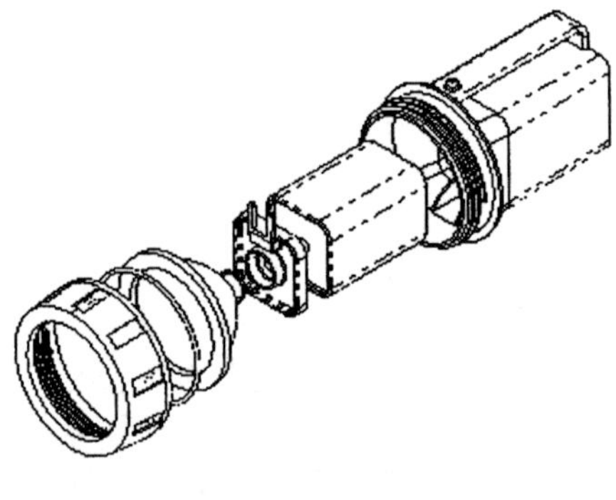

Explode Step 6.

93) Create the sixth step. Click **New Step**. Click the **front vertical edge** of the HOUSING. Enter **1.8**, **[42.7]** for Distance. Click **SWITCH** from the FeatureManager. Click the **Reverse direction** check box. Click the **Apply Check Mark**. Click **OK**.

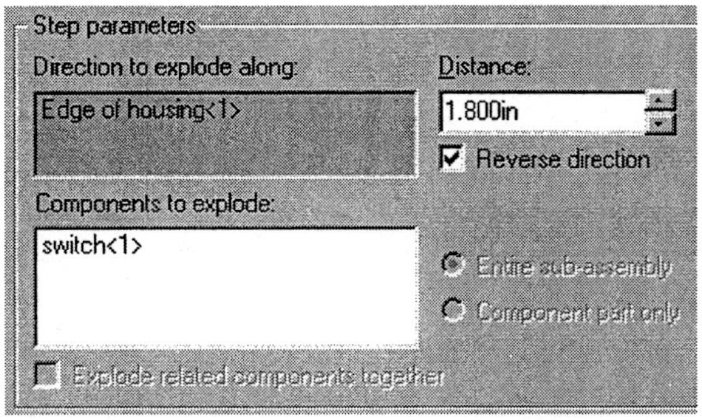

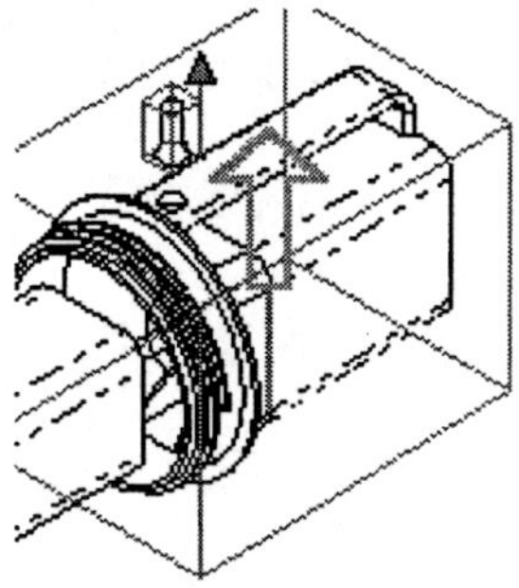

94) Fit the Exploded view to the Graphics window. Press the **f** key.

95) Save the FLASHLIGHT. Click **Save**.

Viewing Exploded State

There are two configurations of the FLASHLIGHT, Exploded and Collapsed.

View the Explode Steps.

96) Click **Configuration Manager** at the bottom of the FeatureManager. Expand Default. Click **Plus** ⊞ to the left of the Default entry. Expand ExpView1. Click **Plus** ⊞ to the left of ExpView1 entry.

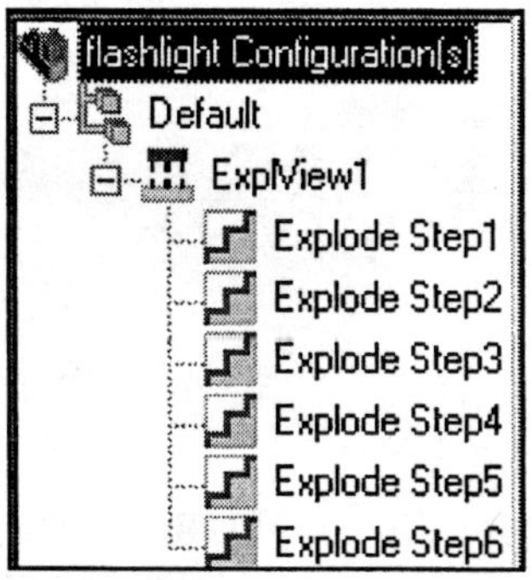

97) Return to the FeatureManager. Click **FeatureManager** at the bottom of the Configuration Manager.

98) Remove the exploded state. Right-click in the **Graphic window**. Click **Collapse** from the Pop-up menu.

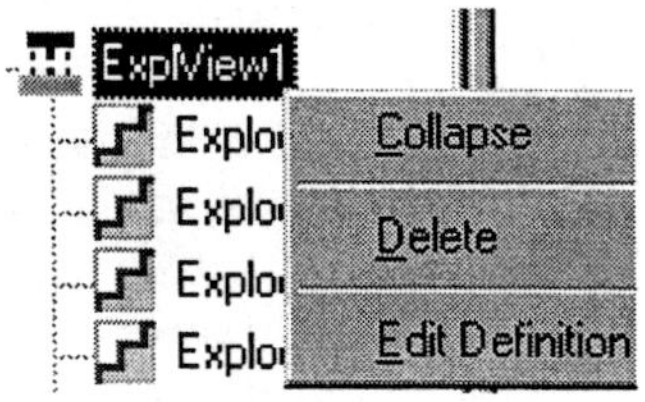

99) Save the FLASHLIGHT. Click **Save**.

Export Files

You receive a call from sales. They inform you that the customer increased the order to 200,000 units. However, the customer requires a prototype to verify the design in six days. What do you do? Answer: Contact a Rapid Prototype supplier.

You export three SolidWorks files:

- HOUSING
- LENSCAP
- BATTERYPLATE

Use the Stereo Lithography (STL) format. Email the three files to a Rapid Prototype supplier.

Example: Paperless Parts Inc. (www.paperlessparts.com).

A Stereolithography (SLA) supplier provides physical models from 3D drawings. 2D drawings are not required. Export the LENSCAP.

100) Open the LENSCAP. Expand the **CAPANDLENS** sub-assembly. Right-click on the **LENSCAP** from the FLASHLIGHT FeatureManager. Click **Open LENSCAP.SLDPRT**.

101) Export the LENSCAP. Click **File**, **SaveAs**. A warning message states that, "Lens-cap.SLDPRT is being referenced by other open documents. "Save AS" will replace these references with the new name. Create a copy of LENSCAP. Click **OK**.

102) Click **STL Files (*.stl)** from the Save as Type drop down list. The Save dialog box displays new options.

103) Click the **Options** button in the lower right corner of the Save dialog box. The STL Export Options dialog is displayed.

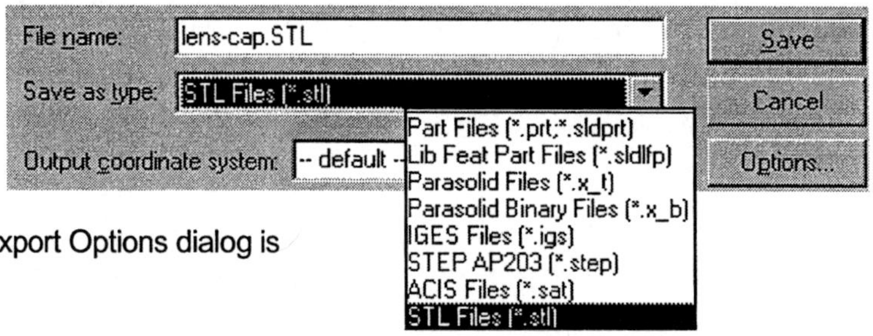

104) Click **Binary** from the Output format box. Click **Course** from the Quality box.

105) Display the STL triangular faceted model. Click **Preview**.

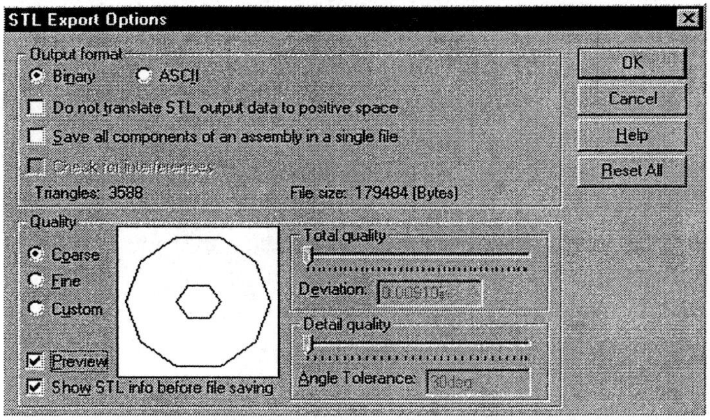

106) Create the binary STL file. Click **OK** from the STL dialog box. Click **Save** from the Save dialog box. A status report is provided. Save the file LENS-CAP.STL. Click **Yes**.

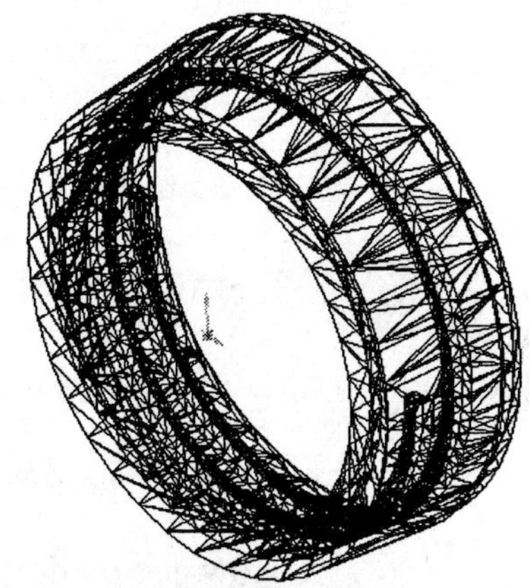

The STL file created in binary format is 179,484 Bytes. The STL file created in ASCII format is 1,015,428 Bytes.

You receive the three SLA physical models for the supplier.

You assemble the rapid prototype models with purchased components.

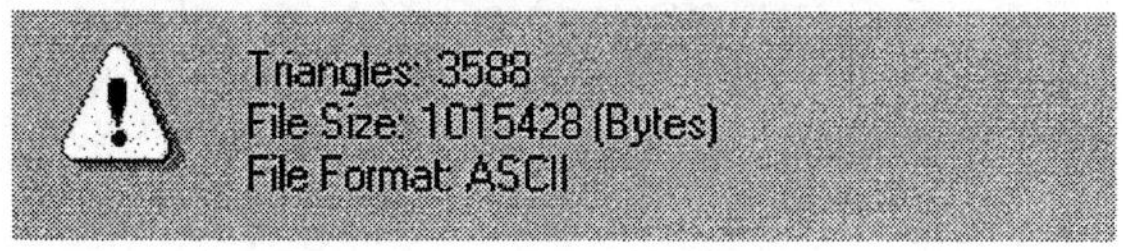

Results:

A flashlight assembly delivered to the customer in six days. Success!

SolidWorks eDrawings provides a facility for you to animate, view and create compressed documents to send to colleagues, customers and vendors.

It is time to go home. The telephone rings. Guess who? The customer is ready to place the order. Tomorrow you will receive the purchase order. You think about the concerns of manufacturing, purchasing and shipping. In the rush to create the prototype, you forgot about the packaging, the part numbers and the silk screen vendor. Congratulation, the project is just beginning! Let's try some more examples.

Questions

1. Describe an assembly.
2. Name 5 helpful assembly techniques.
3. What are Mates and why are they important in assembling components?
4. In an assembly, each component has ______# degrees of freedom. Name them.
5. True of False. A Fixed component cannot move.
6. Name the Mate types.
7. Define Move Component and Rotate Component.
8. Describe the function of Selection Filters.
9. Describe a Section view.
10. True of False. Components in an assembly can be hidden.
11. What are suppressed features and components? Provide examples of each.
12. How do you determine interference between components in an assembly?
13. Why do you create an Exploded View? How do you create a multi step Exploded View?
14. What type of file formats can you save from SolidWorks?

Examples

Create the Assemblies.

Exercise 4.1:

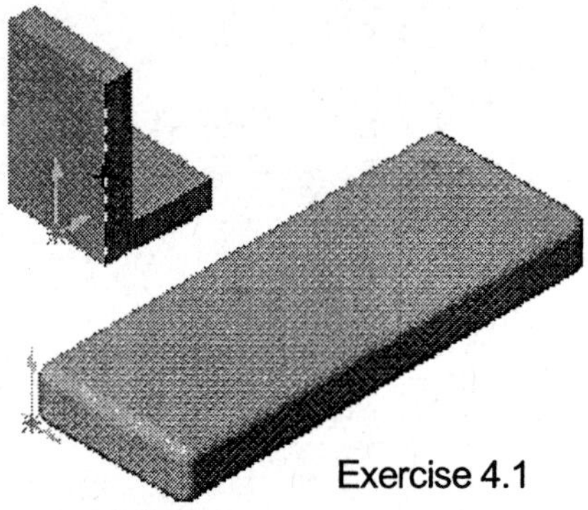

Exercise 4.1

a) Create two parts:

- L-BRACKET
- PLATE

The Base component is the PLATE.

b) Review the following assembly commands before mating the L-BRACKET:

- Move component .
- Rotate component .

c) Complete the assembly. Utilize one PLATE and two L-BRACKETs and two BOLTS. Use on-line help to define the Assembly Mirror function.

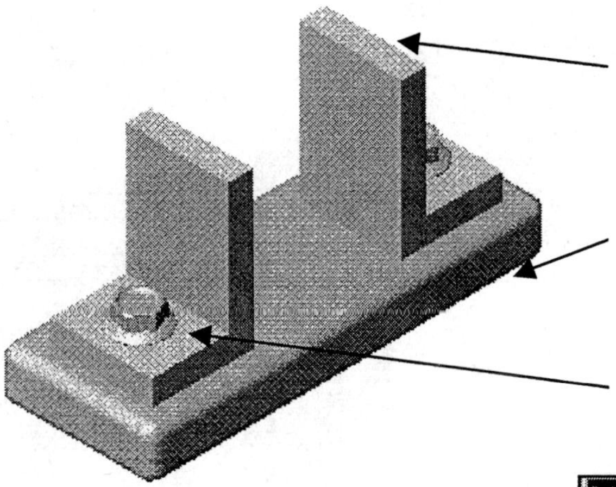

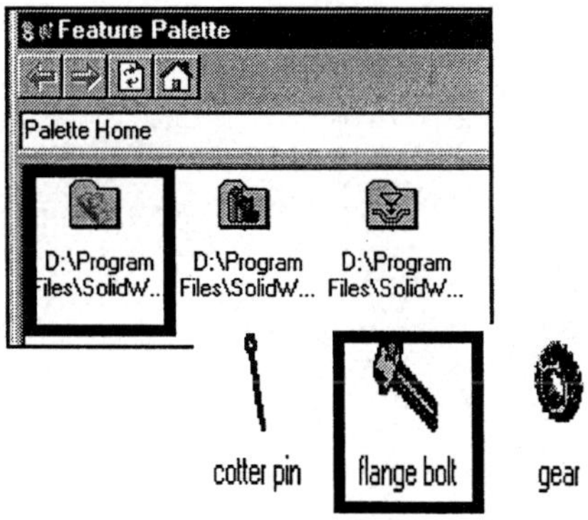

FLANGE BOLTs are found in Tools, Feature Palette menu. Double-click the **hardware folder**  icon. The hardware Feature Library components are displayed.

The flange bolt  icon represents a family of similar shaped components in various configurations.

Exercise 4.2: Create the SUPPORT assembly from the PLATE and STANDOFF.

Exercise 4.3: Create the HANGER assembly from the parts, U-BRACKET and PIN.

Exercise 4.4: Create the FIXTURE assembly from the parts, MOUNTING PLATE and T-SECTION.

SUPPORT assembly

Exercise 4.2

HANGER assembly

Exercise 4.3

FIXTURE assembly

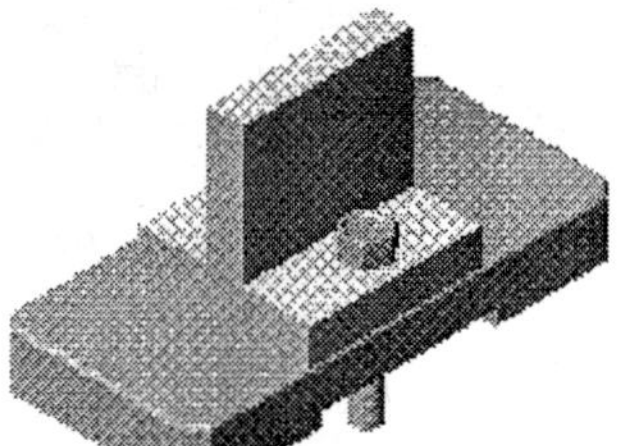

Exercise 4.4

Design Projects.

Exercise 4.5: Create a D-size battery.

Exercise 4.6: Create a battery HOLDER to hold 4-D size batteries.

Exercise 4.5

Exercise 4.6

Exercise 4.7: Create a WHEEL assembly. A SHAFT supports the WHEEL. The SHAFT connects two L-BRACKETS. The L-BRACKETS are mounted to a BASE PLATE. Use purchased parts to save time and cost. The only dimension provided is the WHEEL.

Select a WHEEL diameter:

- 3 inch
- 4 inch
- 100mm

Exercise 4.7

Find a material supplier using the WWW. (Hint: www.globalspec.com)

WHEEL Assembly Parts:

BASE PLATE

BUSHINGS

L-BRACKET

BOLTS

SHAFT

Exercise 4.8: Industry Collaborative Exercise.

Note: SolidWorks Animator Add In is required for this exercise. Select Tools, Add-Ins, SolidWorks/Animator. Review Exercise 2.6.

Create an exploded view of the LENSANDCAP assembly. Create a collapsed animation of the LENSANDCAP assembly. Incorporate the new animation into a PowerPoint presentation.

SolidWorks Animator add-in software application captures motion and animates SolidWorks assemblies. SolidWorks Animator generates Windows-based animations (**.avi** files) that you can support on any Windows-based computer. Combine PhotoWorks and SolidWorks Animator to output photo-realistic animations.

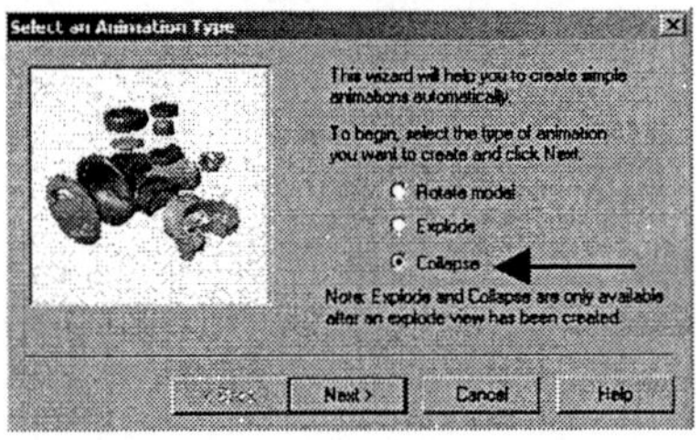

Start with an exploded view of the LENSANDCAP assembly. Click Tools, Add Ins. Check the SolidWorks Animator check box. Create a collapse animation of the LENSANDCAP assembly. Click the Animator Wizard icon. Click Collapse. Click Next. Enter 10 seconds for Duration. Enter 0 for Start Time. Click Finish to view the collapse.

Time = 0 seconds

Time = 5 seconds

Time = 8 seconds

Modify the duration of the ORING component. Modify the start time of the LENSANDBULB assembly. Split the FeatureManager and display the Animation Manager. Right-click ORING from the Animation Manager. Click Edit Path.

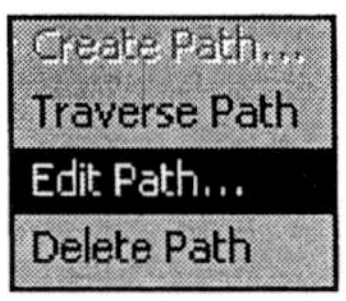

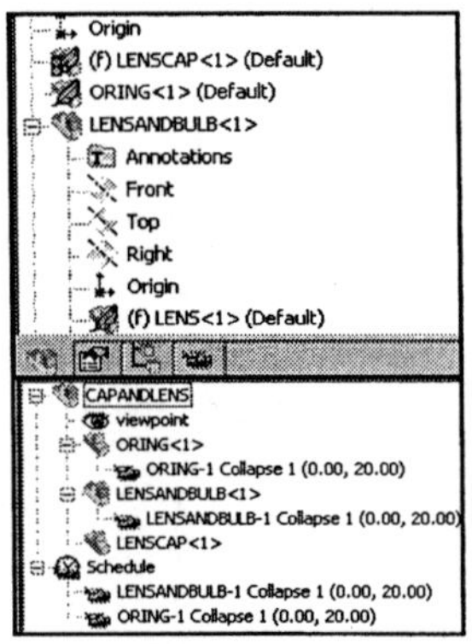

Right-click ORING from the Animation Manager.
Enter 10 for Duration. Click Edit Path.
Right-click the LENSANDBULB.
Enter 10 for start time. Enter 20 for Duration. The ORING will move first, then the LENSANDBULB. The total animation time is 30 seconds.

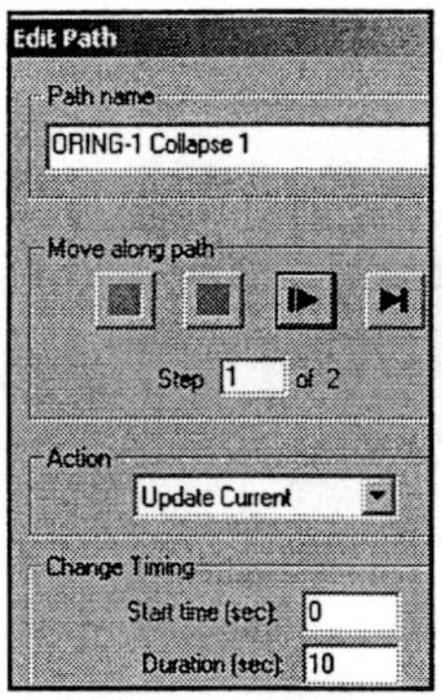

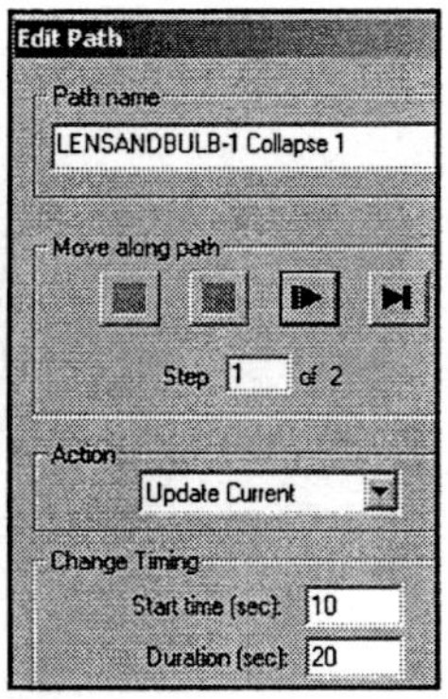

Click Record Animation to File . The Renderer is the SolidWorks screen. Microsoft Video 1 is the default Video Compression. The time required to create the .AVI file is greater than the time required to create the SolidWorks/Animator screen file.

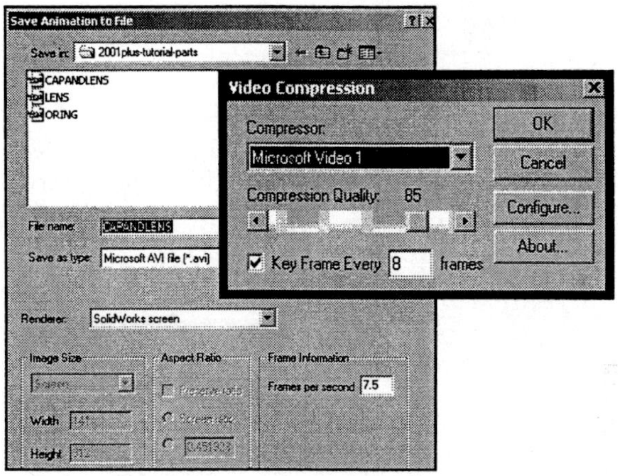

Play the .AVI file with the Windows Media Player.

Insert the .AVI file as a Movie into PowerPoint and create a presentation.

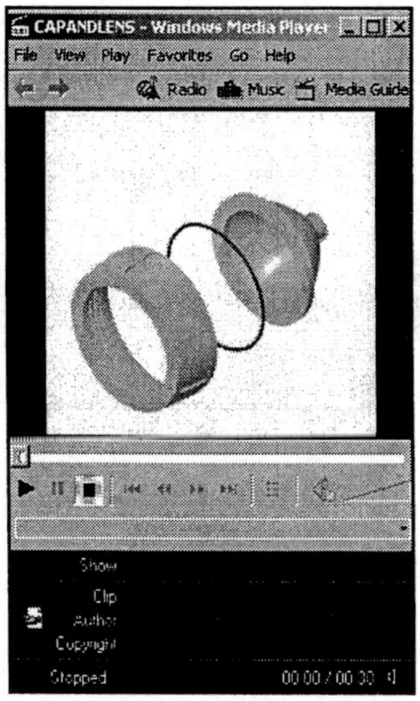

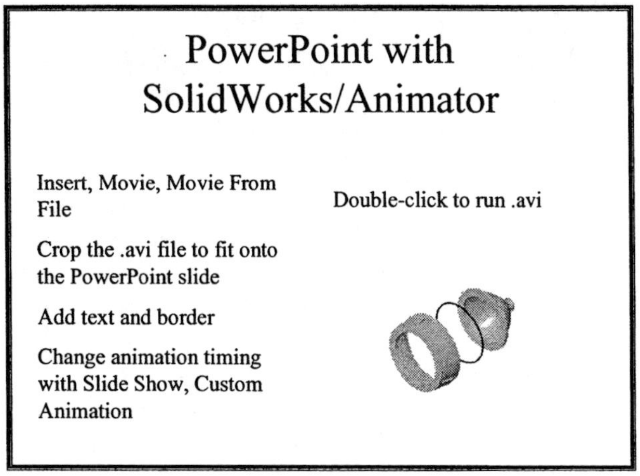

Play the animation in PowerPoint.

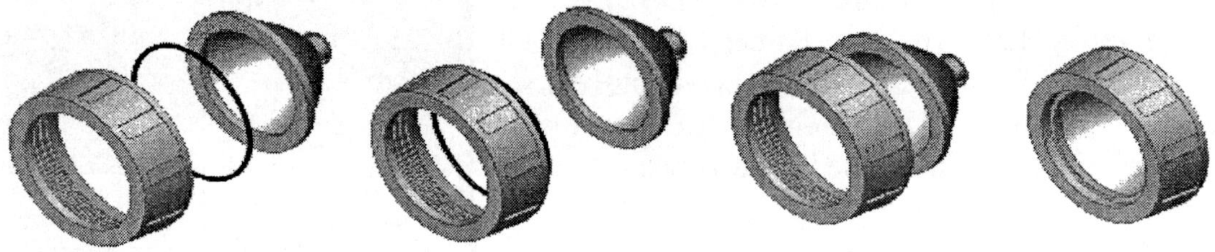

Time = 0 seconds Time = 5 seconds Time = 20 seconds Time = 30 seconds

Note: A manufacturing procedure can be created by combining parts and assemblies from SolidWorks, PhotoWorks, SolidWorks/Animator and PowerPoint. The following example illustrates the process to create the 3 Cylinder Rack assembly.

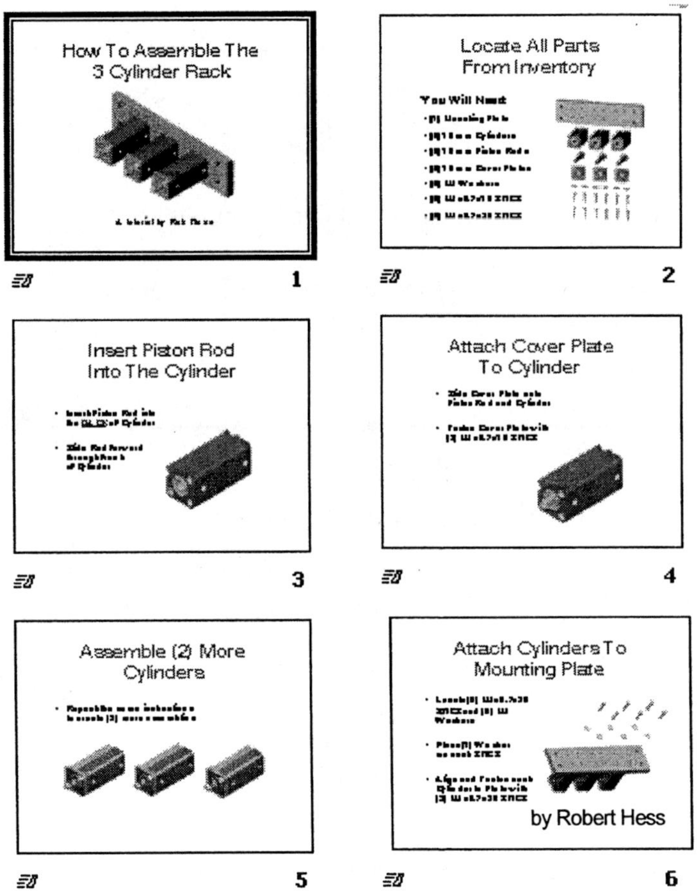

Exercise 4.9: Industry Collaborative Exercise.

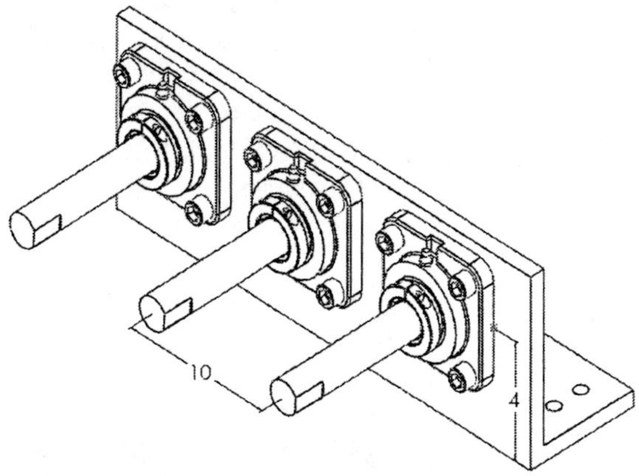

Mounted Bearing Assembly
Model and Images
Courtesy of Emerson Power Transmission Corporation, Ithaca, NY a subsidiary of Emerson.

In a make-buy decision process, you decide on which parts to manufacture and which parts to purchase. In assembly modeling, you decide on which parts to design and which parts to obtain from other sources.

You now work with a team of engineers on a new project. The senior engineer has specified a shaft size of 1 inch, [25.4mm] to be used in conjunction with a mounted ball bearing. Three bearings are mounted to a support plate. Each shaft is separated by a minimum of 10 inches, [254mm]. The center of each shaft is located 4 inches, [100mm] about the bottom face of the support plate. The senior engineer provides two key bearing requirements.

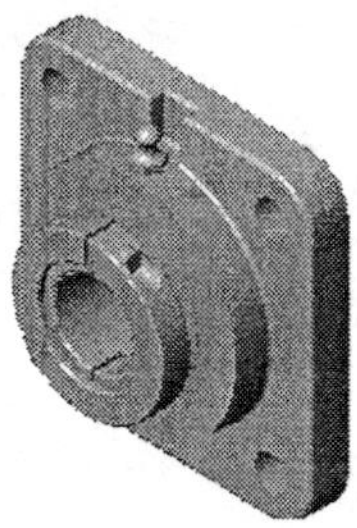

- The bearing contains a concentric collar.
- The bearing must be located in a 4-bolt flange block.

Time is critical. Obtain purchased components directly from the manufacturer, Emerson-EPT, website: http://www.emerson-ept.com.

Design the L-SHAPE SUPPORT PLATE to hold three bearings. Assemble the shaft to the bearing. Utilize SolidWorks Toolbox or create the fasteners for this project.

Note: Access is required to the World Wide Web with a valid email address to complete this exercise.

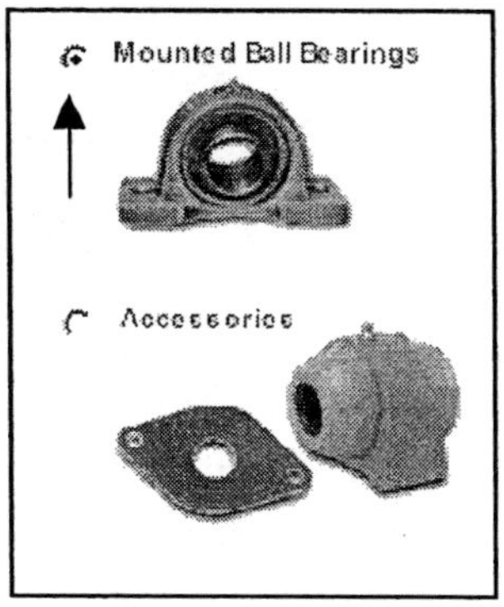

Size and download a component for the CUSTOMER assembly.

1) Obtain the BEARING from EMERSON-EPT. Invoke a web browser. Enter the URL: **www.emerson-ept.com**.

2) Enter your email address and password. If you are a new user to the web site, click the Register button and enter the requested information. You will be emailed with the required information.

3) Select EDGE Online in the left corner of the home web page. Click the Mounted Ball Bearings button.

Select 4 Bolt Flange Block for Housing Type. Select Concentric Collar for Locking Device. Select 1 for Shaft Size.

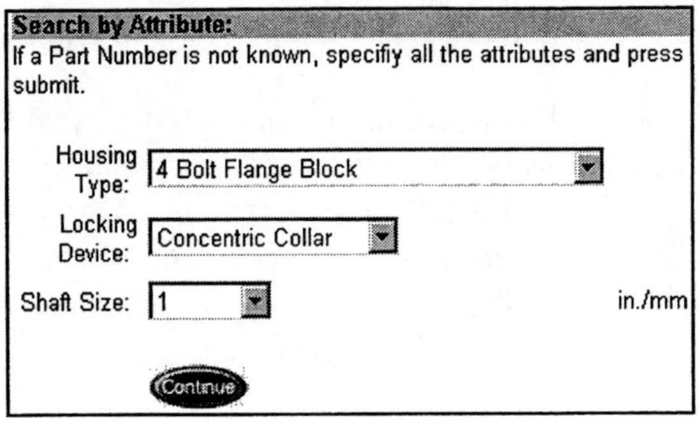

4) The Part Number and Description are displayed. Two part numbers are valid. Double-click VF4B-216.

Search by Attributes with Housing Type: **4 Bolt Flange Block**
Locking Device: **Concentric Collar**
Shaft Size: **1**

Part Number	Housing Type	Locking Device	Shaft Size	Description
VF4B-216 ←	4 Bolt Flange Block	Concentric Collar	1	Browning - Standard Duty - VF4B200 Series - 4 Bolt Flange - Concentric Lock - Contact Seal
VF4B-316	4 Bolt Flange Block	Concentric Collar	1	Browning - Medium Duty - VF4B300 Series - 4 Bolt Flange - Concentric Lock - Contact Seal

5) Display the file formats to download. Click 3D button. Select the SolidWorks Assembly (.sldasm). Select 2001PLUS for Version. Click the Download Files button.

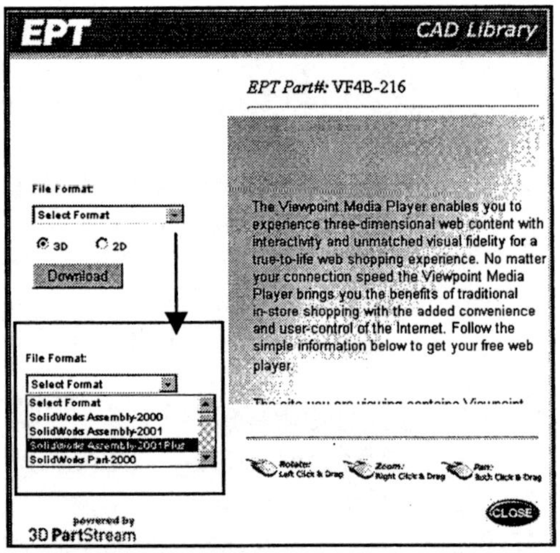

6) The zipped SolidWorks file is downloaded to your computer. Unzip the file.

7) Return to the emerson-ept home page. Click the elink option. Click eCatalog.

8) Enter the part number VF4B-216 for the bearing to obtain the specification sheet. Utilize the bolt size required in your design. Create the 3-BEARING Assembly. The Shaft diameter requires a close running fit.

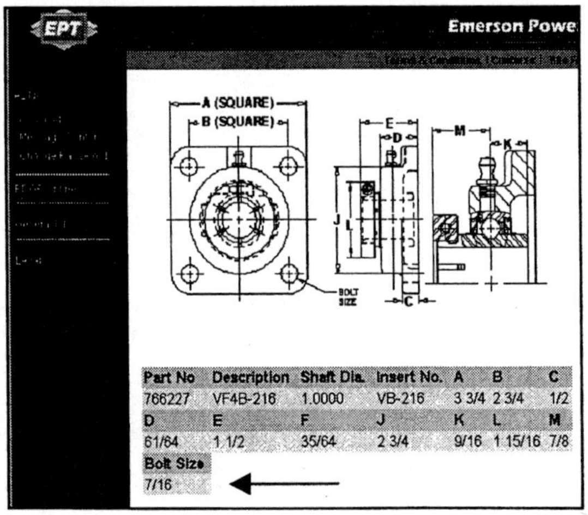

Project 5

Fundamentals of Drawing

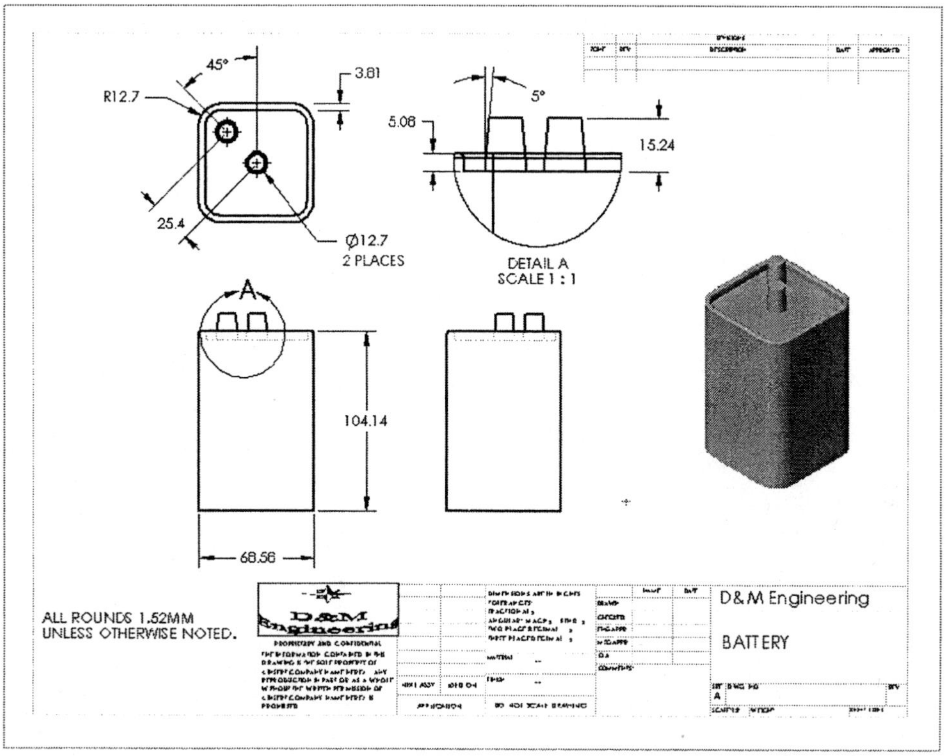

Below are the desired outcomes and usage competencies based upon the completion of Project 5.

Project Desired Outcomes	Usage Competencies
A BATTERY drawing with a customized sheet format containing a company logo, title block and sheet information.	Ability to create a custom sheet format, logo, title block and sheet information.
	An understanding of displaying Standard and Detail views.
	Ability to insert, create and modify dimensions.
A FLASHLIGHT assembly drawing with a Bill of Materials.	Knowledge to develop and incorporate a Bill of Materials into a drawing.

Notes:

Project 5 – Fundamentals of Drawing

Project Objective

Create a BATTERY drawing with a customized sheet format containing a company logo, title block and sheet information. Create a FLASHLIGHT assembly drawing with a Bill of Materials.

Project Situation

The individual parts and assembly are completed. What is the next step? You are required to create drawings for various internal departments, namely: production, purchasing, engineering, inspection and manufacturing. Each drawing may contain unique information and specific footnotes. Example: A manufacturing drawing would require information on assembly, Bill of Materials, fabrication techniques and references to other relative information.

Project Overview

In this project you will create two drawings:

- A BATTERY drawing.
- A FLASHLIGHT assembly drawing.

The BATTERY drawing contains three standard views, a Detailed view and an Isometric view. Do you remember what the three standard views are? They are: Top, Front and Right side, Figure 5.1. Orient the views to fit the drawing sheet. Incorporate the BATTERY dimensions into the drawing.

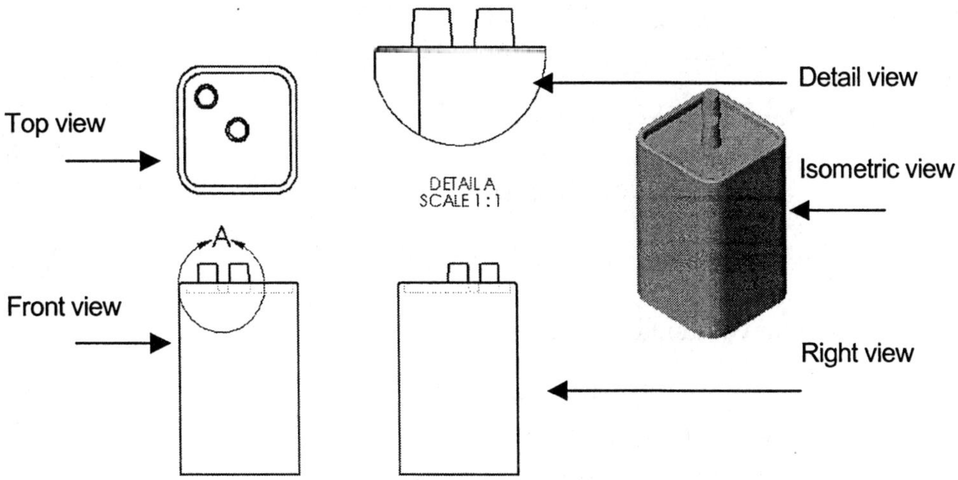

Figure 5.1

The FLASHLIGHT assembly drawing contains an exploded view. The drawing contains a Bill of Materials and balloon text, Figure 5.2.

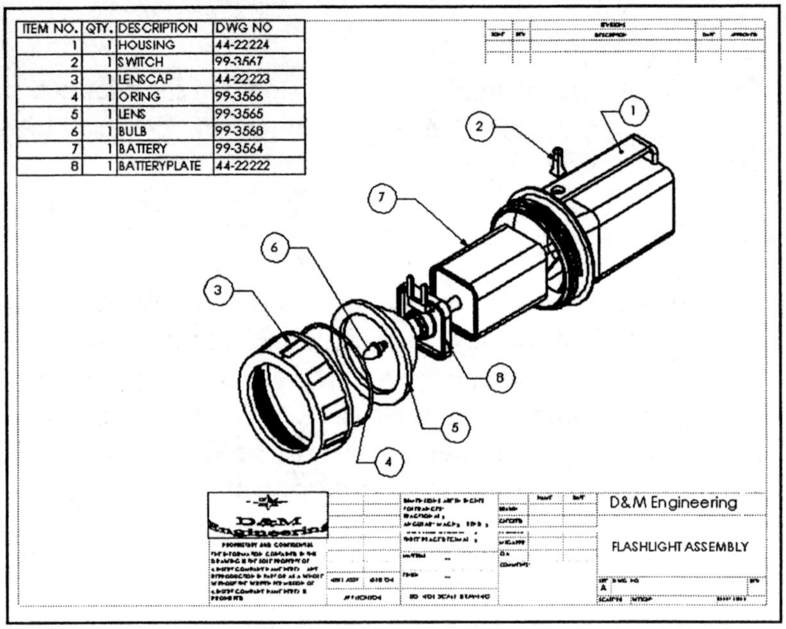

Figure 5.2

Note: Microsoft EXCEL 97 or a later version is required to create the Bill of Materials.

Both drawings utilize a custom sheet format containing a company logo, title block and sheet information.

There are two major design modes used to develop a drawing: Edit Sheet Format and Edit Sheet.

The Edit Sheet Format mode provides the ability to:

- Change the title block size and text headings.
- Incorporate a logo.
- Add drawing, design or company text.

The Edit Sheet mode provides the ability to:

- Add or modify views.
- Add or modify dimensions.
- Add or modify text.

Drawing Template and Sheet Format

The foundation of a SolidWorks drawing is the Drawing Template. Drawing size, drawing standards, company information, manufacturing and or assembly requirements, units and other properties are defined in the Drawing Template.

The Sheet Format is incorporated into the Drawing Template. The Sheet Format can contain border, title block and revision block information, company name and or logo information, Custom Properties and or SolidWorks Properties.

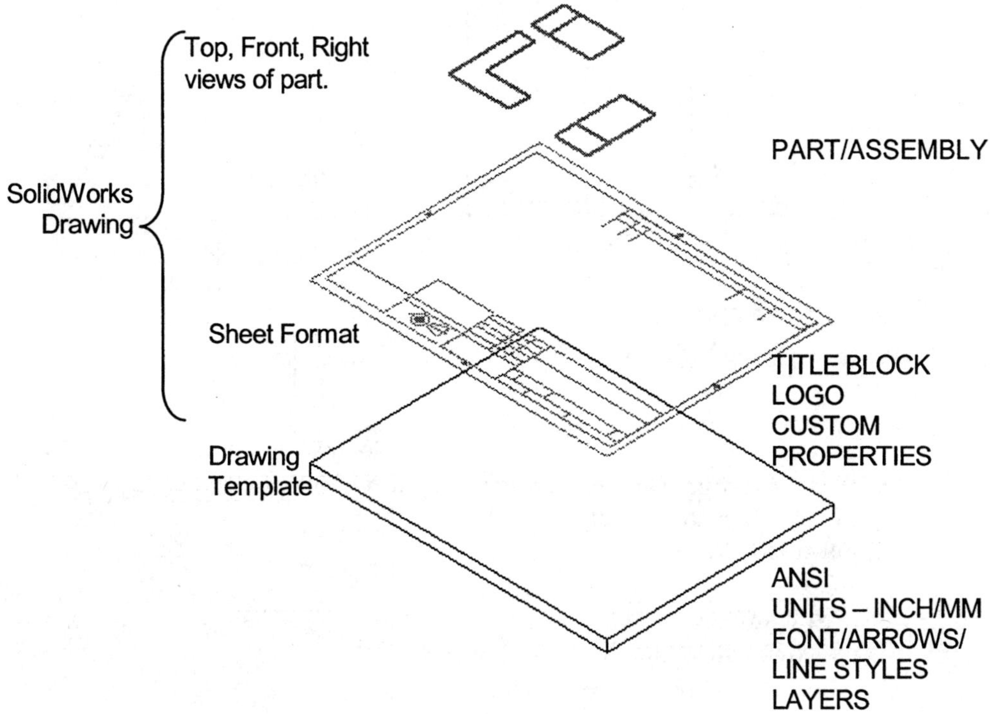

Views from the part or assembly are inserted into the SolidWorks Drawing.

There are various sheet format options in SolidWorks:

- Standard Sheet Format.
- Custom Sheet Format.
- No Sheet Format.

Create a custom Sheet Format and a custom Drawing Template.

Note: For companies supporting a First Angle projection scheme, views in this project are placed in various locations.

For non-ANSI dimension standards, the dimensioning techniques are the same, even if the displayed arrows and text size are different. For printers supporting millimeter paper sizes, select A4-Landscape, [297mm x 210mm]. Notes and dimensions for millimeter drawings are provided in brackets [x].

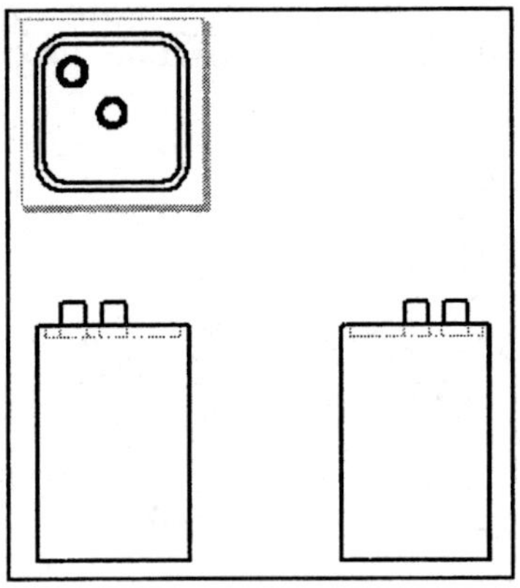

Third Angle Projection

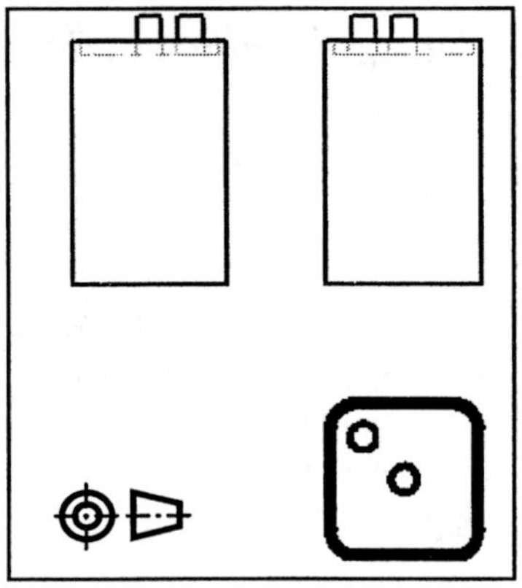

First Angle Projection

Create the BATTERY drawing.

1) Click **New**. Click **Drawing**. Click **OK**.

2) Select **A-Landscape**, **[A4 Landscape]** from the Standard Sheet Format drop down list. Click **OK**.

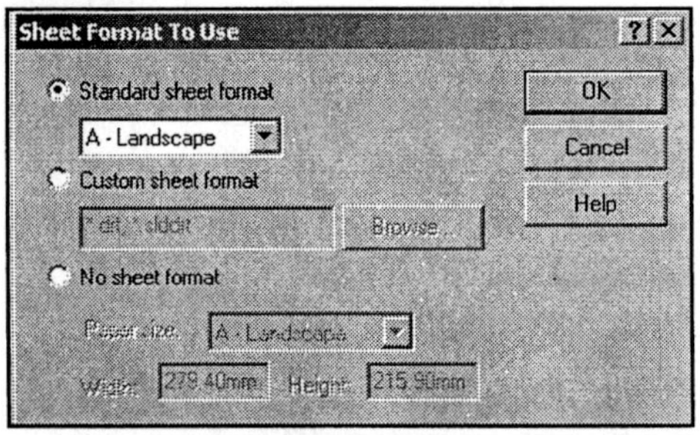

The A-Landscape sheet format is displayed in a new Graphics window. The sheet format border defines the drawing size, 11in. x 8.5in. Not: The A4-Landscape sheet format is [297mm x 210mm].

The Drawing and Line Format toolbars are displayed left of the Graphics window. The Drawing toolbar is displayed on the left side of the PropertyManager. The Title Block is in the lower left corner.

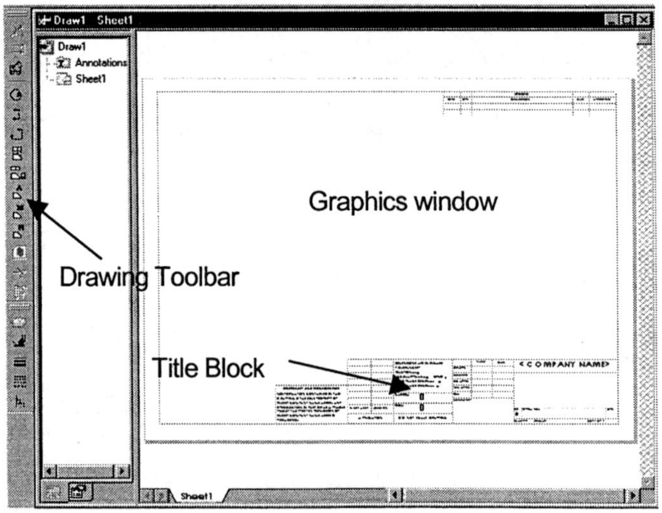

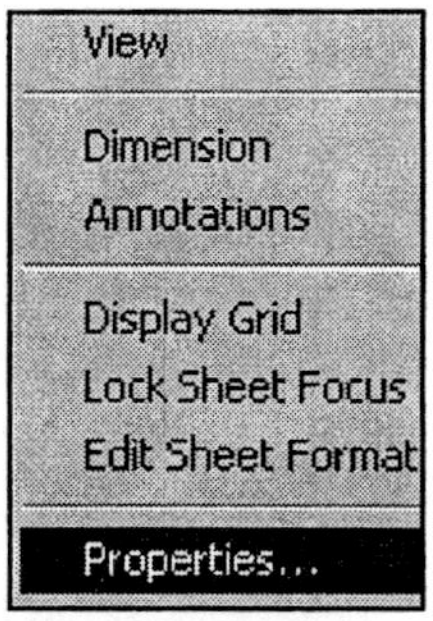

3) Right-click in the **Graphics window**. Click **Properties**. The Sheet Setup Properties are displayed.

4) Enter **1:2** for sheet Scale. Select **Third Angle** for Type of Projection. Click **OK**.

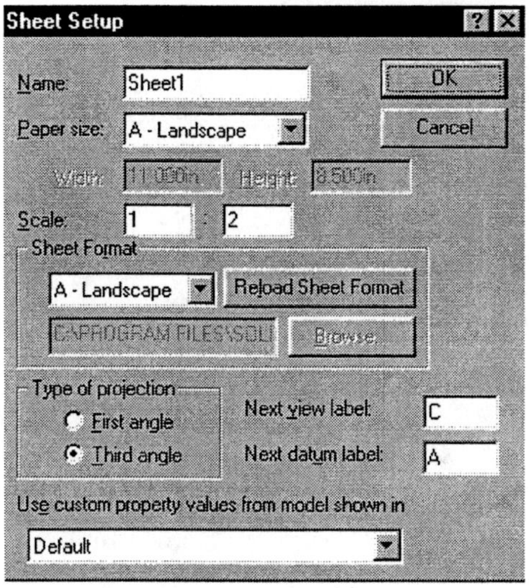

5) Set ANSI dimension standard. Click **Tools, Options, Document Properties.** Click **ANSI**. Click **Units, Inches, [Millimeters]**. Select **3**, **[2]** for the number of decimal places to display.

The Detailing options provide the ability to address: dimensioning standards, text style, center marks, witness lines, arrow styles, tolerance and precision.

There are numerous text styles and sizes available in SolidWorks. Companies develop drawing format standards and use specific text height for Metric and English drawings. Many engineering drawings use the following format:

- Font: Century Gothic – All capital letters.
- Text height: .125in. or 3mm for drawings up to B Size, 17in. x 22in.
- Text height: .156in. or 5mm for drawings larger than B Size, 17in. x 22in.
- Arrow heads: Solid filled with a 1:3 ratio of arrow width to arrow height.

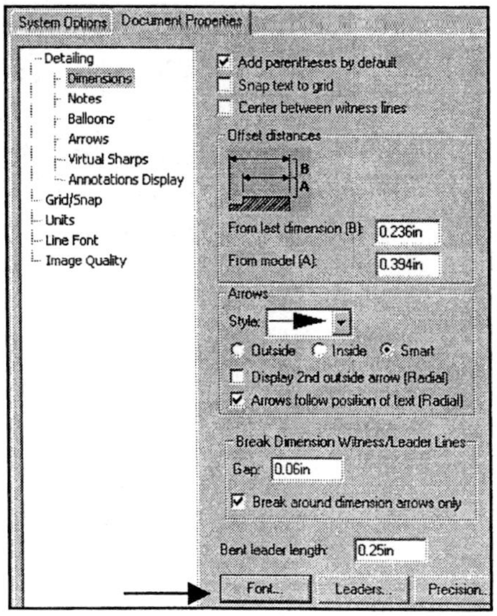

Set Dimensions Font.

6) Display the Detailing options dialog box. Click **Dimensions** from the Document Properties tab.

7) Change the default text height. Click the **Font** button from the Dimensions box. The Choose Font dialog box is displayed.

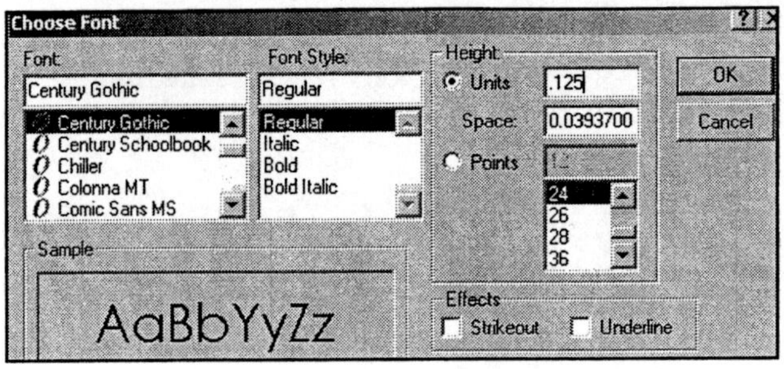

8) Set the dimension text. Click **Century Gothic** from the Font list box. Click the **Units** button. Enter **.125**, **[3.0]** for Height. Click **OK** from the Choose Font dialog box.

9) Click **OK** from the Document Properties dialog box.

Title Block

The title block contains vital part or assembly information. Each company may have a unique version of a title block. Most title blocks contain the following type of information:

Company Name/Logo:	Part number:
Part name:	Drawing number:
Drawing description:	Revision number:
Sheet number:	Material & finish:
Tolerance:	Drawing scale:
Sheet size:	Revision block:
CAD file name:	Engineering Change Orders:
Quantity required:	Drawn by:
Checked by:	Approved by:

The title block is located in the lower right hand corner of the drawing. Note: You cannot modify the Sheet Format text, lines or title block information unless you are in the Edit Sheet Format mode. You cannot insert views unless you are in the Edit Sheet mode.

Edit Sheet Format - Title Block.

10) Edit the company name. Right-click **Edit Sheet Format** from the Pop-up menu. The drawing title block lines turn blue.

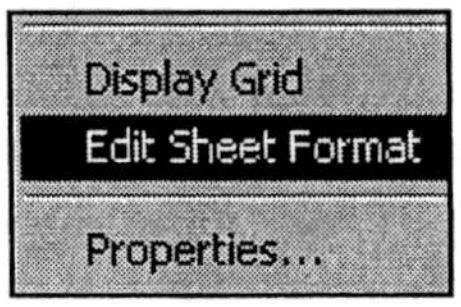

11) View the right side of the title block. Click **Zoom to Area** on the sheet format title block. Double-click on the **<COMPANY NAME>** text.

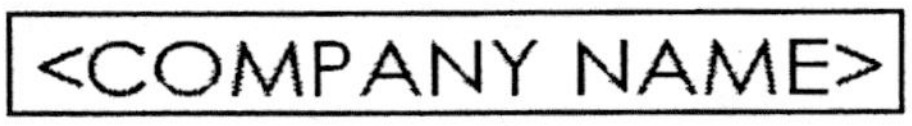

12) Enter a company name. Click a **position** inside the Note text box. Example: Enter **D & M Engineering, Inc.** Click **OK**.

13) Uncheck the **Use document font** check box. Click the **Font** button to change the font size. Accept the text. Click **OK**. The text is displayed in the title block.

Note: Use your own company name.

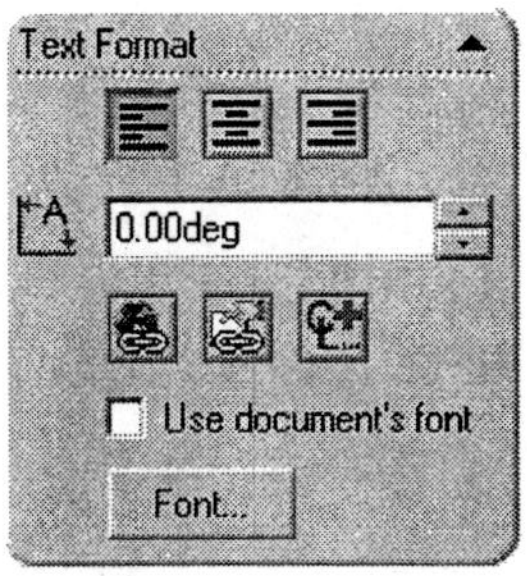

Company Logo

A company logo is normally located in the title block. Create a company logo by copying a picture file from Microsoft ClipArt using Microsoft Word. Copy/Paste the logo into the SolidWorks drawing.

Note: The following logo example was created in Microsoft Word 2000 using the COMPASS.wmf and WordArt text. Any ClipArt picture, scanned image or bitmap can be used.

Create a logo.

14) Create a New Microsoft Word Document. Click **New** from the Standard toolbar in MS Word. Click **ClipArt** from the Draw toolbar.

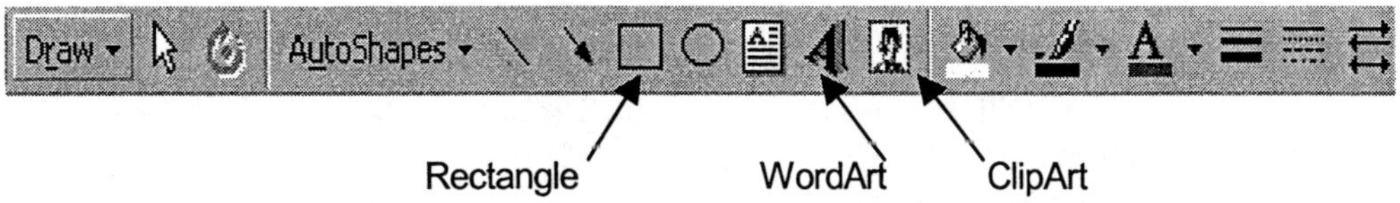

15) Double-click on the **COMPASS.wmf** file. Insert the **compass.wmf** picture file from the ClipArt menu. Click **Insert Clip**.

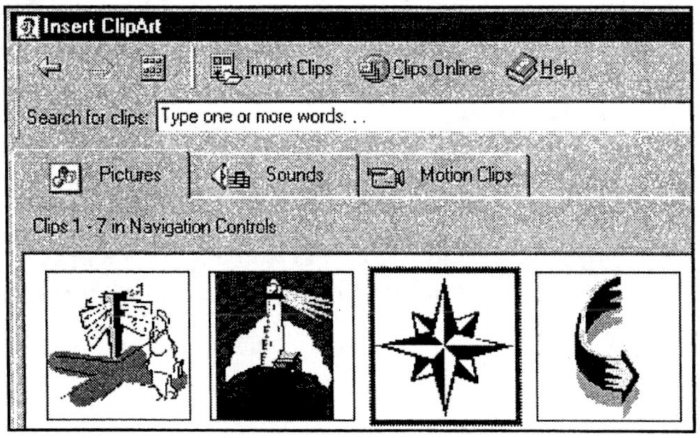

16) Redefine the picture layout. Right click **Format Picture, Layout, Square** to display the drag handles.

Add text to the logo picture. Click **Insert Word Art** from the Draw toolbar. Click a **WordArt** style. Click **OK**.

17) Enter **D & M Engineering** in the text box. Click **24** from the Size drop down list. Click **OK**.

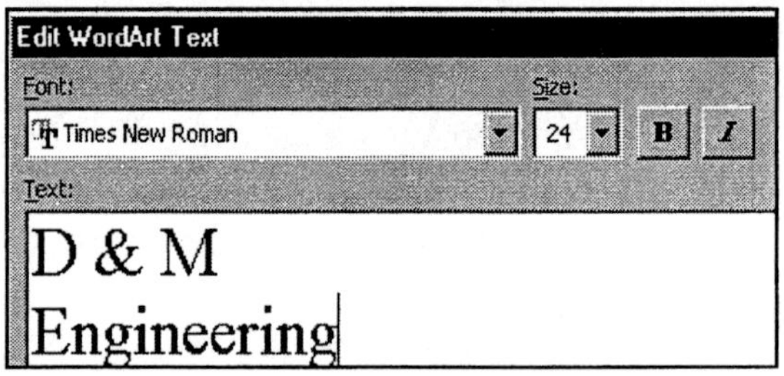

18) Drag the **Word Art text** under the picture. Size the **Word Art text** by dragging the picture handles.

19) Select the geometry. Press the **Shift** key. Select the **compass picture**. Select the **WordArt text**. Right click **Group**. Click **Copy**.

20) The logo is placed into the Clipboard. Close Microsoft Word. Click **File, Exit**.

21) Place the logo into the title block. Click inside the SolidWorks **Graphics window**, on the left-hand corner of the title block.

22) Click **Edit**, **Paste**. Size the **logo** to the SolidWorks title block by dragging the picture handles.

23) Fit the drawing to the screen. Press the **f** key.

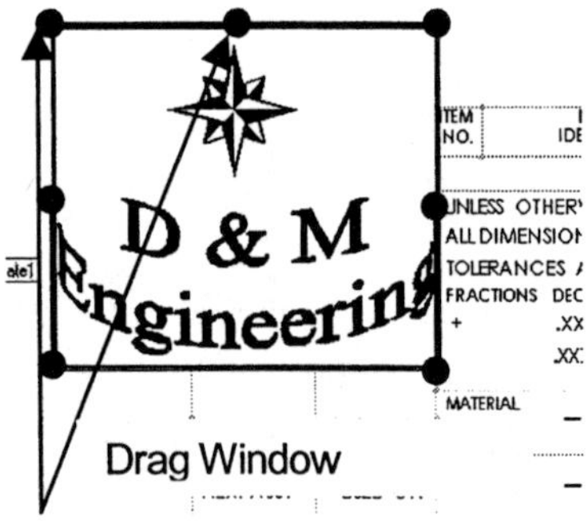

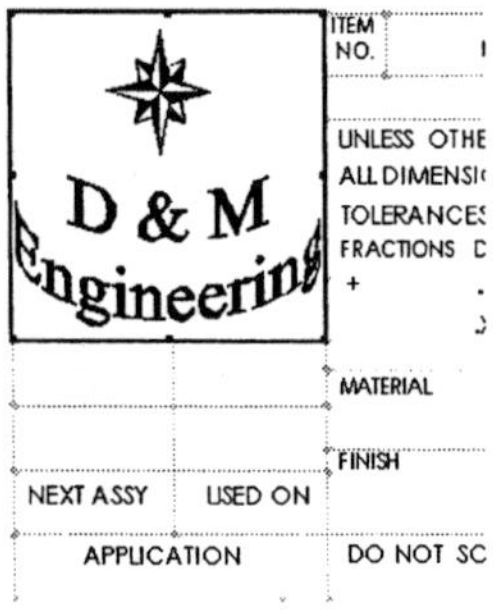

24) Return to Edit Sheet. Right-click in the **Graphics window**. Click **Edit Sheet**. The title block is displayed in gray.

Save the Sheet Format.

25) Click **File, Save Sheet Format**. The Save Sheet Format dialog box appears.

Save the Sheet Format as a custom Sheet Format. The custom Sheet Format is used for all drawings in this Project. Combine the Sheet Format with the standard Drawing Template to create a custom Drawing Template.

26) Click the **Custom Sheet Format** button. Click the **Browse** button from the Save Sheet Format dialog box.

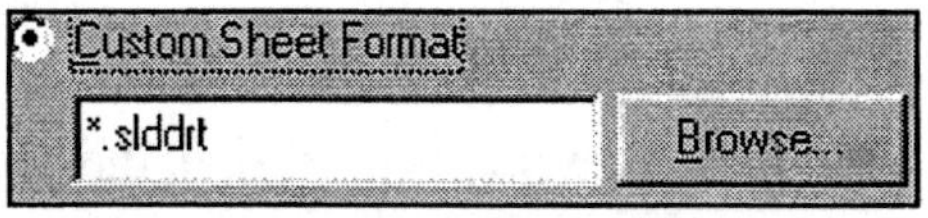

27) The default Sheet Format folder is called data. The file extension for Sheet Format is .slddrt. Enter File name. Enter **CUSTOM-A**, [**CUSTOM-A4**]. Click **Save** from the Save Sheet Format dialog box. The Sheet Format is used with a Drawing Template. Click **OK**.

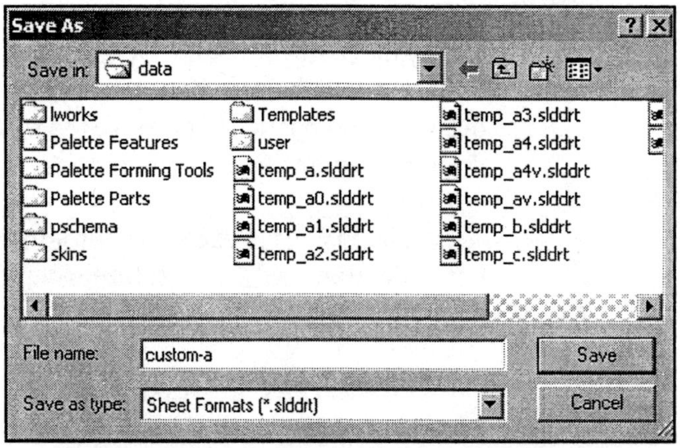

28) Save the Drawing Template. Click **File, Save As**. Click **Drawing Template (*.drwdot)** from the Save as type list. Enter **CUSTOM-A-DRAWINGTEMPLATE, [CUSTOM-A4-DRAWINGTEMPLATE]** for File name. Click **Save**.

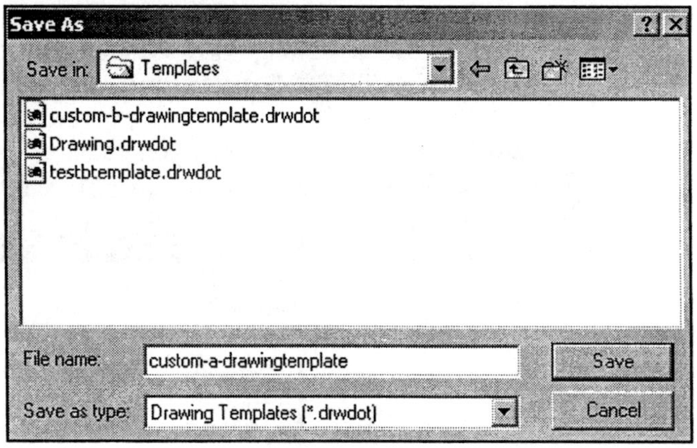

29) Close All parts and drawings. Click **Windows, Close All**.

30) Utilize the new custom Drawing Template. Click **File, New**. Click **CUSTOM-A-DRAWINGTEMPLATE, [CUSTOM-A4-DRAWINGTEMPLATE]**. The custom Sheet Format 1 is contained in the Drawing Template. Click **OK**.

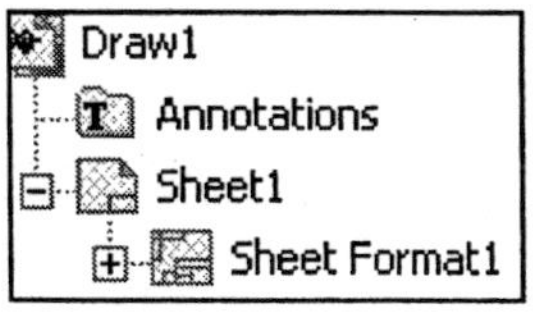

31) Save the drawing. Click **Save**. Select **My Documents** file folder in the Save In text box. Enter **BATTERY** for filename. Click **Save**.

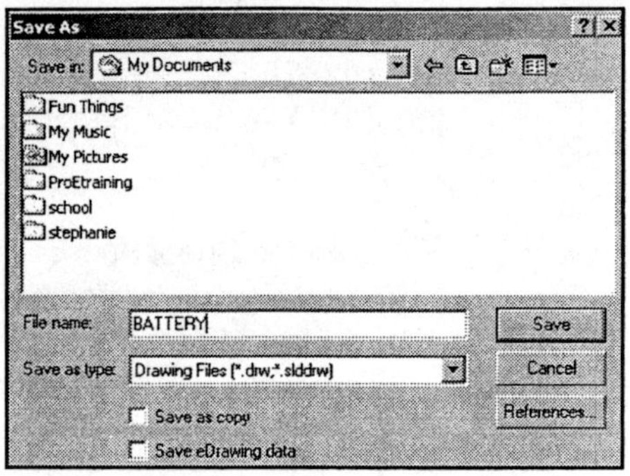

Create the Drawing from a Part

A drawing contains part views, geometric dimensioning and tolerances, notes and other related design information. When a part is modified, the drawing automatically updates. When a dimension in the drawing is modified, the part is automatically updated. The following tasks are recommended before starting the BATTERY drawing:

- Verify the part: Click File, Open. Select BATTERY. Note: This is the part used to create the first drawing. The part shares the same database as the drawing. Do not delete or move the part. The drawing will not be valid. The drawing requires the associated part.

- View dimensions in each part. From the FeatureManager, slide the Rollback bar to the Extruded-Base feature. Step through each feature of the part and review all dimensions. Modify features if required.

32) Open the BATTERY part. Click **Open**. Enter **BATTERY.sldprt.**

Insert three standard views.

33) Display the drawing. Click **Window** from the Main menu. Click **Tile Horizontal**. Click **BATTERY-Sheet1** Graphics window.

34) Drag the **BATTERY Part** icon from the Feature Manager into the center of the drawing Graphics window. Click the **Maximize** icon for the BATTERY-Sheet1 drawing window.

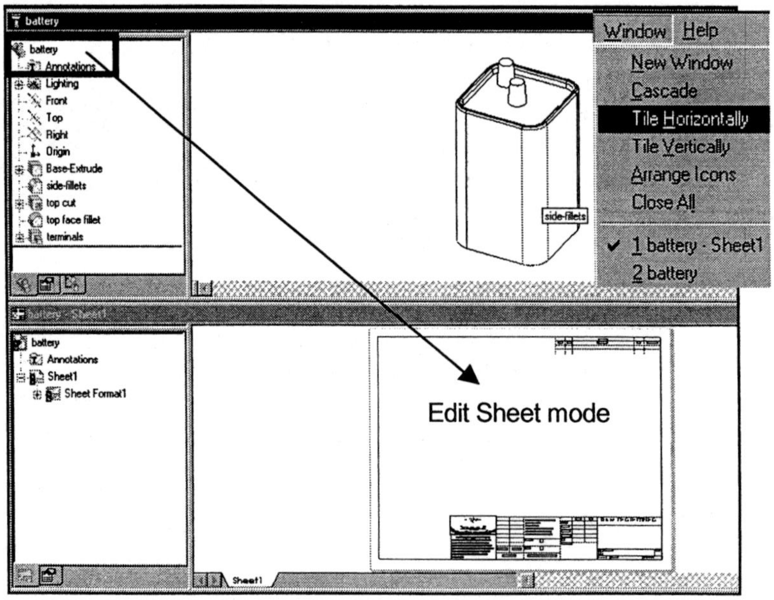

You cannot insert a part into a drawing when Edit Sheet Format is selected. Edit Sheet Format displays all lines in blue. The 3 standard drawing views flash once and then disappear. Right-click in the Drawing Graphics window. Click Edit Sheet.

The three standard views are displayed in the Graphic window. It may take a few seconds to create the standard views depending upon the configuration of your computer. The drawing views may be positioned too close to the title block. Views can be added, deleted and moved. The default Display Model is Shaded.

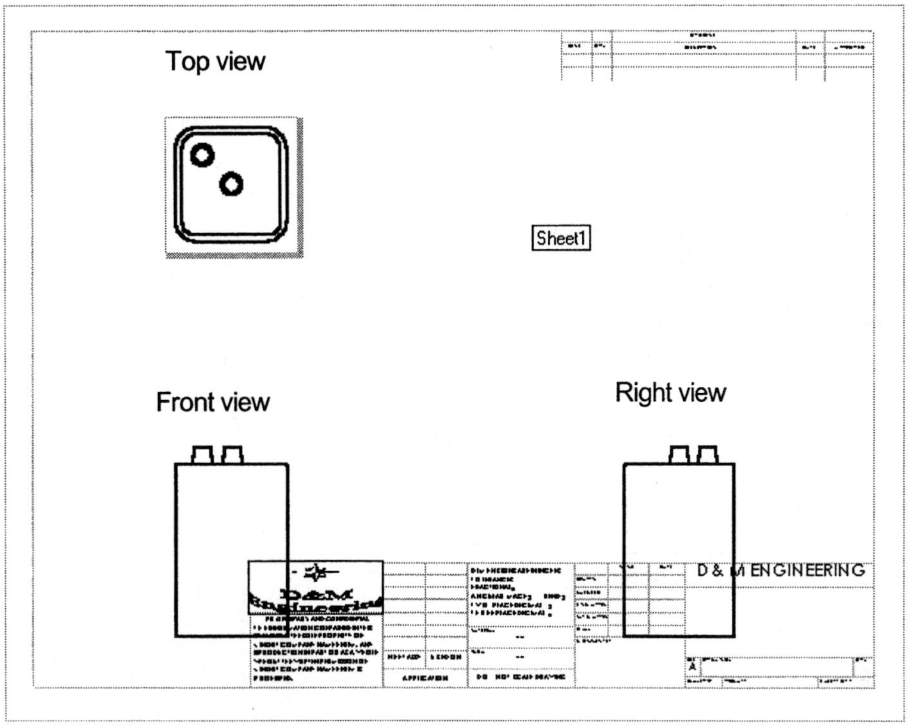

The mouse pointer provides feedback in both the Drawing Sheet  and Drawing View  modes. The mouse pointer displays the Drawing Sheet  icon when the Sheet properties and commands are executed. The mouse pointer displays the Drawing View  icon when the View properties and commands are executed.

Each drawing has a unique file name. In SolidWorks, drawing file names end with an .slddrw suffix. Part file names end with an .sldprt suffix.

A drawing or part file can have the same prefix. A drawing or part file cannot have the same suffix. Example: Drawing file name: BATTERY.slddrw. Part file name: BATTERY.sldprt.

Move Views

Reposition the view on a drawing. Provide approximately 1in. - 2in., (25mm – 50mm) between each view for dimension placement. Views are displayed in Hidden Lines Removed for clarity.

Hide the Origins.

35) Click **View**, **Origins** from the Main toolbar.

Move Views.

36) Click the view boundary of **Drawing View1** (Front). The mouse pointer displays the Drawing View  icon. The view boundary is displayed in green.

37) Position the **mouse pointer** on the edge of the view until the Drawing Move View  icon is displayed.

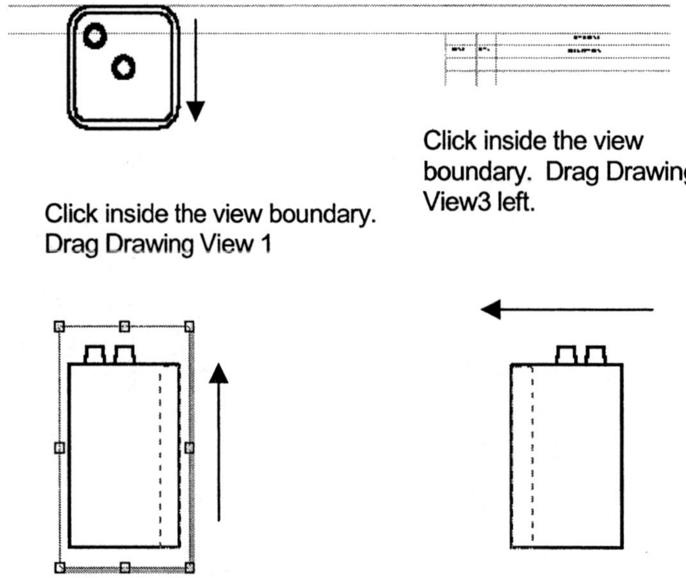

38) Drag **Drawing View1** in an upward vertical direction. Drawing View2 (Top) and Drawing View3 (Right) move aligned to Drawing View1 (Front).

39) Hidden lines are displayed with thin dashed lines. Click **Hidden in Gray**  to display the hidden lines.

40) Move Drawing View3 in a right to left direction. Click the **Drawing View3** view boundary. Position the **mouse pointer** on the edge of the view until the Drawing Move View  icon is displayed. Click **Hidden in Gray** .

41) Drag **Drawing View3** in a right to left direction towards Drawing View1.

42) Move Drawing View2 in a downward vertical direction. Click the **view boundary**. Drag **Drawing View2** in a downward direction towards Drawing View1. Click **Hidden in Gray** .

Named View

Named views display the part or assembly in various orientations. Add named views to the drawing at anytime.

Add an Isometric view to the drawing.

43) Click **Drawing View1** (Front) **view boundary.** The view boarder is displayed in green. Click **Named View**

 from the Drawing toolbar. The named views for the BATTERY are displayed. Click **Isometric** from the View Orientation text box. Isometric view is placed on the mouse pointer.

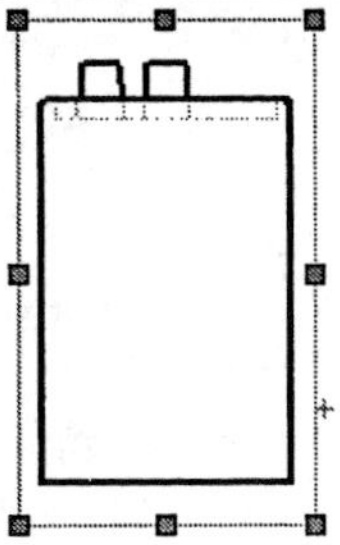

44) Position the Isometric view. Click the **Graphics window** to the right of Drawing View2.

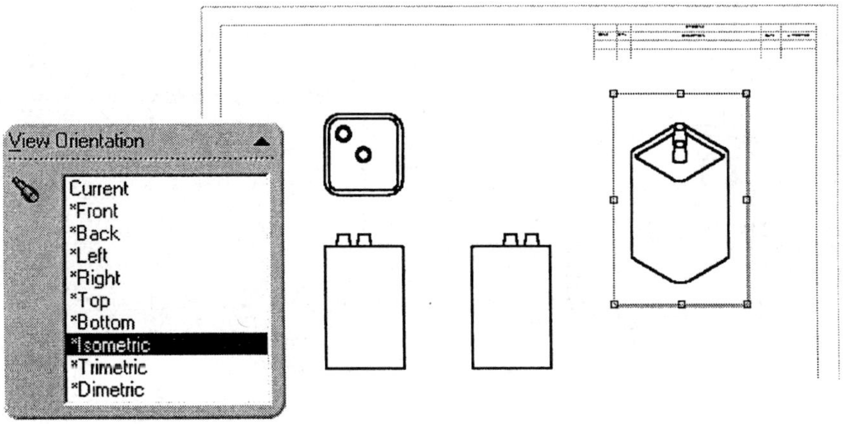

45) Save the BATTERY drawing. Click **Save.**

Detail View

A Detailed view enlarges an area of an existing view. You need to specify location, shape and scale. Create a Detail view using a 1:1 scale.

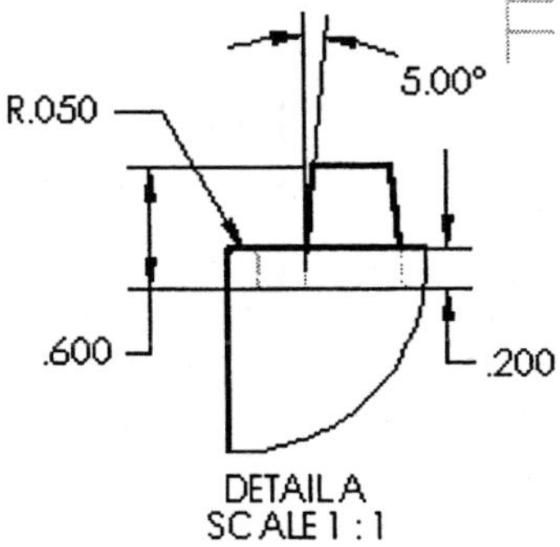

Add a Detail view to the drawing.

46) Click **Detail View**  from the Drawing toolbar. The circle sketch tool is selected. Sketch a **circle** with the center point located between the two terminals in Drawing View1(Front). Click the Detail view **location** to the right of the Drawing View2(Top). Change the view Scale. Select the **Custom Scale** check box. Enter **1:1** in the Custom Scale text box. Click **OK** from the PropertyManager.

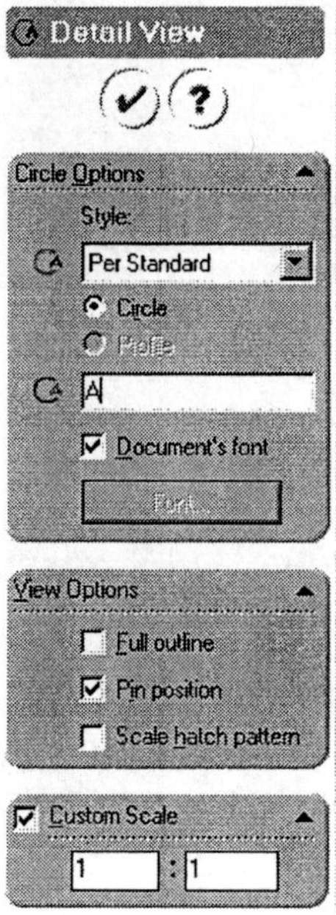

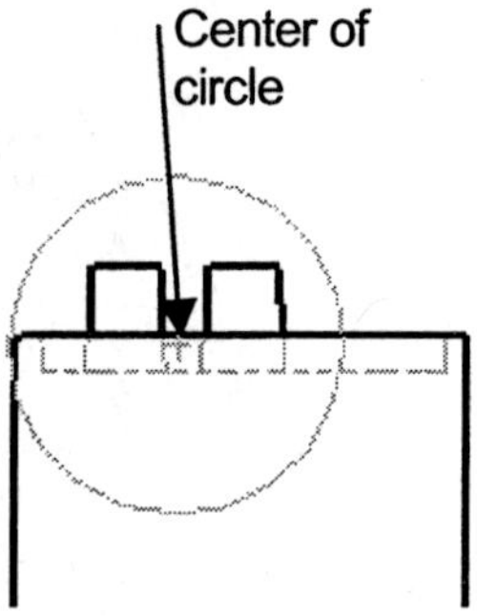

47) Fit the drawing to the Graphics window. Press the **f** key.

48) Save the BATTERY drawing. Click **Save**.

The Drawing Views are complete. Move the views to allow for spacing of the dimensions and notes.

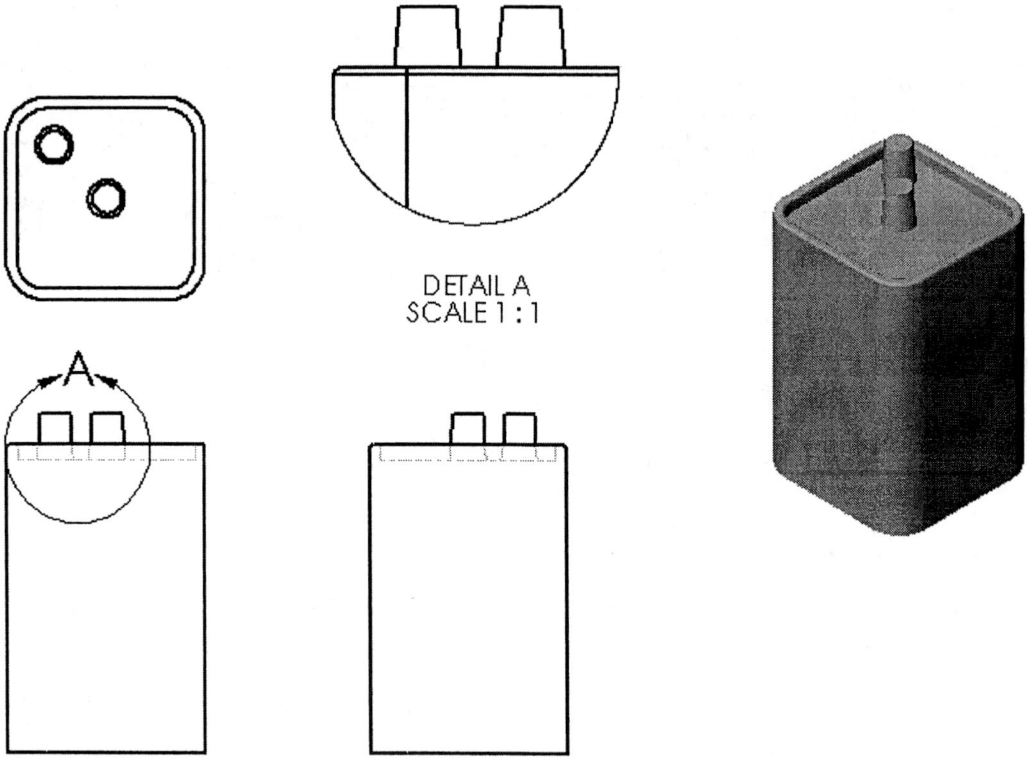

View Display

Drawing views are displayed in Wireframe, Hidden in Gray, Hidden Lines Removed and Shaded 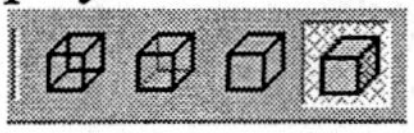 modes.

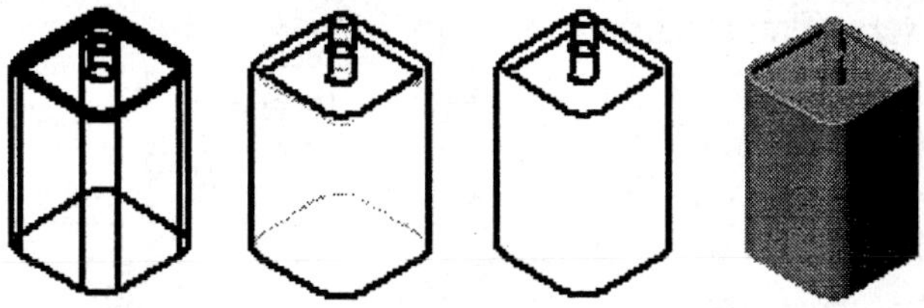

Tangent edges are displayed either in the Visible, With Font or Removed modes.

Display hidden lines, profile lines and tangent edges in various view modes.

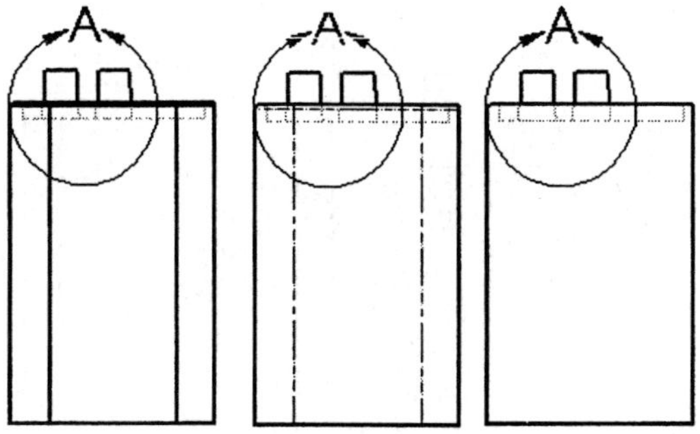

49) Display hidden lines in the Detail view. Click **Detail view boundary**. Click **Wire frame**  from the View toolbar.

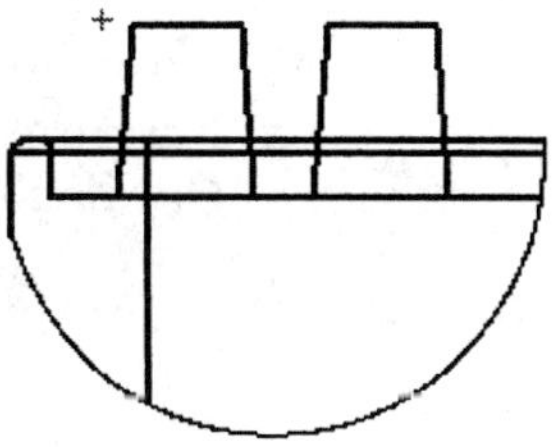

50) Remove Tangent Edges. Right-click **Drawing View1**(Front) from the FeatureManager. Click **Tangent Edge**. Click **Tangent Edges Removed**.

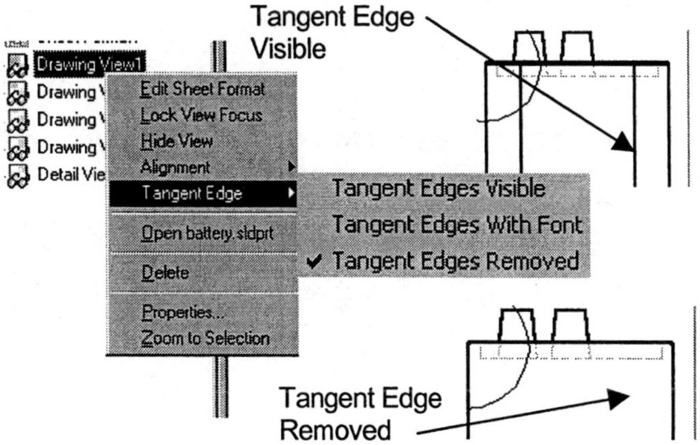

Insert Dimensions from the Part

Dimensions created for each feature are inserted into the drawing. Display the first dimensions in the Front view.

Illustrations for this project are provided in inches and millimeters. The Detailing option, Dual Dimensions Display produces both inches and millimeters for each dimension. The primary units are set to inches and the secondary units are set to millimeters.

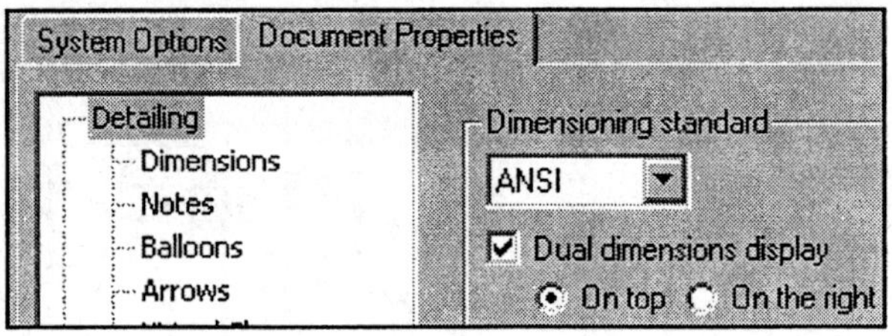

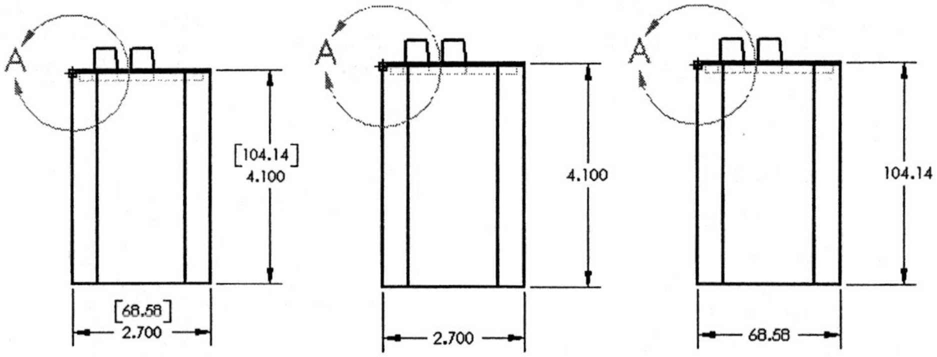

Dual Dimensions Display | Units = Inches | Units = Millimeters

Insert dimensions.

51) Insert dimensions into the Drawing View1(Front) view. Click **Drawing View1 view boundary**.

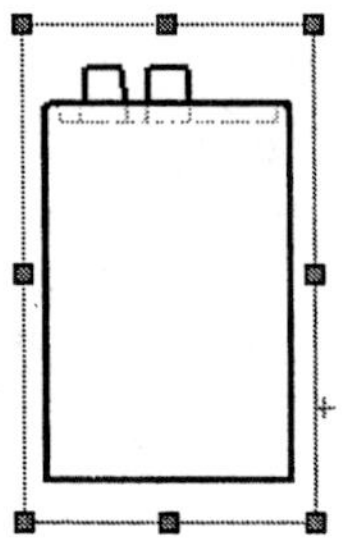

52) Click **Insert Model Items**  from the Annotations toolbar. The Insert Model Items dialog box is displayed.

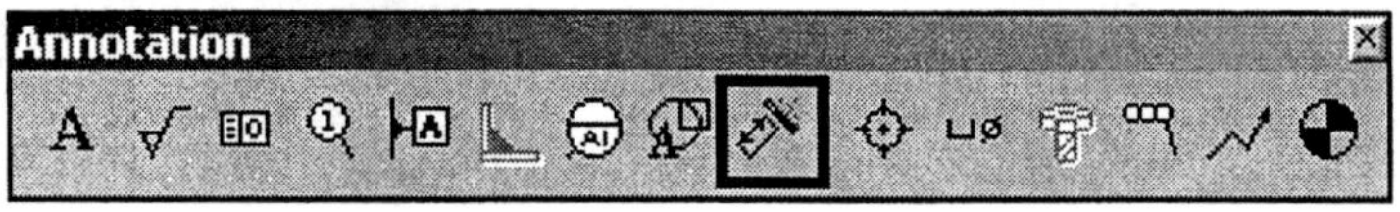

53) Uncheck **Include items from hidden features**. Uncheck **Import items into all views**. Click **Drawing View1** in the Import into views list box. Click **OK**.

Note: Dimensions may appear in different locations.

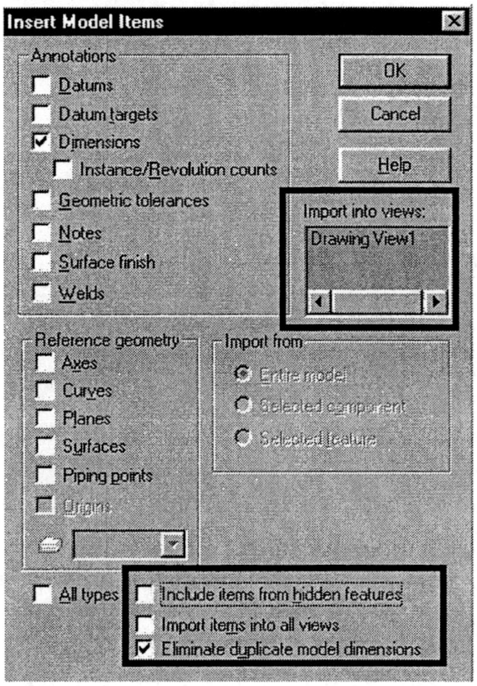

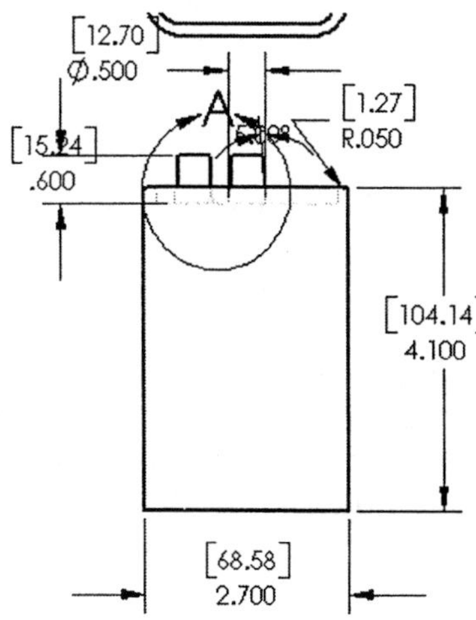

The dimensions are located too far from the profile lines and on top of the profile. You will move them later in this section.

Drawing dimension location is dependent upon:

- Feature dimension creation.
- Selected drawing views.

Move Dimensions in the Same View

Move dimensions within the same view. Use the mouse pointer to drag dimensions and leader lines to a new location. Leader lines reference the size of the profile. A gap must exist between the profile lines and the leader lines. Shorten the leader lines to maintain a drawing standard. Use the Arrow buttons in the PropertyManager to flip dimension arrows.

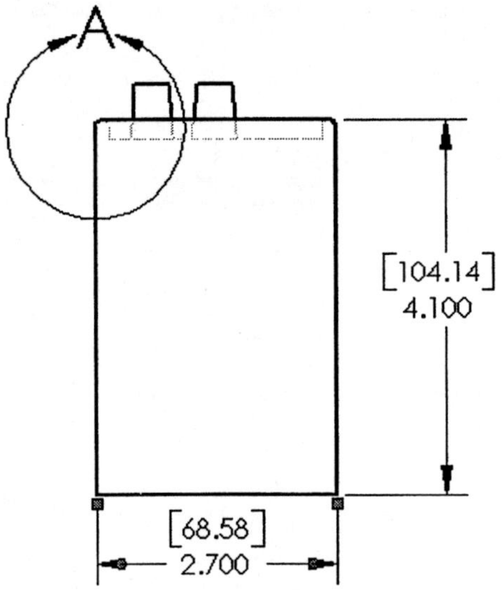

Move linear dimensions in Drawing View1.

54) Click the vertical dimension text **2.700**, **[68.58]**. A green box is drawn around the text. The mouse pointer displays the Linear Dimension icon. Drag the **dimension text** to approximately 10mm below the horizontal line of the BATTERY. Center the **text** between the two arrowheads. Click the **circular green point** on the arrowhead to flip the arrow head to the inside.

55) Click the vertical dimension text **4.100**, **[104.14]**. Drag the **text** to the right of the BATTERY.

56) Move the leader line endpoints. The end points of the leader line are displayed as green squares. Drag each **square green end point** to the left until it is off the profile line.

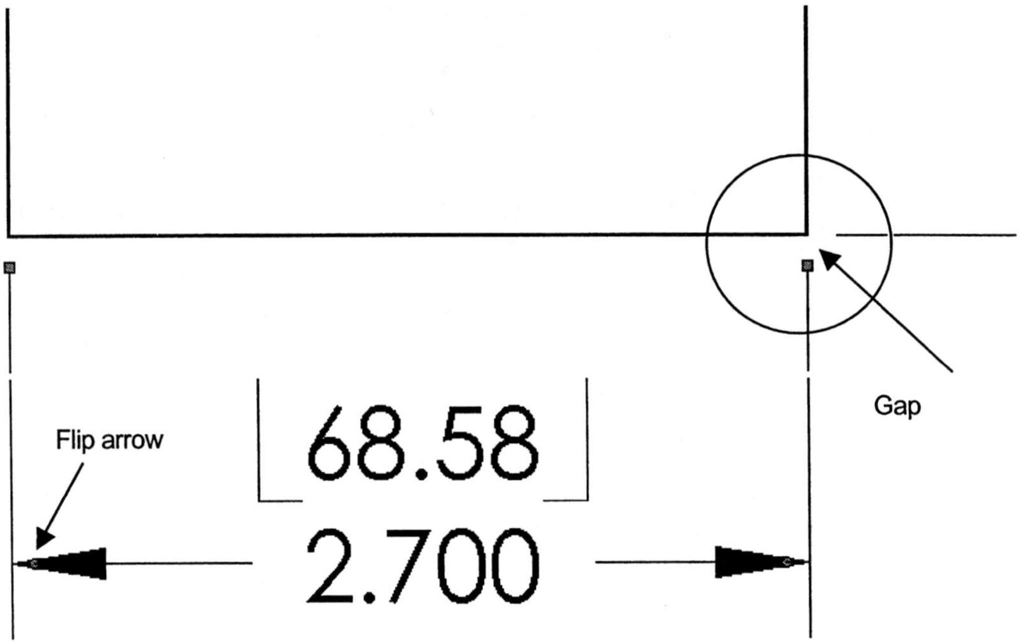

A gap exists between the profile line and the leader lines.

Move Dimensions to a Different View

Move the dimensions from Drawing View1 to Drawing View4 (Detail View) and Drawing View2 (Top). Note: When moving dimensions from one view to another, drag only the dimension text. Release the dimension text inside the new drawing view boundary.

Move dimensions to a different drawing view.

57) Hold the **Shift** key down. Click the vertical dimension **.600**, **[15.24]**. Drag the **dimension text** from Drawing View1 to Drawing View4. Release the **mouse button** and the **Shift** key when the pointer is inside the Drawing View4 boundary.

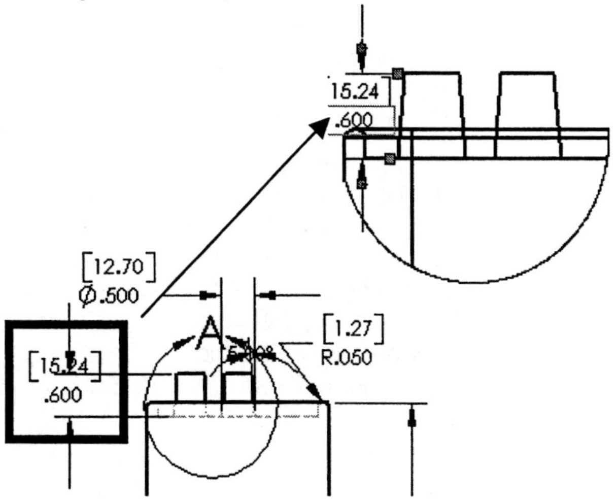

58) Drag the **dimension text** to the right of the profile lines. Drag the **extension lines** off the profile.

59) Move the draft angle dimension **5.00°** into Drawing View4. Click **5.00°**. Drag the **text** outward approximately 10mm above the profile line.

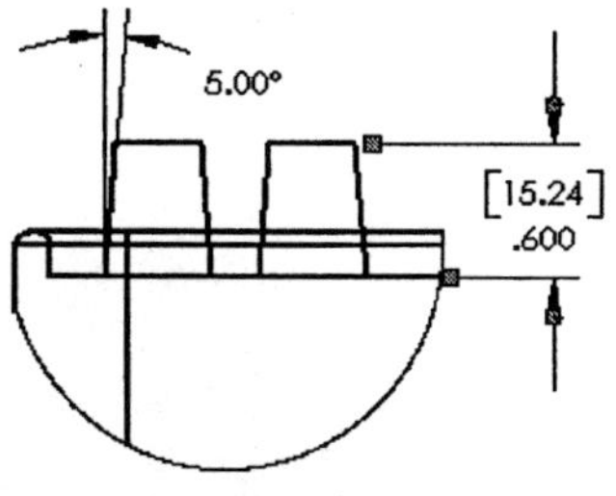

DETAIL A
SCALE 1 : 1

60) Drag the vertical dimension **.600**, **[15.24]** to the right to avoid crossing the leader lines. Drag the leader line **green end points** off the profile lines.

61) Position the diameter text. Click the diameter dimension **Ø.500, [Ø12.70]** in Drawing View1. Drag the **dimension text** from Drawing View1 to Drawing View2. Position the dimensions off the top profile. Drag the **text** downward.

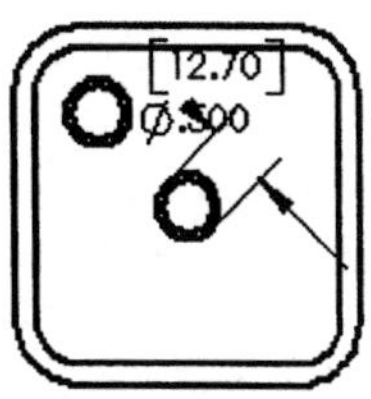

62) Position the Detail label A text. Click **A** and drag upward approximately 10mm from the profile line.

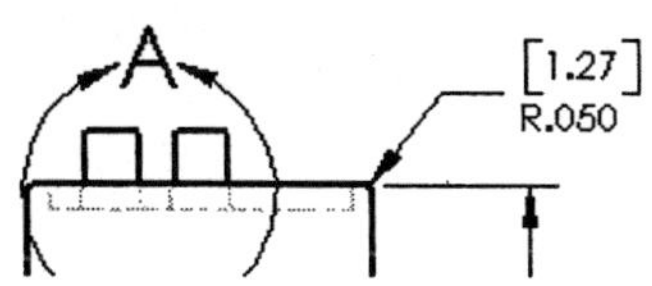

63) Position the Radial text in Drawing View1. Click **R.050**, **[R1.27]** text. Drag the **text** outward approximately 10mm from the profile line. Flip the **arrowhead** if required.

64) Insert remaining dimension into all views. Click **Insert** from the Main menu. Click **Model Items.** The Insert Model Items dialog box is displayed. Uncheck **Include items from hidden features.** Check Import items into all views. Click **OK.**

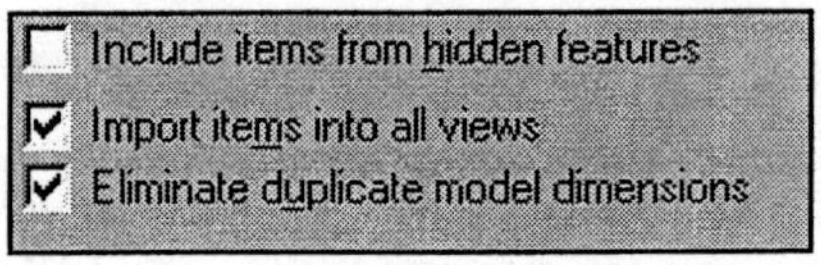

Drawing View2 displays crowded dimensions. Move the overall dimensions. Place dimensions in the view where they display the most detail.

Position the dimensions in Drawing View2.

65) Right-click on the dimension, **Ø.500,** **[Ø12.70]**. Click **Display Options**. Check the **Display as Diameter.** Click **OK.** Drag the **diameter dimension** to the lower right corner of the view. Drag the **other dimensions** outside the profile lines of Drawing View2. Drag the 45.00° **extension line** to the center circle.

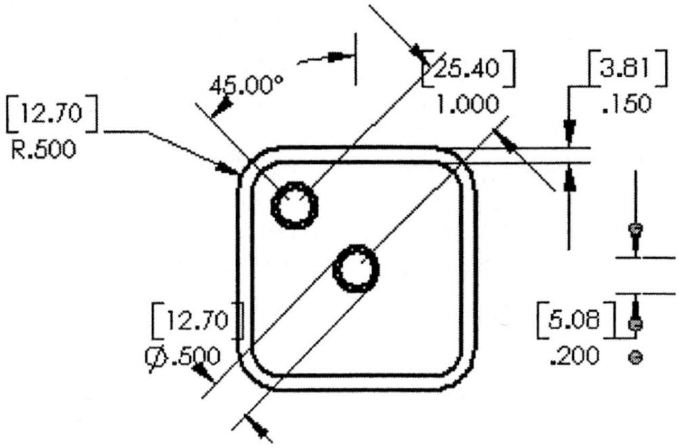

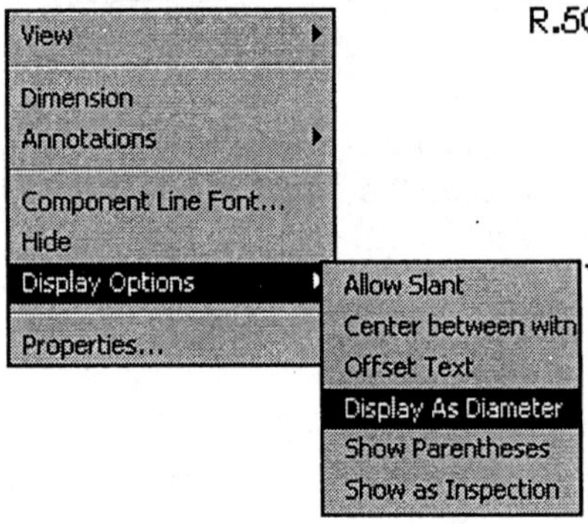

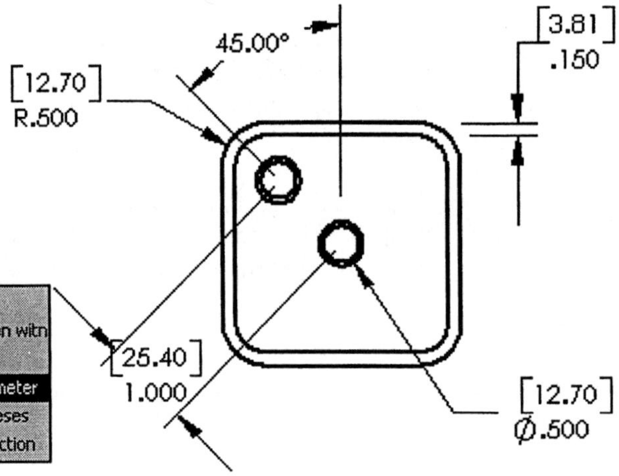

Modify the diameter text.

66) There are two battery terminals. Click the **Ø.500,** **[Ø12.70]** text. Enter **2 PLACES** in the text box below. Click **OK.**

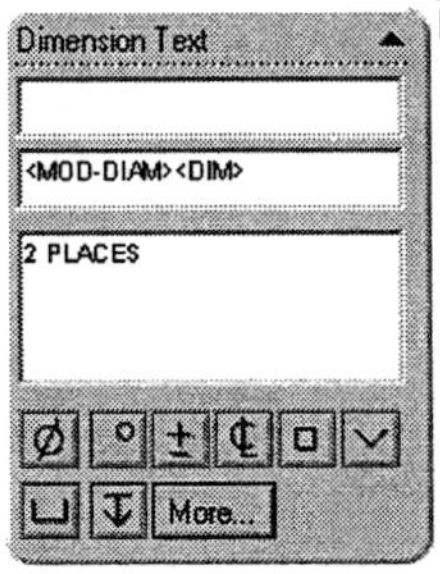

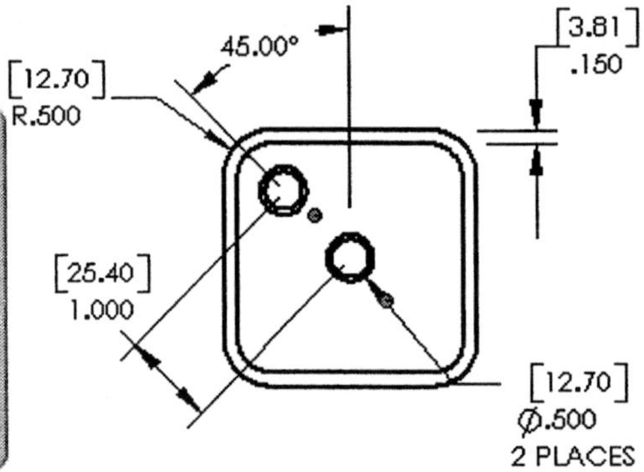

Move dimensions in the Detail view.

67) Drag the **.200**, **[5.08]** dimension to the left of the profile.

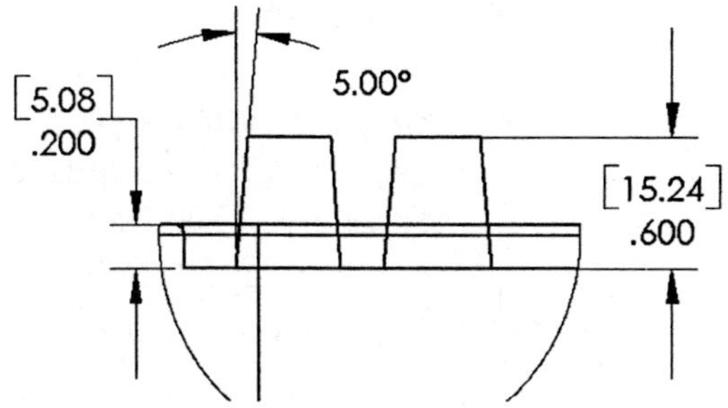

68) Notes: Center Marks, Hole Callouts and other Dimension Annotations are accessed through the Annotations toolbar menu. Right-click in the gray **toolbar region**. Click **Annotations**.

Create Center Marks

Centerlines are composed of alternating long and short dash lines. The lines identify the center of a circle, axes or cylindrical geometry. The battery terminals in Drawing View2 require Center Marks. Center Marks represents two perpendicular intersecting centerlines.

Create the Center Marks.

69) Place Center Marks on the hole in Drawing View2 (Top). Click **Center Mark** from the Annotation menu. The mouse pointer displays the Center Mark icon. Click the **circumference** of the center Terminal.

Resize the Center Marks.

70) The Center Marks are too large. Uncheck the **Use document's default** box. Enter **.05**, **[1]** for Mark Size.

71) Click the **circumference** of the second Terminal. Click **OK**. The new Center Marks are displayed.

72) A gap is required between the Center Mark and the end points of the leader lines. Drag the **leader lines** from the linear dimensions off the Center Marks.

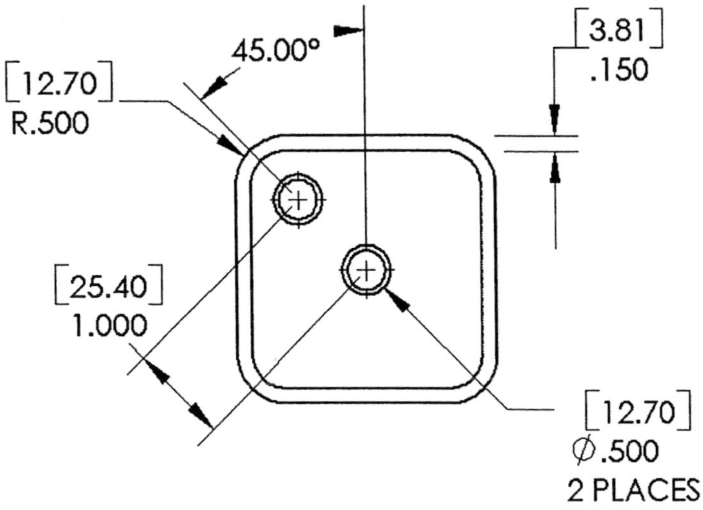

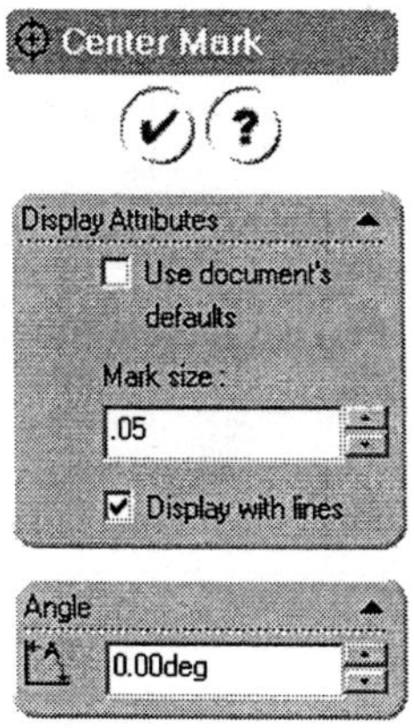

Trailing Zero Display

Your company decides to contract a manufacturer in Korea to procure the battery. The drawing dimensions must be supplied in millimeters. Modify the Document Property settings. The ANSI trailing zero display standard is different for inch and millimeter drawings. Trailing Zero Display is set to the Smart option for inch dimensions. Trailing Zero Display is set to Removed option for millimeter dimensions.

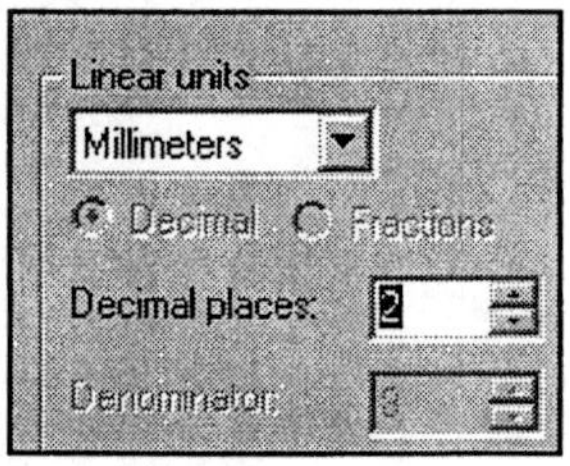

Modify the Document Properties.

73) Click **Tools, Options, Document Properties**. Click **Units**. Select **Millimeter** for primary units. Select **2** for number of decimal places to display.

74) Select **Remove** for Trailing zeros. Uncheck **Dual dimension display**.

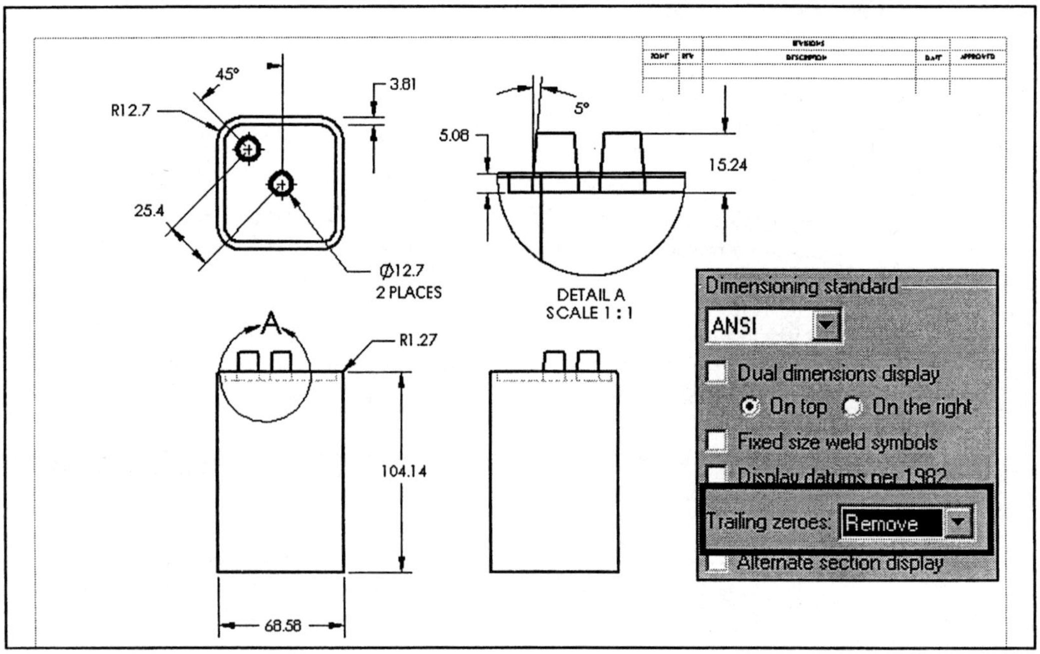

General Notes

Plan ahead for general drawing notes. Notes provide relative part or assembly information. Example: Material type, material finish, special manufacturing procedure or considerations, preferred supplier, etc.

Below are a few guidelines when creating general notes:

- Use capitol letters.
- Use left text justification.
- Font size should be the same as the dimension text.

Create parametric notes by selecting dimensions in the drawing. Example: Specify the Fillet Radius of the BATTERY as a note in the drawing. If the Radius is modified, the corresponding note is also modified.

Create general notes in the Revision section of the drawing. The documentation control manager issues a revision number when a drawing is initiated or changed. Use the Edit Sheet Format mode. Enter the Revision, Description and Date. Align the text.

Hide superfluous feature dimensions. Do not delete the dimensions. Recall hidden dimension with the View, Show Hidden command.

Create a Parametric drawing note.

75) Click **Note** from the Annotations toolbar. Click the **start point** in the lower left corner of the Graphics window, away from the title block.

76) Type two lines of notes in the Note text box. For line 1: Enter **ALL ROUNDS.** Press the **Space** key. Select the **R1.27** text in Drawing View1. The variable name "D1@Top Face Fillet@battery@DrawingView1" is added to the text box. Press the **Space** key. Enter **MM.**

77) For line 2: Enter **UNLESS OTHERWISE NOTED.** on line 2. Display the Note. Click **OK.**

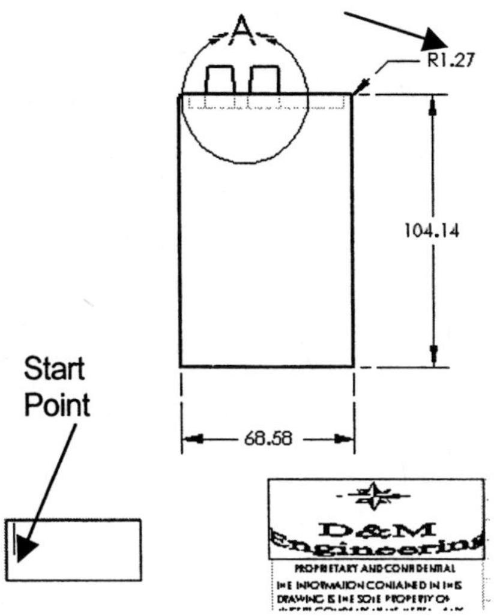

ALL ROUNDS "D1@Top Face Fillet@BATTERY-1@Drawing View1" MM
UNLESS OTHERWISE NOTED.

The parametric note specifies that the radius of all Rounds is 1.27 MM.

Do not double dimension a drawing with a note and the corresponding dimension. Hide the radial dimension.

78) Remove the R1.27 dimension. Right-click the **R1.27** dimension in Drawing View1. Click **Hide.**

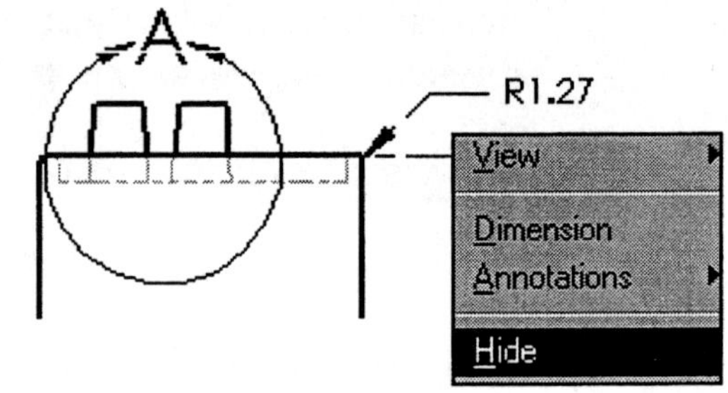

79) Modify the parametric note. Right-click **Open BATTERY.sldprt** in Drawing View1. The BATTERY part is displayed. Click the **Top Face Fillet radius.** Enter **.060**, [**1.52**]. Click **Rebuild.** Click **Save.**

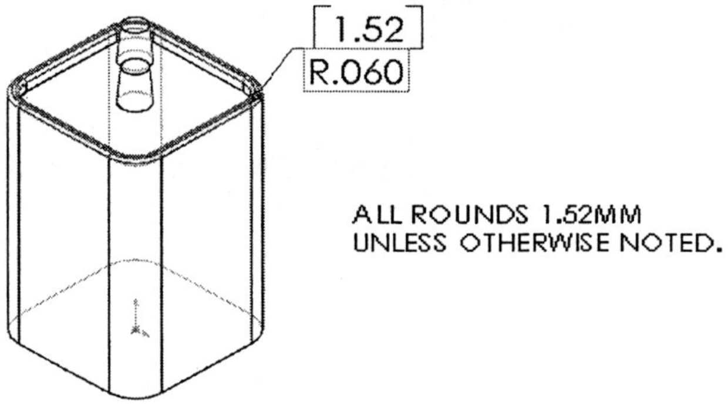

80) Right-click on the **BATTERY part icon.** Click **Open Drawing.** The BATTERY drawing is displayed. The parametric note, ALL ROUNDS 1.52 MM UNLESS OTHERWISE NOTED is displayed. The updated views reflect the new Fillet radius.

81) Save the BATTERY part and BATTERY drawing. Click **Save.**

Note: Unit conversion from inch to millimeters is automatically calculated when the part units are set to inches and the drawing units are set to millimeters.

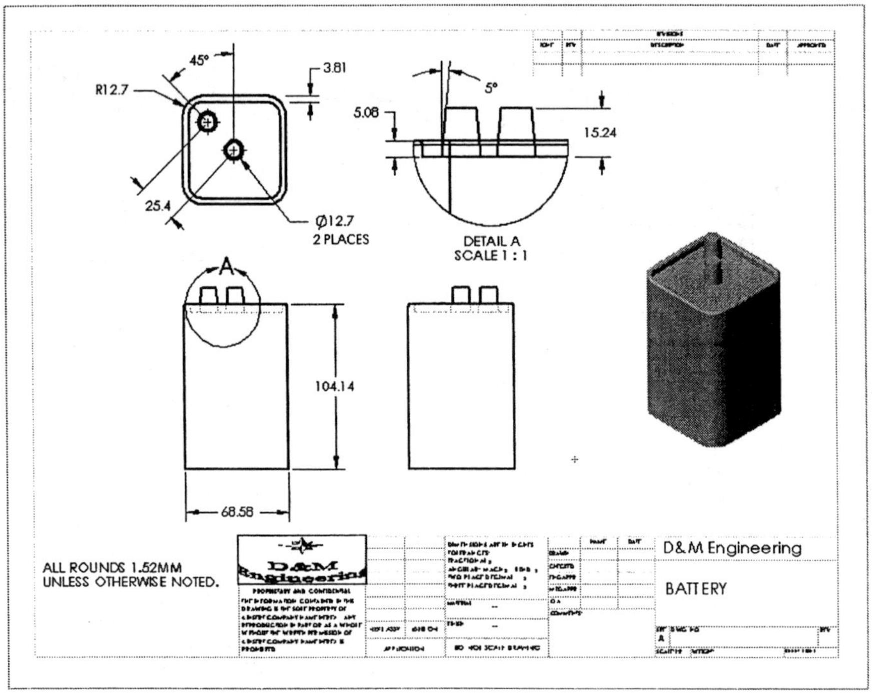

Note: When both the part units and drawing units are set to inches, set Trailing Zeros to the Smart option before creating Parametric Notes.

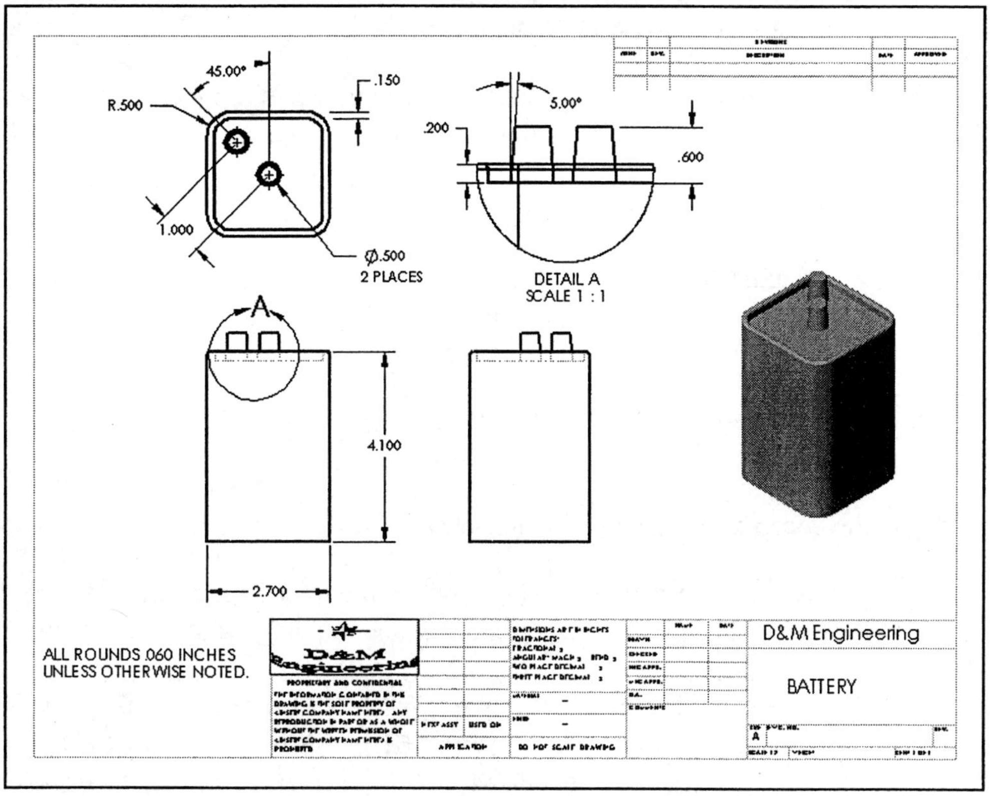

Revise Text and Align Text.

82) Right-click **Edit Sheet Format**. Copy/Paste existing notes. **Zoom in** on the upper right corner of the drawing. Click on the **REV** text. Click **Edit**, **Copy** from the Main menu. Locate the text. Click a **point** in the cell below the REV text. Click **Edit**, **Paste**. The REV text is displayed in the second row. Modify the text. Double-click on the **REV** text. Enter **1**. Click **OK**.

REVISIONS			
ZONE	REV.	DESCRIPTION	DATE
	1	ECO 32510 RELEASE TO MANUFACTURING	10-10-02

83) Repeat **Copy/Paste** for **Description** and **Date**. **Enter ECO 32510 RELEASE TO MANUFACTURING** for DESCRIPTION. Enter **10-10-02** for date or the current date. update the date for the book and picture?

84) Align the Date text. Hold the **Ctrl** key down. Click the note, **DATE**. Click the note **10-10-02**. Right-click in the **Graphics window**. Click **Align**. Click **Leftmost**. The notes are leftmost aligned.

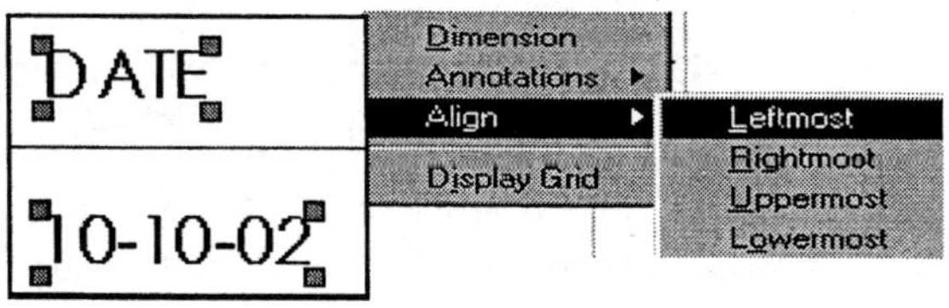

Part Number and Part Name

A part number is a numeric representation of the part. Each part has a unique number. Each drawing has a unique number. Drawings can incorporate numerous part numbers or assembly numbers.

Note: Software applications incorporate unique part numbers to create and perform the following activities:

- Bill of Materials.
- Manufacturing procedures.
- Cost analysis.
- Inventory control, Just in Time, etc.

As the designer, you are required to procure the part and drawing numbers from the documentation control manager.

In this project, use the following prefix codes to categories created parts and drawings. The part name, part number and drawing numbers are as follows:

Category:	Prefix:	Part Name:	Part Number:	Drawing Number:
Molded Parts	44-	BATTERYPLATE	44-A26	44-22222
		LENSCAP	44-A27	44-22223
		HOUSING	44-A28	44-22224
Purchased Parts	99-	BATTERY		99-3564
		LENS		99-3565
		O-RING		99-3566
		SWITCH		99-3567
		BULB		99-3568
Assemblies	10-	FLASHLIGHT	10-F123	10-350123

Link notes in the title block to the SolidWorks Properties. Create a note for the title of the drawing that is linked to the SolidWorks file name, BATTERY. Create additional notes in the title block that completes the drawing.

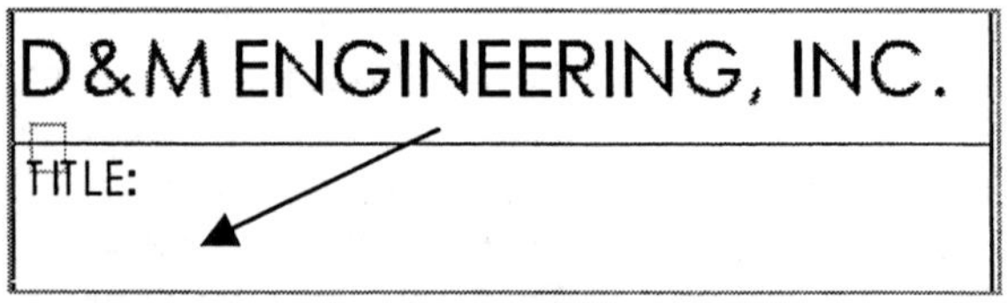

Create a linked note.

85) Zoom in on the lower right corner of the drawing. Click **Note** from the Annotations toolbar. Click a **start point** in the TITLE section of the title block.

The Property dialog box is displayed. Click **Link to Property**. The Link to Property dialog box is displayed. Select **SW-FileName** from the drop down list. The variable $PRP"SW-File Name" is displayed in the Note text box. Display the linked text in the Title. Click **OK**.

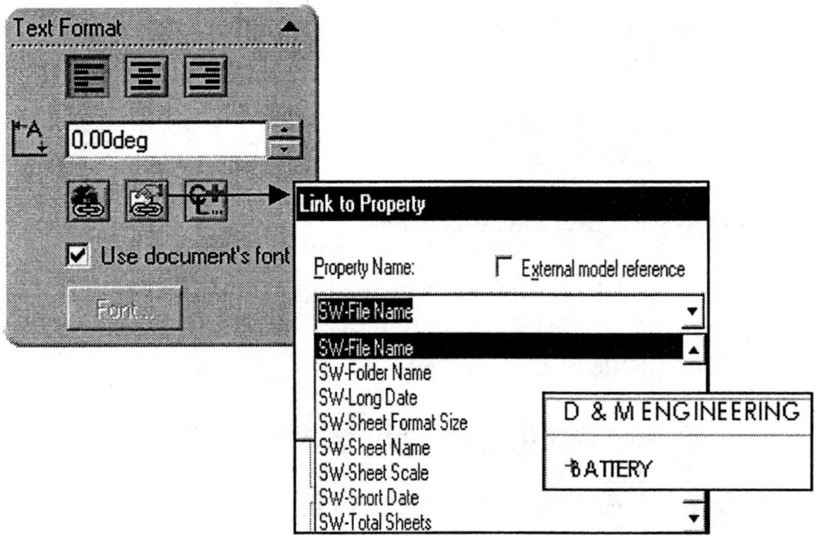

Notes in the Title Block

Additional notes are required in the title block. The text box headings: SIZE B, DWG. NO., REV., SCALE, WEIGHT and SHEET OF are entered in the SolidWorks default Sheet Format.

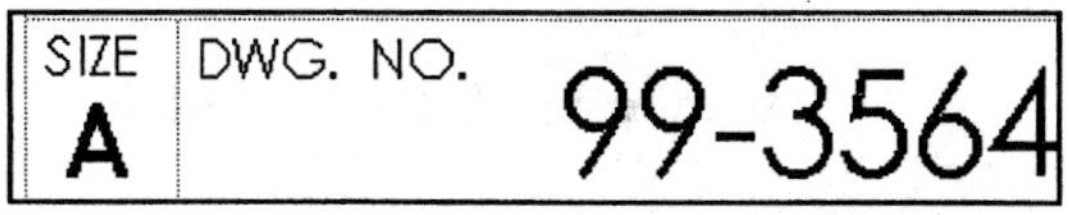

86) Create a note in the DWG. NO. text box. Click **Note**. Click a **start point** in the upper left hand corner below the DWG. NO. text. Enter the drawing number. Enter **99-3564**. Click **OK**.

Note: Modify the font size to address smaller text boxes. Remove the check mark from the Use document's font. Select the Font button for the different text size.

87) Enter each **note** listed in the Title Block Entry table.

Title Block Entry:	Description:	Notes:
DWG. NO.	Unique number to identify and track the drawing.	99-3564
REV.	Indicates the number of official changes that were made to the original drawing.	1
SCALE	Size to which the drawing is created.	1:1
SHEET	Number of sheets that make up the drawing.	1 OF 1

Complete the notes in the title block.

88) Enter **name** and **date** in the DRAWN box. Enter **MILLIMETERS** for units.

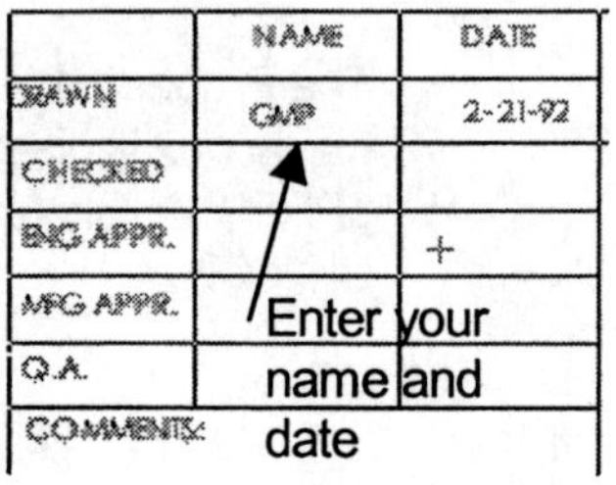

89) Right-click in the **Graphics window**. Click **Edit Sheet**.

90) Fit the drawing to the Graphics window. Press the **f** key.

91) Save the BATTERY drawing. Click **Save**.

DIMENSIONS ARE IN MILLIMETERS
TOLERANCES:
FRACTIONAL<MOD-PM>

Exploded View and Bill of Materials

Add an Exploded view and Bill of Materials to the drawing. Add the FLASHLIGHT assembly Exploded view. The Bill of Materials reflects the components of the FLASHLIGHT assembly.

Create a drawing with a Bill of Materials. Perform the following steps:

- Create a new drawing from the sheet format.
- Display the Exploded view of the assembly.
- Insert the Exploded view of the assembly into the drawing.
- Label each component with Balloon text.
- Create a Bill of Materials.

Let's start.

92) **Close** all parts and drawings.

93) Create a new drawing. Use the A size sheet format. Click **New** from the Standard toolbar. Click **CUSTOM-A Drawing Template.** Click **OK.** Save the drawing. Click **Save.** Select **My Documents** folder in the Save In text box. Enter the assembly drawing name. Enter **FLASHLIGHT**. Click **Save.**

94) Display the Exploded view of the assembly. Open the FLASHLIGHT assembly. Click **Open** . Click **FLASHLIGHT**.

95) Display the Exploded view. Click **Configuration** at the bottom of the ViewManager. Click the **Plus Sign** to the left of the Default entry. Right-click **ExpView1**. Click **Explode**.

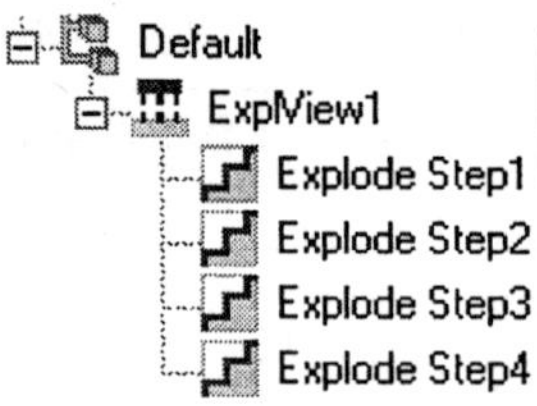

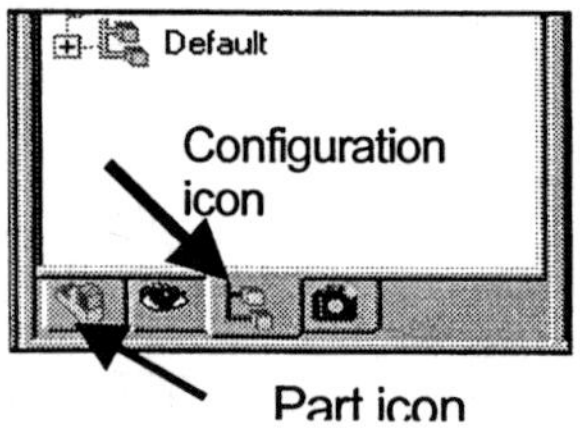

96) Split the FeatureManager. Position the **mouse pointer** on the top of the Split Bar. Drag the **Split Bar** downward to display the Configuration FeatureManager and the Part FeatureManager.

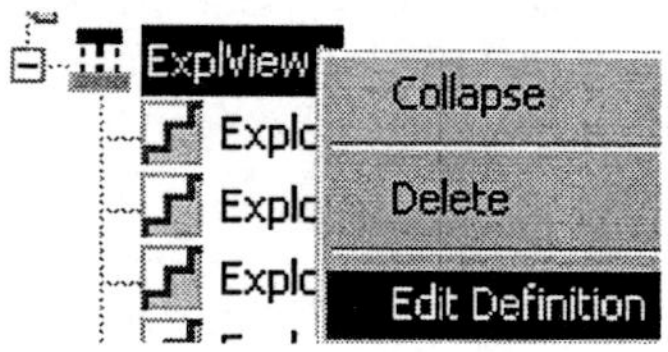

97) The BULB component is contained inside the LENS. Explode the BULB. Right-click **ExplView1**. Click **Edit Definition**.

98) Create Explode Step 7. Click **New Step**. Select **BULB** from the FeatureManager. Click the **edge** of the HOUSING for Direction to Explode. Select **Component Part Only**. Position the BULB in front of the LENS. Enter a **value** for distance. Click **OK**.

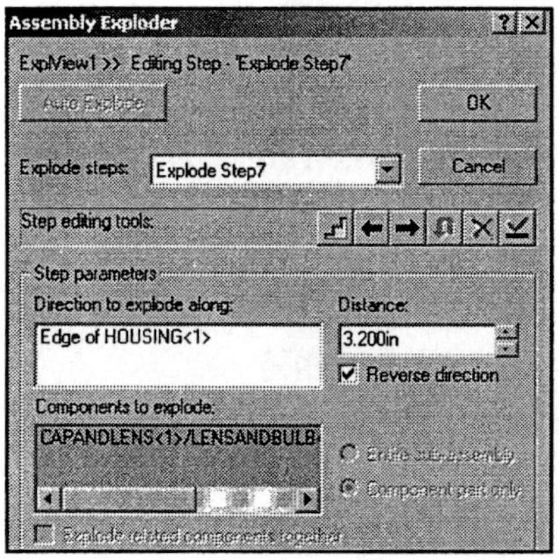

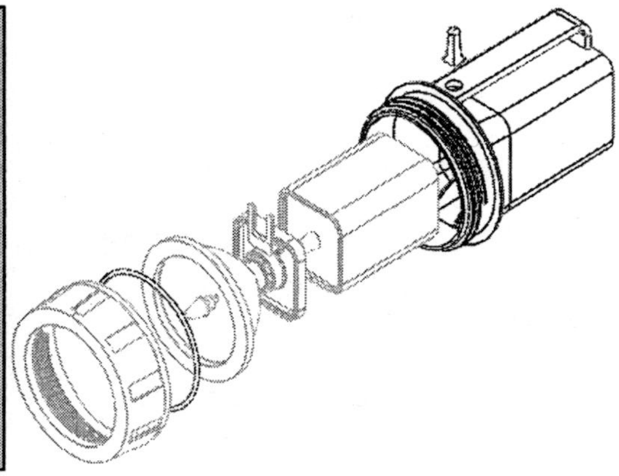

99) Fit the Exploded view to the Graphics window. Press the **f** key.

100) Tile the Graphics window to display both the FLASHLIGHT assembly and FLASHLIGHT-Sheet1 drawing. Click **Window, Tile Horizontal**. Click the **FLASHLIGHT-Sheet1** Graphics window.

101) Insert the Exploded view into the drawing. Click inside the drawing **Graphics window**. Click **Named Views** from the Drawing toolbar. Click the **FLASHLIGHT assembly** icon in the Assembly window. Click **Isometric** from the Named View dialog box. The mouse pointer displays a + symbol.

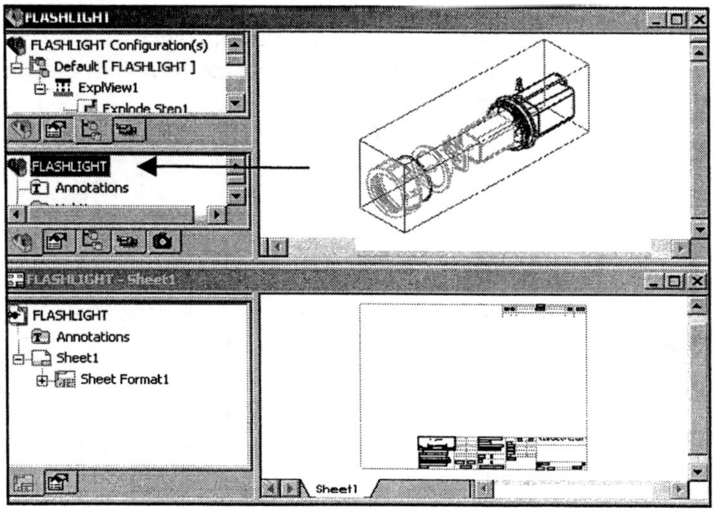

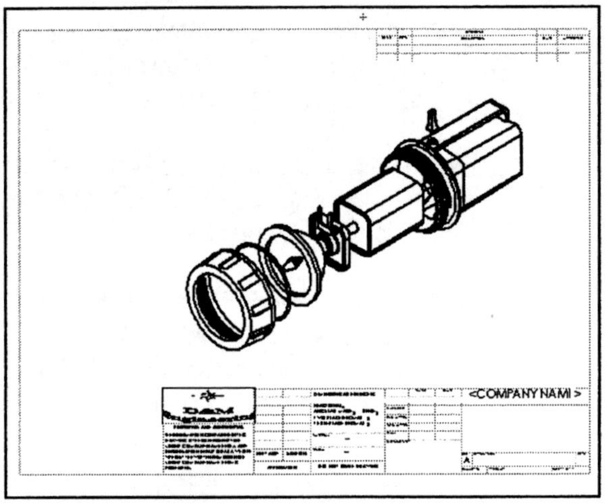

102) Position the view on the right side of the drawing. Click the **inside location** of the drawing sheet. Modify the drawing sheet scale. Click **Custom Scale** in the Property Manager. Enter **1:3**. Click **OK**.

Select the **FLASHLIGHT assembly icon** to obtain the entire exploded assembly.

103)Close the Assembly window. **Maximize** the Drawing window.

104) Position the view and resize the view boundary. Drag the **view** below the Revision block. Resize the view boundary. Drag the **green square handles** approximately 10mm from the profile lines.

105) Turn the Origins off. Click **View**, **Origins**. Click **Hidden Lines Removed**. Right-click **Tangent Edge**. Click **Tangent Edge Visible**.

106) Save the FLASHLIGHT assembly drawing. Click **Save**.

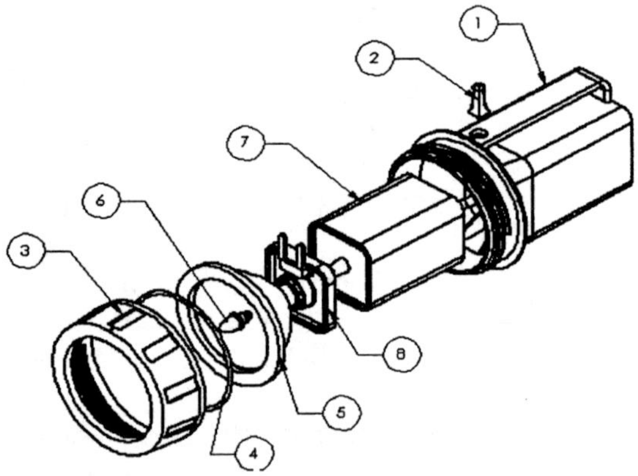

Label each component.

107)A unique item number is placed inside a circle. The circle is called Balloon text. Create the Balloon text. Click **Balloon**  from the Annotations toolbar.

108) Place the first Balloon. Click the **edge** of the HOUSING handle. A Balloon appears with a leader line and the number 1 inside a circle . Place the second Balloon. Click the **edge** of the BATTERYPLATE. Repeat for the other **components** in the assembly.

109) The Balloon text mode is active until it is deactivated. Deactivate the Balloon text mode. Click **Balloon**  from the Annotations toolbar.

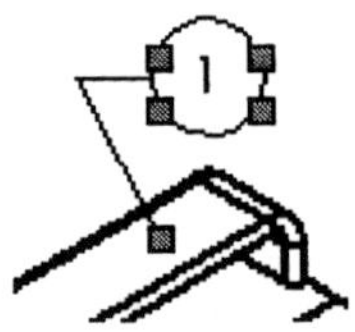

Drag handles to move arrow or text

110) Reposition the Balloon text. Click the **Balloon** text. Drag the **text** by its handles to the desired location.

Note: When the Drawing Standard is set to ANSI, the Balloon note displays arrowhead on a selected edge and a filled dot on a selected face.

Balloon Circular Split Line Style settings contain both item number and quantity. Each Balloon references a part with a SolidWorks File Name. The SolidWorks File Name is linked to the Part Name in the Bill of Materials by default. Customize both the Balloon text and the Bill of Materials according to your company's requirements.

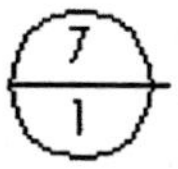

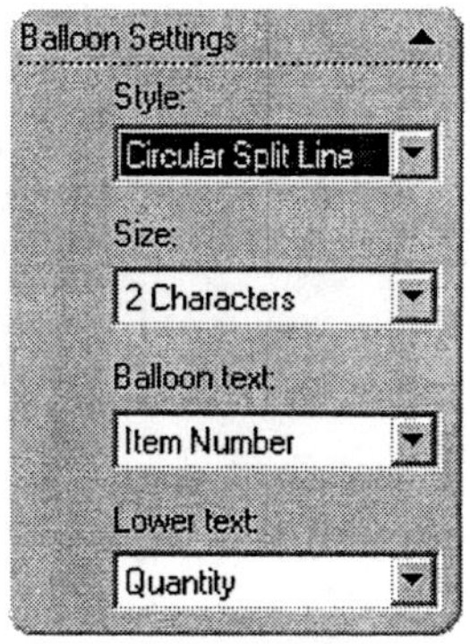

Create a Bill of Materials.

111) Select the **view boundary**. Click **Insert** from the Main menu. Click **Bill of Materials**. Accept the default Bill of Material template, BOMTEMP.XLS. Click **Open**.

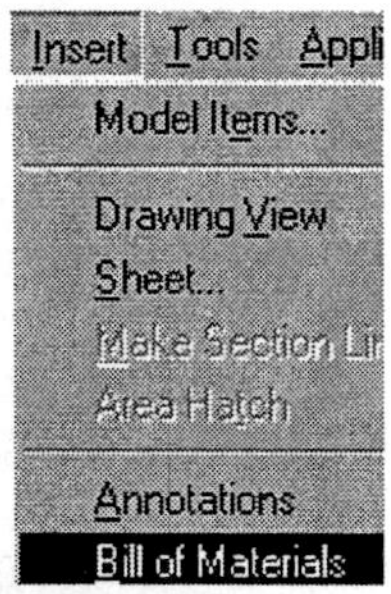

112) The Bill of Materials Properties dialog box is displayed. Click the **Use documents note font when creating the table** checkbox. Uncheck the **Use table anchor point** check box. Click **OK**.

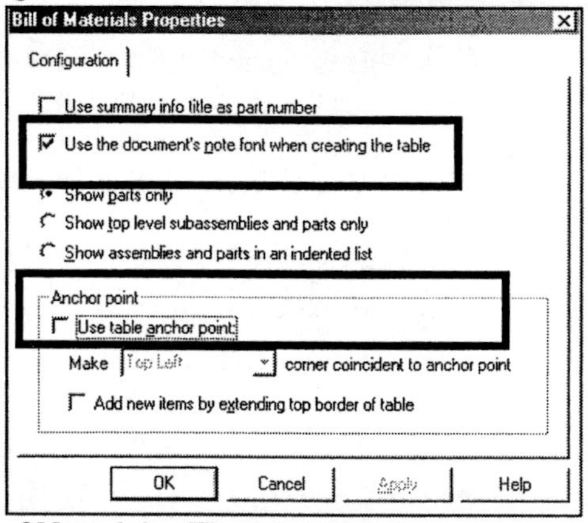

113) Move the Bill of Materials. Click the **Bill of Materials**. The grip points turn green. Drag the **Bill of Materials** below the upper left corner of the Drawing.

The custom Bill of Material incorporates:

- ITEM NO.
- QUANTITY.
- DESCRIPTION.
- PART NUMBER.

The Bill of Materials is created with an EXCEL spreadsheet.

Note: MS EXCEL 97 or later is required. Utilize the default Bill of Material by changing the column heading from PART NO. to DESCRIPTION. Change the column heading from DESCRIPTION to DWG NO., "Drawing Number".

Start Microsoft EXCEL.

114) Double-click the inside **Bill of Material** table. The EXCEL spreadsheet and toolbars are displayed. EXCEL defines entries in the table by cells. Each cell is labeled by a row letter and a column number. Example: ITEM NO. is in cell A1.

ITEM NO.	QTY.	PART NO.	DWG NO
1	1	HOUSING	
2	1	SWITCH	

115) Edit the text. Click the **DESCRIPTION** text in cell D1. Enter **DWG NO**. Click **cell D2**. Enter **44-22224** in the DWG NO. column for the HOUSING. Click **cell D3**. Enter **99-3567** in the DWG NO. column for the SWITCH. Enter the corresponding remaining **DRAWING NUMBERS**.

	A	B	C	D
1	ITEM NO.	QTY.	PART NO.	DWG NO
2	1	1	HOUSING	44-22224
3	2	1	SWITCH	99-3567
4	3	1	LENSCAP	44-22223
5	4	1	ORING	99-3566
6	5	1	LENS	99-3565
7	6	1	BULB	99-3566
8	7	1	BATTERY	99-3564
9	8	1	BATTERYPLATE	44-22222

Sheet1

116) Edit the text. Click the **PART NO.** text in cell C1. Enter **DESCRIPTION** in cell C1.

ITEM NO	QTY.	DESCRIPTION	DWG NO
1	1	HOUSING	44-22224
2	1	SWITCH	99-3567

117) Exit EXCEL. Click the **drawing sheet**.

118) Save the FLASHLIGHT drawing. Click **Save**. The Bill of Materials is saved with the drawing.

Create a linked note for the drawing Title.

119) Click **Note** from the Annotations toolbar. Click a **start point** in the TITLE section of the title block.

D&M ENGINEERING, INC.

TITLE:

Click **Link to Property**. The Link to Property dialog box is displayed. Select **SW-FileName** from the drop down list. The variable $PRP"SW-File Name" is displayed in the Note text box. Enter **ASSEMBLY**. Display the linked text, FLASHLIGHT ASSEMBLY in the Title box. Click **OK**.

$PRP:"SW-File Name" ASSEMBLY

FLASHLIGHT ASSEMBLY

120) Save the FLASHLIGHT assembly drawing. Click **Save**.

The FLASHLIGHT assembly drawing is complete.

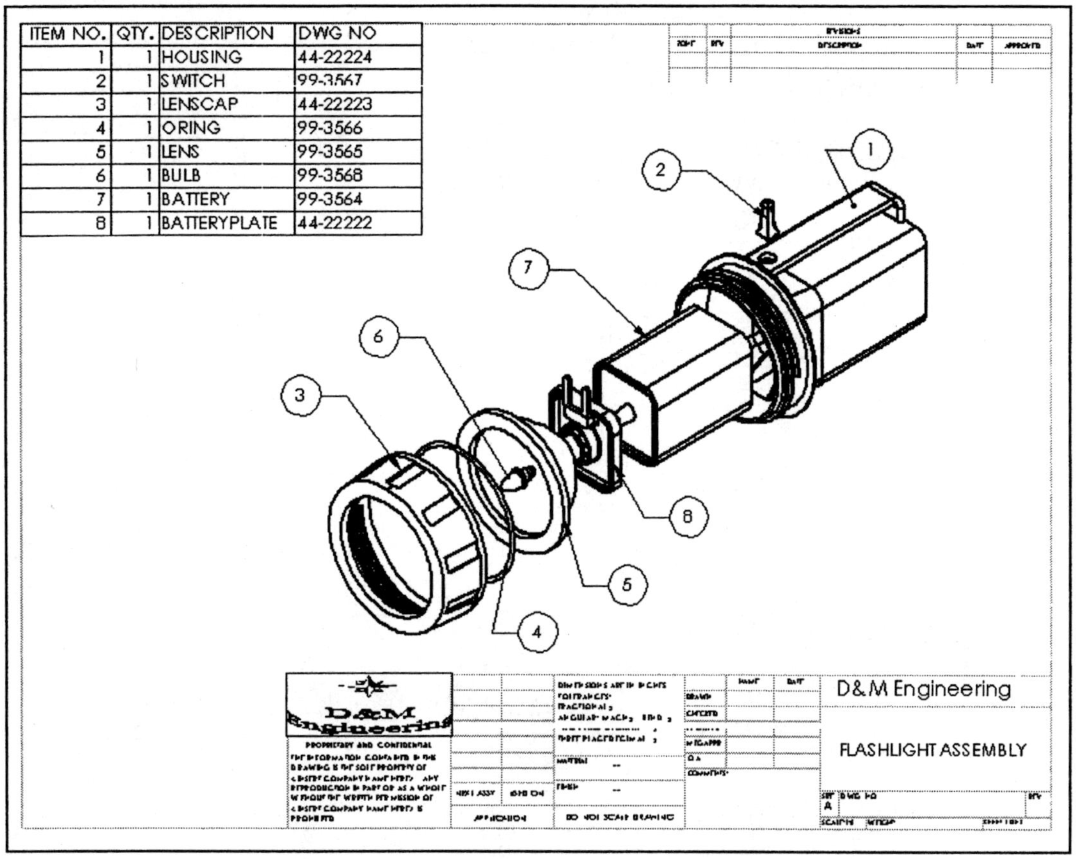

Associative Part, Assembly and Drawing

The associative part, assembly and drawing share a common database. Verify the association between the part, assembly and drawing.

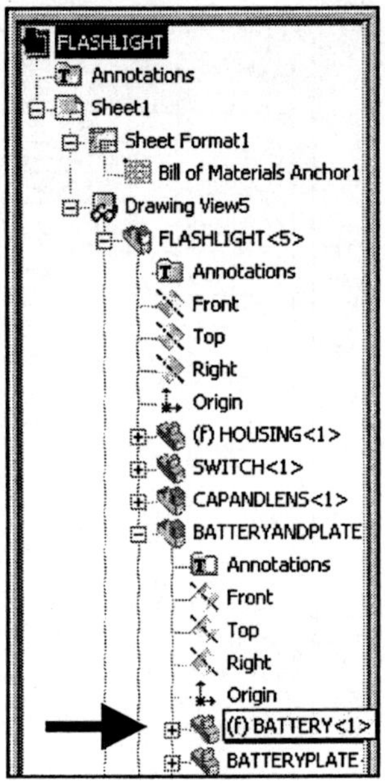

121) Modify a dimension on the drawing. Expand the Isometric Drawing View in the FLASHLIGHT FeatureManager. Click the **Plus** icon. Expand the BATTERYANDPLATE. Click the **Plus** icon. Right-click **BATTERY**. Select **Open BATTERY**. The BATTERY part opens.

122) Open the BATTERY drawing. Right-click on the **BATTERY part icon**. Click **Open Drawing**.

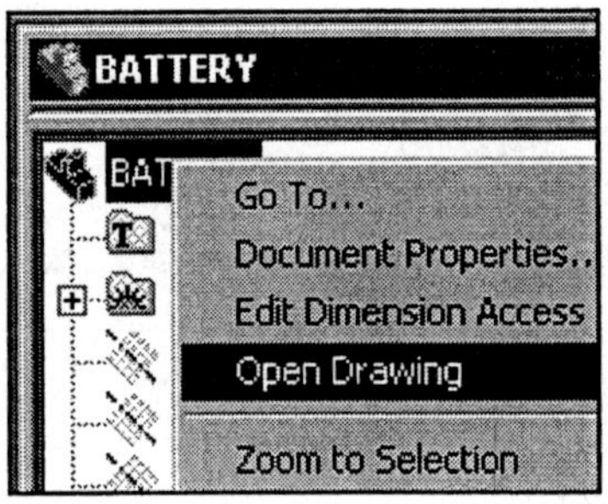

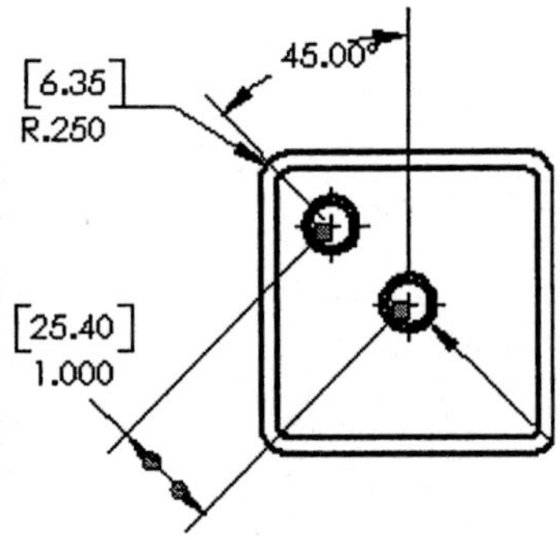

123) Double-click on the radial fillet dimension **.500**, **[12.7]** in Drawing View2 (Top). Enter **.250**, **[6.35]** in the Modify dialog box.

124) Click **Rebuild**. All views are updated.

125) View the updated part. **Open** the BATTERY part. The system provides a prompt. The file is changed and it requires an update. Click **OK**. View the updated assembly. **Open** the FLASHLIGHT assembly.

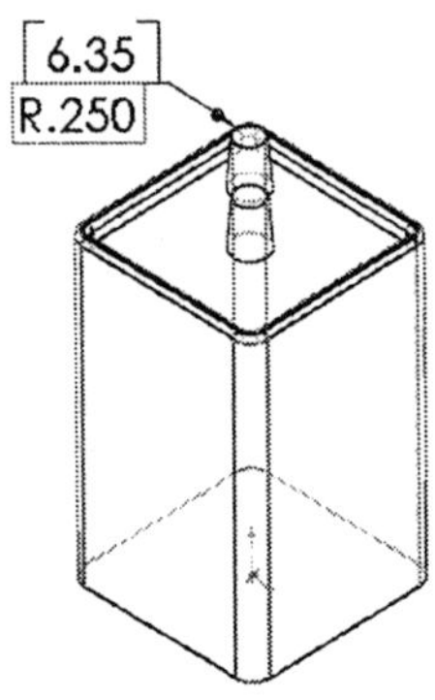

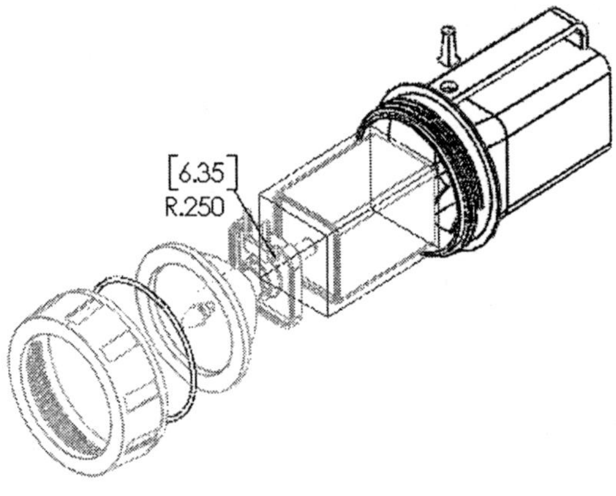

Drawings are an integral part of the design process. Parts, assemblies and drawings all work together. From the initial design concepts, you created parts and drawings that fulfill the design requirements of your customer.

Additional SolidWorks examples, see below are provided in the text **Drawing and Detailing with SolidWorks 2001/2001Plus**, Planchard & Planchard, SDC Publications 2002.

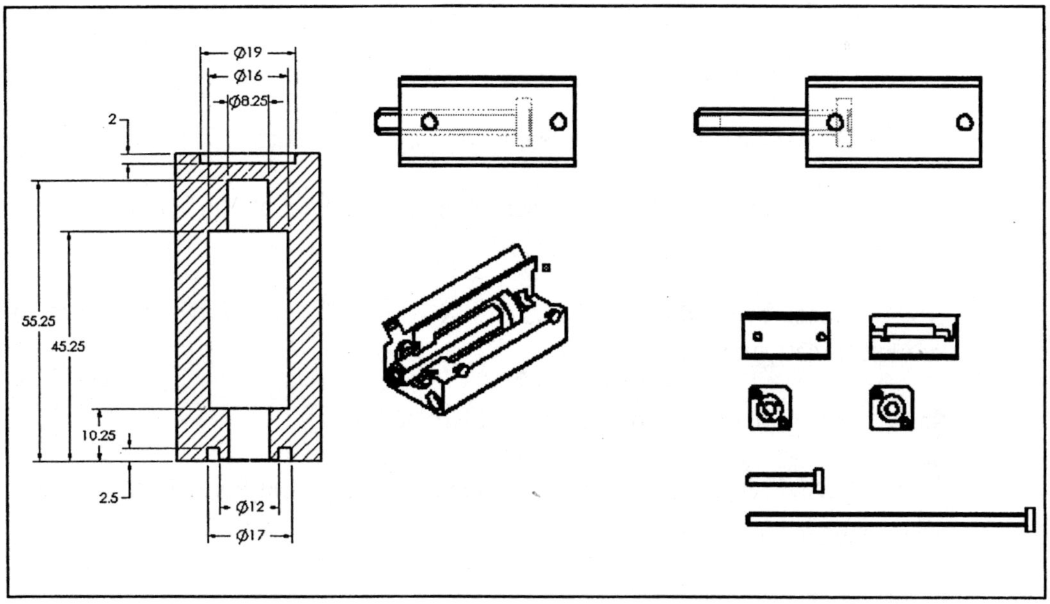

Questions

1. What is a Bill of Materials?

2. Name the two major design modes used to develop a drawing in SolidWorks?

3. What are the three various sheet format options in SolidWorks?

4. Name seven components that are commonly found in a title block.

5. How do you add an Isometric view to the drawing?

6. In SolidWorks, drawing file names end with a ______ suffix?

7. True or False. In SolidWorks, if a part is modified, the drawing is automatically updated.

8. True or False. In SolidWorks, when a dimension in the drawing is modified, the part is automatically updated.

9. Name three guidelines to create General Notes.

10. True or False. Most engineering drawings us the following font: Time New Roman – All small letters.

11. What are Leader lines?

12. What are Center Marks?

13. How do you add an Exploded view from an assembly to a drawing?

14. How do you label components in an Exploded view on an assembly drawing?

Exercises

Exercise 5.1: Create a drawing for the U-CHANNEL. The U-CHANNEL part was created in Project 1, Exercise 1.4.

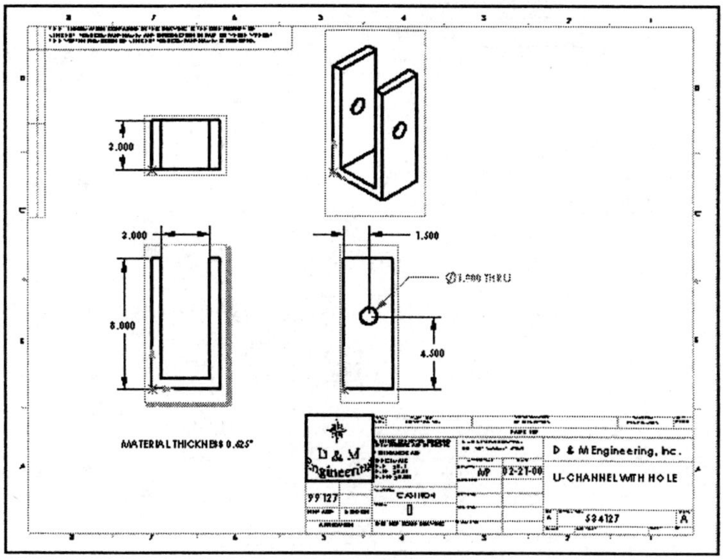

Exercise 5.2: Create an assembly drawing and Bill of Materials for Exercise 1.15 in Project 1.

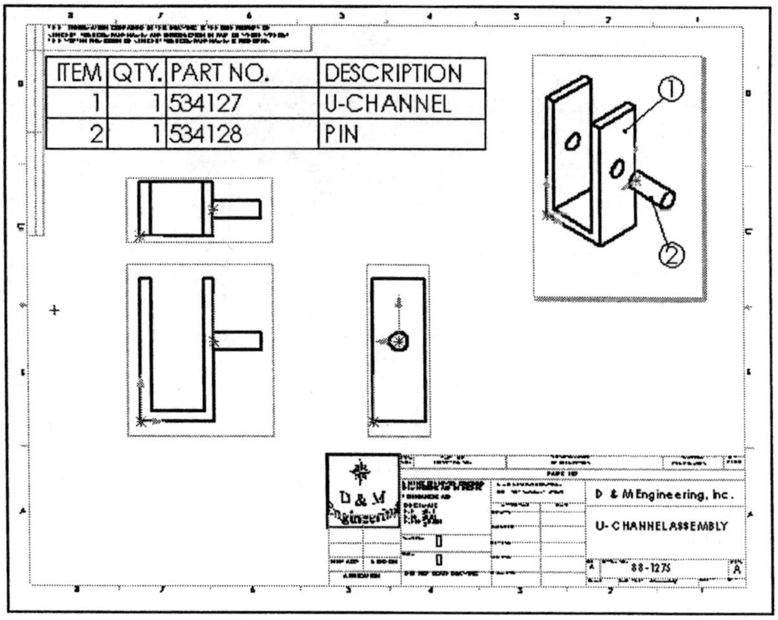

Exercise 5.3: Create an assembly drawing and Bill of Materials for two PLATEs and four STANDOFFS in the below drawing from Exercise 1.14 in Project 1.

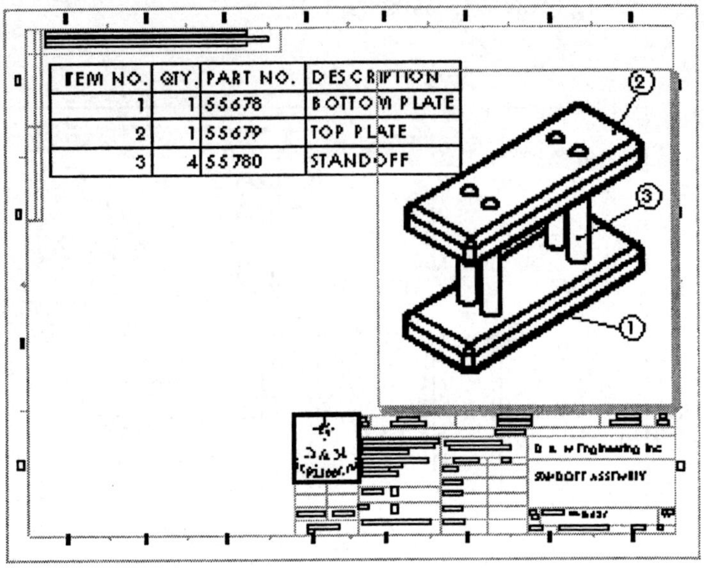

Exercise 5.4: Create an assembly drawing and Bill of Materials for the BATTERYANDPLATE assembly.

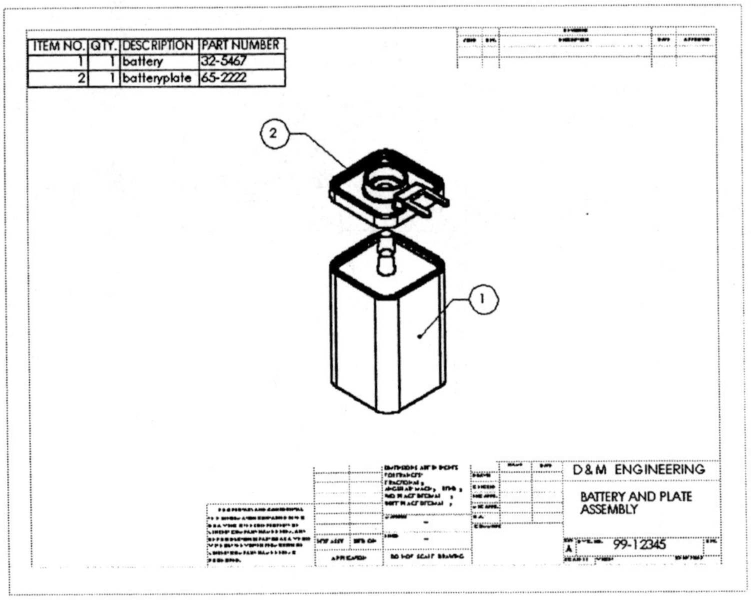

Exercise 5.5: Industry Collaborative Exercise.

Create a new part called FLAT-PLATE. FLAT-PLATE is (100mm x 100mm x 30mm). Add 4 - M10 x1.5 Thru holes. Create the first hole with the HoleWizard. Create the other three holes with a Linear Pattern. Save the FLAT-PLATE. Review the design with your teammates.

An engineer on your team expresses concerns about the standard M10x1.5 screw fastener. The engineer specifies a threaded insert to prevent the screws from loosening under vibration. The threaded insert must adhere to the following design considerations:

- M10 x 1.5 Thread Size (Nominal Diameter = 10 mm, Thread Pitch = 1.5).
- Free Running.
- Tapped depth less than 30mm.

Your tasks on this project are as follows:

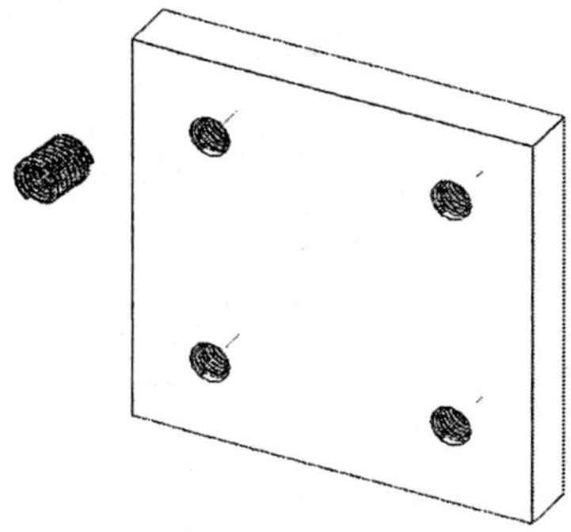

- Find a threaded insert and record the part number.
- Determine the tapped hole specification based upon the part number.
- Download the SolidWorks part file.
- Create the SolidWorks assembly that contains the FLATPLATE and the 4 threaded inserts.
- Create a drawing for the FLATPLATE with a Hole Call Out. Create the Drawing Notes required to document the drill and tap required for the threaded insert.

Note: Emhart Fastening Teknologies, A Black & Decker Company, manufactures Heli-Coil® Screw Thread Inserts.

Obtain Heli-Coil® Screw Thread Insert information from the following URL: http:// www.emhart.com. Select the Heli-Coil® Selector button.

Select the Innovations button. Click the Emhart Mentor® SmartCAT button. Register and enter your email address.

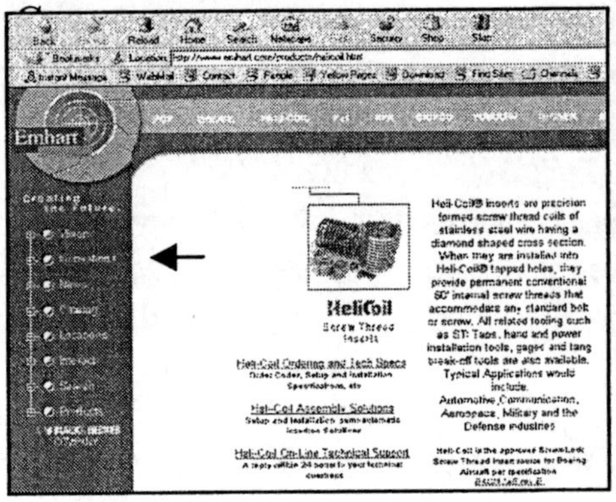

-

Coil®. Click the Search Selector.

Enter the following product requirements:

- Thread size: M10x1.5
- Type: Free
- Length: 15mm
- Packaging: Bulk

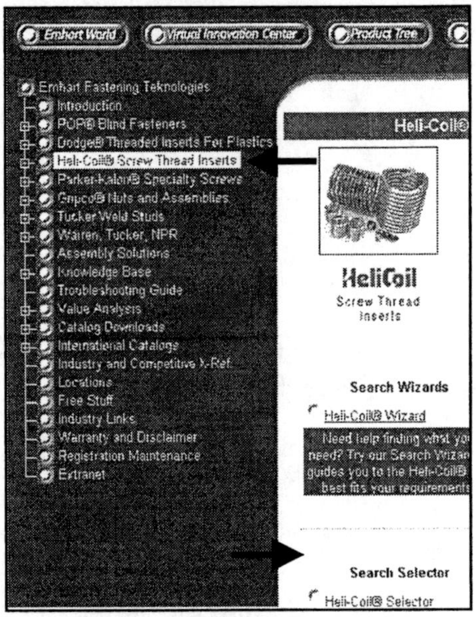

Obtain the part number and min/max outside diameter. Select the Spec button.

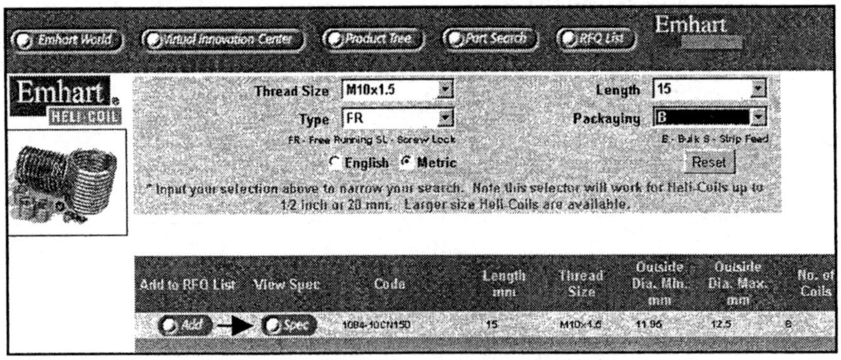

Click the Product Tools button to obtain the drawing notes required for manufacturing. Click the CAD Drawings button to view and download the component.

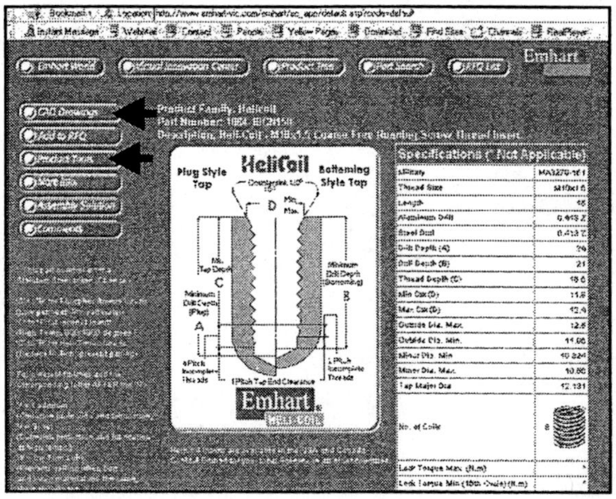

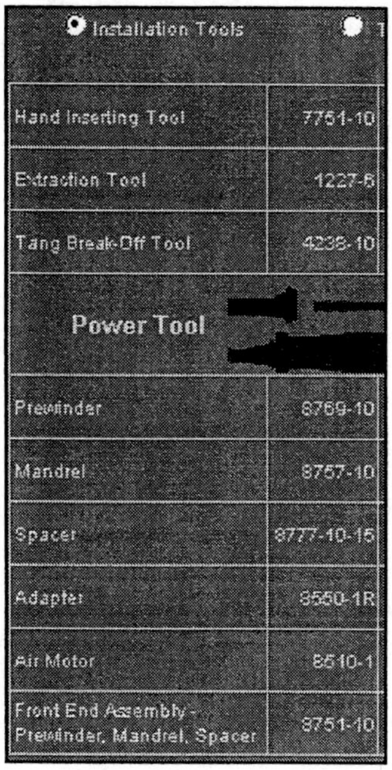

Modify the FLATPLATE. Edit the Thru Hole. Change the Standard to HeliCoil® Metric Tapped Hole. Enter 21 for Tap Drill Type & Depth.

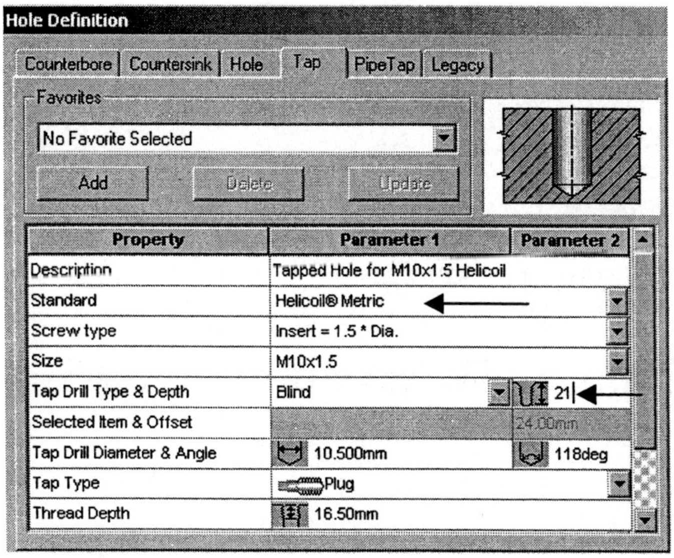

Components Courtesy of
Emhart Teknologies
New Haven, CT USA
Emhart® and Heli-Coil® are registered trademarks of the Black & Decker Company

Create a new assembly with the Heli-Coil® threaded insert and the FLATPLATE. P is the Pitch of the thread. P = 1.5.

Create a Distance Mate between the two components. Position the Heli-Coil® threaded insert between 0.75P to 1.5 P from the top face of the FLATPLATE.

Create a Coincident Mate between the Origin of the Heli-Coil® and Temporary Axis of the Tapped Hole.

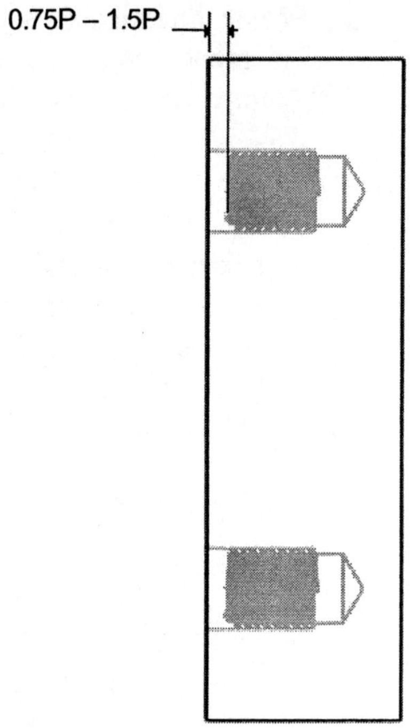

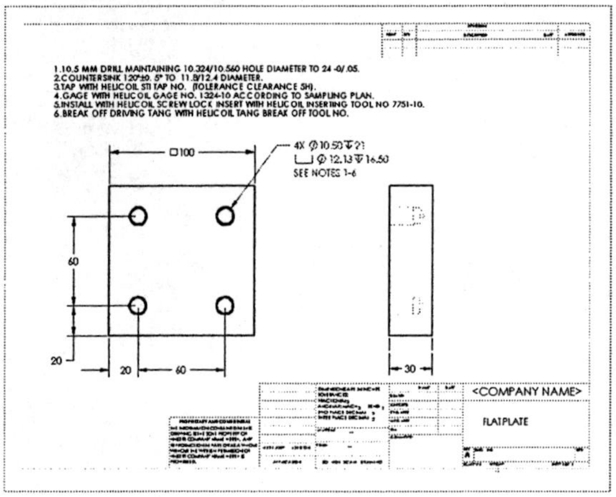

Create a drawing for the FLATPLATE. Add the following Notes to complete the drawing:

NOTES:

1. 10.5 MM DRILL MAINTAINING 10.324/10.560 HOLE DIAMETER TO 24 -0/.05.

2. COUNTERSINK 120°±0. 5° TO 11.8/12.4 DIAMETER.

3. TAP WITH HELICOIL STI TAP NO. 2087-10 (TOLERANCE CLEARANCE 5H).

4. GAGE WITH HELICOIL GAGE NO. 1324-10 ACCORDING TO SAMPLING PLAN.

5. INSTALL WITH HELICOIL SCREW LOCK INSERT 4184-10CN WITH HELICOIL INSERTING TOOL NO 7751-10.

6. BREAK OFF DRIVING TANG WITH HELICOIL TANG BREAK OFF TOOL NO. 4238-10.

Store Drawing notes for threaded inserts in a Notepad.txt document to conserve time.

Exercise 5-6: Industry Collaborative Exercise.

Parts, assemblies and drawings are utilized over the World Wide Web in many different formats. Example: Animated parts and assemblies are applied to the web site created by Mike J. Wilson.

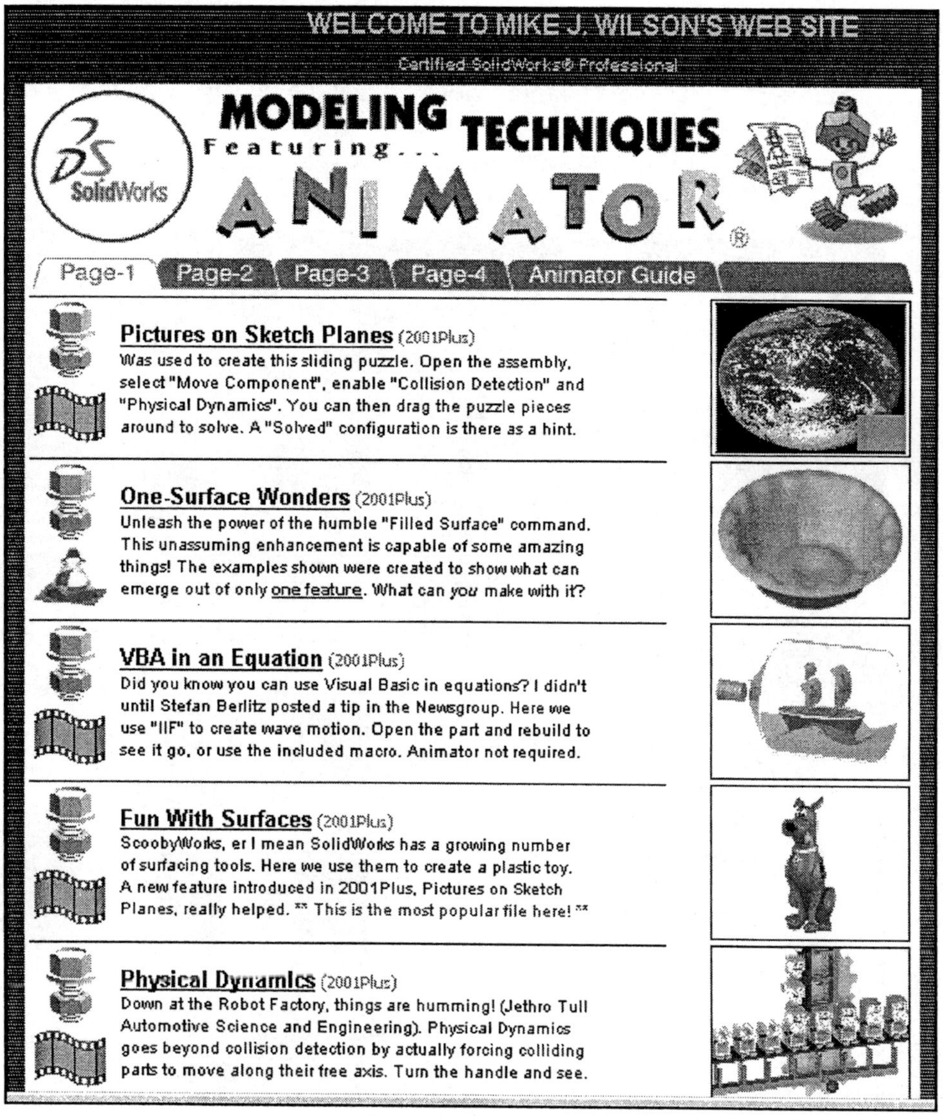

Models and images courtesy of Mike J. Wilson
www.mikejwilson.com

A SolidWorks eDrawing is a compressed document that does not require the corresponding part or assembly. SolidWorks eDrawing is animated to display multiple views and dimensions.

Create an eDrawing of the BATTERY drawing. Select Tools, Add-Ins, eDrawings. The eDrawing toolbar contains the Publish eDrawing icon.

Review the eDrawing on-line Help for additional functionality. Click Publish eDrawing from the eDrawing toolbar.

Click the Animate button.

The Front view of the BATTERY drawing is displayed. Click the Next button. Display the remaining views. Click the Play button.

Save the BATTERY eDrawing. Email the drawing to a friend.

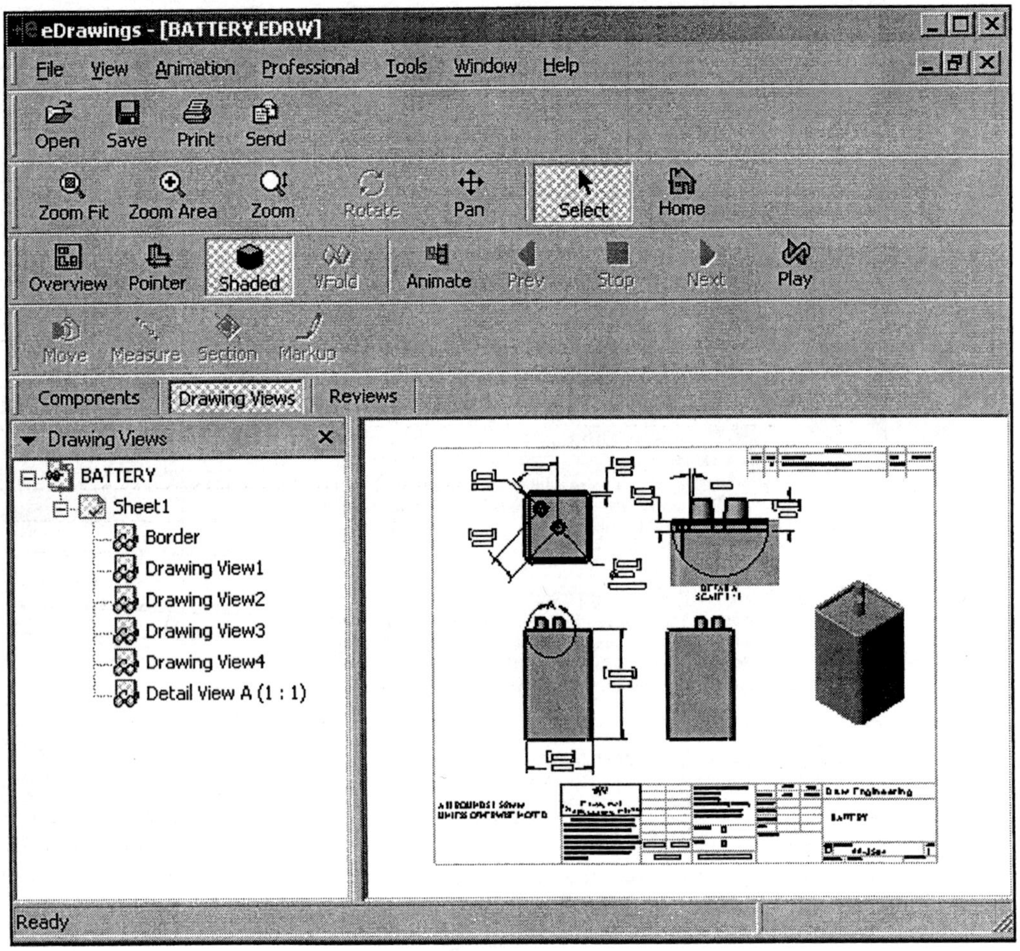

Appendix

Engineering Changer Order (ECO)

D&M

Engineering Change Order

ECO # ________
Page 1 of __

Product Line				Author
	☐	Hardware		Date
	☐	Software		Authorized Mgr.
	☐	Quality		Date
	☐	Tech Pubs		
Change Tested By				

Reason for ECO(Describe the existing problem, symptom and impact on field)

D&M Part No.	Rev From/To	Part Description	Description	Owner

ECO Implementation/Class			Departments	Approvals	Date	
All in Field	☐		Engineering			
All in Test	☐		Manufacturing			
All in Assembly	☐		Technical Support			
All in Stock	☐		Marketing			
All on Order	☐		DOC Control			
All Future	☐					
Material Disposition			ECO Cost			
Rework	☐		DO NOT WRITE BELOW THIS LINE(ECO BOARD ONLY)			
Scrap	☐		Effective Date			
Use as is	☐		Incorporated Date			
None	☐		Board Approval			
See Attached	☐		Board Date			

The SolidWorks URL: http://www.solidworks.com contains information on local resellers, Gold Partners, Solutions Partners and SolidWorks users groups.

The SolidWorks URL: http://www.3DpartStream.net contains additional engineering electronic catalog information.

Helpful on line SolidWorks information is available from the following URLs:

http://www.nhcad.com

- Configuration information and other tips and tricks.

http://mechengineer.com/snug/

- News group access and local user group information.

http://www.solidworktips.com

- Helpful tips, tricks on SolidWorks and API.

INDEX

SolidWorks Tutorial provides the location of a feature with the Part name or Assembly name from each project. Do not skip steps in this tutorial-based project book. Our goal is to show how multiple design situations and steps are combined to produce a successful project. Use SolidWorks On-Line Help and Glossary for additional options and commands.